ISBN 0-486-21769-8
Dover
3.50

Music, Physics and Engineering

HARRY F. OLSON, E.E., PH.D., D.SC., HON.

Staff Vice President
Acoustical and Electromechanical Research
RCA Laboratories, Princeton, New Jersey

SECOND EDITION

DOVER PUBLICATIONS, INC.
New York

Published in Canada by General Publishing Company, Ltd., 30 Lesmill Road, Don Mills, Toronto, Ontario.
Published in the United Kingdom by Constable and Company, Ltd., 10 Orange Street, London WC 2.

This Dover edition, first published in 1967, is a revised and enlarged version of the work first published by the McGraw-Hill Book Company, Inc., in 1952 under the title *Musical Engineering*.

Standard Book Number: 486-21769-8
Library of Congress Catalog Card Number: 66-28730

Manufactured in the United States of America
Dover Publications, Inc.
180 Varick Street
New York, N.Y. 10014

Preface to the Second Edition

Many and varied advances have been made in musical engineering in the 13 years since the appearance of the first edition. The most important developments have been made in the field of sound reproduction, subjective acoustics, and electronic music.

The most recent, significant, and important advance in the field of sound reproduction is the large-scale commercialization of stereophonic sound in the consumer or mass market complex. Stereophonic sound provides auditory perspective of the reproduced sound and thereby preserves a subjective illusion of the spatial distribution of the original sound sources. Three mediums for stereophonic sound reproduction have been commercialized on a large scale for the consumer's use in the home, namely, magnetic tape, disc and frequency-modulation radio.

The main purpose of sound reproduction is to provide the listener with the highest order of artistic and subjective resemblance to the condition of the live rendition. To achieve this objective requires the utmost degree of excellence of the physical performance of the equipment as directed by the psychological factors involved. The state of the art in sound reproduction has advanced to a stage where a high order of physical performance can be obtained. Up until very recent times, the application of the psychological characteristics as related to sound reproduction have lagged behind the physical considerations. The commercialization of stereophonic sound in the consumer complex has hastened the work in the subjective aspects of sound reproduction.

The properties of a musical tone described in the first edition of *Musical Engineering* remain the same. Once a tone has been described it is possible to produce this tone by purely electronic means. Thus, it will be seen that any tone produced by a voice or musical instrument can be produced by an electronic system. In addition, electronic synthesizers provide the capability of creating musical tones which cannot be produced by the voice or conventional instruments.   Furthermore, the electronic system can generate any sound or combination of sounds, which have or have not been produced before, that may have any musical significance.

New theories in the field of communication provide powerful tools for use by the composer in the composition of music.   Random probability systems which operate under the guidance of certain rules and limits lead to the composition of music.   Computers or specially designed computers have also been employed for the composition of music.

The following subject matter has not changed since the first edition, namely, the fundamentals of sound and music, the construction and characteristics of musical instruments, and the properties of music, and therefore requires no revision.   Accordingly, the major revision of this book involves sound reproduction, subjective acoustics and electronic music.

HARRY F. OLSON

Princeton, N. J.
November, 1966

Preface to the First Edition

The following book on *Musical Engineering* is intended to serve as a textbook, as a handbook, or as an instruction book for students, teachers, musicians, engineers, laymen, and enthusiasts interested in the subjects of speech, music, musical instruments, acoustics, sound reproduction, and hearing.

The major portion of the book employs simple physical explanations, illustrations, and descriptions which can be read without any special training in music, engineering, physics, or mathematics. The objective is to provide a useful and informative book for the progressive musician, the music teacher, the musically inclined layman, the avid music listener, and the sound-reproduction enthusiast. The serious-minded individuals in these vocations and avocations cannot afford to be unacquainted with the applied aspects of speech, music, musical instruments, acoustics, and sound reproduction, because the scientific applications in these subjects have produced a tremendous change in the art, culture, industry, education, and entertainment in this country. Furthermore, with the ever-increasing technological advances, even more dramatic and profound changes will occur in the future. The remainder of the book is devoted to technical descriptions of the action, performance, and characteristics of musical instruments, the vibrating systems and the analogies of musical instruments, the acoustics of music and the processes of sound reproduction. Therefore, it should be of value to the engineer and applied scientist concerned with these subjects.

Particular efforts have been directed toward writing a book that will provide useful information to the reader. As an aid in attaining this objective, the book is well illustrated with 303 figures and 28 tables. In many instances each figure contains several parts, so that a complete theme is presented in a single illustration. A further aid to a clear presentation is obtained by the use of specific definitions of musical and technical terms.

Original or reproduced speech and music are the important and significant sounds for the dissemination of information, art, and culture. A knowledge of the mechanisms for producing these sounds and the characteristics of these sounds is important for studies in speech, music, musical instruments, hearing, and the reproduction of sound. The scientific aspects of speech, music, musical instruments, and hearing have advanced along with acoustics, electronics, communications, and sound reproduction to the point where the practical aspects of these subjects are in the domain of applied science. Accordingly, musical engineering is used to designate a practical consideration of the interrelated subjects of speech, music, musical instruments, acoustics, sound reproduction, and hearing. In view of the importance of the engineering aspects of music and its relation to other fields, it seems timely and logical to devote a book to the subject of musical engineering.

The presentation of musical engineering in this book includes the following subjects: The nature of sound waves. The fundamental generators for the production of a sound wave. Musical terminology. Musical scales. Resonating and radiating systems used in musical instruments. String, wind, reed, brass, percussion, and electrical musical instruments and the human voice mechanism. The acoustic spectrums, power output, directivity patterns, growth, decay, and duration characteristics of musical instruments and the human voice. The fundamental properties of speech and music. The human hearing mechanism. The acoustics of theaters, studios, and rooms. Sound-pickup techniques in recording, sound motion pictures, radio broadcasting, and television. Sound-reproducing systems.

A few words about the subject matter of this book in relation to other books by the author may be of interest to the reader. Each book covers a separate and distinct subject as follows: In the book *Dynamical Analogies* the objective is to present the subject of analogies in vibrating systems. The use of analogies makes it possible to solve problems in sound reproduction, underwater sound, ultrasonics, servos, motors, and engines by electrical circuit theory. In the book *Elements of Acoustical Engineering* the objective is to present the subject of modern acoustics in the applied-science domain. *Elements of Acoustical Engineering* is devoted to the subject of modern acoustics in all phases and aspects which includes the fundamentals of acoustics, the theory, design, and measurement of all electroacoustic transducers used in sound reproduction, ultrasonics, and underwater sound, the applications in these fields and the environments. In this book, *Musical Engineering*, the objective is to present a description and exposition of the processes, instruments, and characteristics involved in all the steps from the musical notation on paper to the ultimate useful destination of original or reproduced sound, the human hearing mechanism.

The author wishes to acknowledge the assistance of Mr. E. G. May in supplying the theoretical and experimental characteristics of pipes and horns in Chapter 4.   He wishes to express his appreciation to Mrs. E. C. Fechter for typing the manuscript and to Mrs. G. T. Dennis for assistance in reading the manuscript and preparing the index.

HARRY F. OLSON

Princeton, N.J.
January, 1952

Contents

CHAPTER ONE

Sound Waves

1.1 INTRODUCTION

Music is the art of producing pleasing, expressive, or intelligible combinations of tones. The sounds of original music are produced by the human voice or instruments actuated by musicians. Most music is recorded and translated into sound from a symbolic notation on paper. The ultimate objective destination of all music is the human hearing mechanism. Thus the production of music consists of the following processes: the symbolic notation on paper by the composer, the translation of the notation into musical sounds by a musician employing his own voice or an instrument or both, and the actuation of the human hearing mechanism by the musical sounds.

The evolution and production of combinations of tones by composers and musicians, which have been accepted by the listeners as pleasant and expressive, have gone on through all the ages of man. Some of the musical developments have withstood the rigors of time and are in evidence at the present time. Others enjoyed a short-lived popularity and, as a consequence, were lost in the oblivion of the past. The evolution, development, and production of music through all the past epochs has been a very slow process, because the number of persons that could listen first hand to music as it was rendered was obviously limited. The advent of sound reproduction changed all this and made it possible for people in the millions to hear famous actors, artists, and musical aggregations where only the order of a thousand had been able to hear them first hand. As a result of sound reproduction, music entered upon a new era which made the musical past seem rather insignificant in comparison.

The reproduction of sound is the process of picking up sound at one point and reproducing it either at the same point or at some other point either at the same time or at some subsequent time. The most common sound-reproducing systems are the telephone, the phonograph, the radio, the sound motion picture, and television.

1

The telephone is the oldest sound-reproducing system. There is an average of more than one telephone instrument for each family in this country. This means that any person can talk to any other person in the matter of seconds.

The phonograph was the first sound-reproducing system which made it possible for all the people of the world to hear statesmen, orators, actors, orchestras, and bands when, previously, only a relatively few could hear them at first hand. The phonograph is used in every country and clime. There is an average of at least one phonograph per family in this country. For the past decade the sale of records has averaged three per year for every man, woman and child. The popularity of the phonograph is due to the fact that the individual can select any type of information or entertainment and reproduce it whenever he wants it.

The radio, like the phonograph, is a consumer-type instrument. Practically every family owns several radio receivers from the personal to the high-quality types. More than half of the automobiles are equipped with radio receivers. As a result, practically every person can select almost any desired program for listening.

The addition of sound to motion pictures made this type of expression complete. This was the first system in which picture and sound were synchronized and reproduced at the same time. Practically one-half of the population sees a motion picture once a week.

Television is the latest system in which picture and sound are reproduced at the same time. Sound is, of course, important to television because without it the result would be the same as the silent motion picture. On the average, every family owns a television receiver and has the opportunity to select from a myriad of varied programs.

The radio, phonograph, sound motion picture, and television have made it possible for all the people of the world to hear famous statesmen, artists, actors, and musical aggregations where only a relatively small number had been able to hear them at first hand. It is evident that the reproduction of sound has produced in a relatively short time a great change in the education and entertainment of this and other countries. The impact of the telephone, phonograph, radio broadcasting, sound motion pictures, and television upon the dissemination of information, art, and culture has been tremendous. The reproduction of sound in these fields has been as important to the advancement of knowledge as the invention of the printing press.

The science of music has been advanced along with acoustics, mathematics, electrical engineering, electronics, psychology, and physiology. The findings in these fields have made it possible to use a scientific approach in the study of speech and music. As a result, great advances have been made in reducing speech and music to a scientific basis. These advances have been hastened by the advent of the reproduction of sound, because the maximum exploitation of this medium requires a scientific understanding of speech and music. The studies in speech, music, hearing, and sound reproduction have advanced to the point where the fundamental aspects dealing with speech, music, musical instruments, hearing, and sound reproduction are reasonably well established. This implies that the science of music and the related subjects have advanced to the stage where they can be classed as engineering. In view of the importance and status of the science of speech, music, and the related subjects, it seems logical and timely to devote a book to the exposition of the scientific aspects of these subjects. Accordingly, the book presented herewith was written. In keeping with the treatment outlined above, the book is given the title *Musical Engineering*. Musical engineering is the theory and practice of the subjects of speech, music, hearing, acoustics, and electronics in the applied-science domain. More specifically, musical engineering involves the following subjects: the nature of an audio sound wave; the fundamental generators for the production of sound waves; musical terminology; musical scales; resonating and radiating systems used in musical instruments; string, wind, percussion, and electrical musical instruments and the human voice; the frequency spectrums, power output, directivity patterns, growth and decay and duration characteristics of musical instruments; the fundamental properties of speech and music; the human hearing mechanism; the acoustics of rooms; and sound-reproducing systems. It is the purpose of this book to present the subject of musical engineering.

1.2 SOUND

Sound is an alteration in pressure, particle displacement, or particle velocity which is propagated in an elastic medium, or the superposition of such propagated alterations.

Sound is also the auditory sensation produced through the ear by the alterations described above.

From these definitions it will be seen that sound is produced when the air or other medium is set into motion by any means whatsoever. Sound may be produced by a vibrating body as, for example, the sounding board of a piano, the body of a violin, or the diaphragm of a loudspeaker. Sound may be produced by the intermittent throttling of an air stream as, for example, the siren, the human voice, the trumpet and other lip-reed instruments, and the clarinet and other reed instruments. Sound may also be produced by the explosion of an inflammable-gas mixture or by the sudden release of a compressed gas from bursting tanks or balloons.

Sound may be produced by the impact of the wind against certain objects in which the nonlinear properties of the medium convert a steady air stream into a pulsating one.

The properties of sound waves and the most common ways of producing sound waves will be described in this chapter in the sections which follow.

1.3   NATURE OF A SOUND WAVE

An explosion of a small balloon of compressed air produces one of the simplest forms of sound wave. A small balloon filled with compressed air is shown in Fig. 1.1A. The air surrounding the balloon is in repose. In Fig. 1.1B, the balloon has burst and the air which has been confined under pressure is transmitted outward in all directions as a pulse of pressure. In equalization, the pressure or condensation pulse is followed by a rarefaction pulse. In the rarefaction pulse the pressure is below the normal undisturbed atmospheric pressure, and in the condensation pulse the pressure is above the normal undisturbed atmospheric pressure, as shown in C, D, and E of Fig. 1.1. Following the definition of a sound wave, given in Sec. 1.2, it will be seen from the foregoing description that the disturbance produced by the bursting balloon constitutes a sound wave, consisting of a condensation or high-pressure pulse followed by a rarefaction or low-pressure pulse. The sound wave travels outward in all directions at the velocity of sound, that is, 1,100 feet per second. The magnitude of the condensation and corresponding rarefaction falls off inversely as the distance from the point of explosion of the balloon. The sound wave depicted in Fig. 1.1 is one of the simplest types. More complex sound waves consist of more than one condensation and rarefaction, usually of different values. These sound waves are produced by a vibrating body or a throttled air stream.

The sound wave described above is termed a spherical wave. The wavefront in a spherical wave is a continuous spherical surface in which the sound variations for all parts of the surface have the same phase. Phase is the fraction of the whole period or time interval which has elapsed with reference to some fixed origin of time. If a relatively small volume, at a large distance from the source, in a spherical sound wave is examined, the waves will be found to be plane. The wavefront in a plane wave is a continuous plane surface in which the sound vibrations for all parts of the surface have the same phase. For example, in Fig. 1.1E, the simple sound wave approximates a plane wave. This sound wave is shown in greater detail in Fig. 1.2. The sound pressure and particle velocity are shown for equal time intervals over a complete cycle of events. The pressure p is depicted by the pointer of the small pressure gauge. The magnitude and direction of the flow of the air molecules,

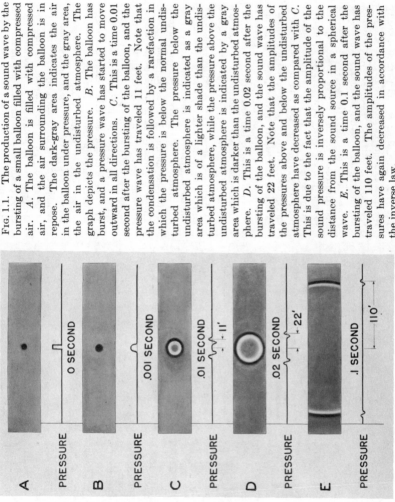

FIG. 1.1. The production of a sound wave by the bursting of a small balloon filled with compressed air. *A.* The balloon is filled with compressed air, and the air surrounding the balloon is in repose. The dark-gray area indicates the air in the balloon under pressure, and the gray area, the air in the undisturbed atmosphere. The graph depicts the pressure. *B.* The balloon has burst, and a pressure wave has started to move outward in all directions. *C.* This is a time 0.01 second after the bursting of the balloon, and the pressure wave has traveled 11 feet. Note that the condensation is followed by a rarefaction in which the pressure is below the normal undisturbed atmosphere. The pressure below the undisturbed atmosphere is indicated as a gray area which is of a lighter shade than the undisturbed atmosphere, while the pressure above the undisturbed atmosphere is indicated by a gray area which is darker than the undisturbed atmosphere. *D.* This is a time 0.02 second after the bursting of the balloon, and the sound wave has traveled 22 feet. Note that the amplitudes of the pressures above and below the undisturbed atmosphere have decreased as compared with *C.* This is due to the fact that the amplitude of the sound pressure is inversely proportional to the distance from the sound source in a spherical wave. *E.* This is a time 0.1 second after the bursting of the balloon, and the sound wave has traveled 110 feet. The amplitudes of the pressures have again decreased in accordance with the inverse law.

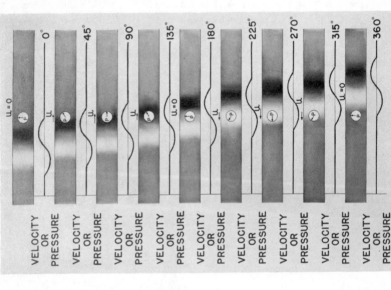

FIG. 1.2.   The pressure and particle-velocity relations in a simple plane sound wave of one cycle. A cycle is any set of variations starting at one condition and returning once to the same condition. A cycle covers an angle of 360 degrees. In the above example, the sound wave is shown at intervals of 45 degrees during a complete cycle. The pointer of the pressure gauge indicates the pressure above and below the normal undisturbed atmosphere as the sound wave travels past the gauge. When the pointer of the pressure gauge is directed upward, the pressure is that of the normal undisturbed atmosphere. When the pointer is displaced in a clockwise direction from directly upward the pressure is above the normal undisturbed atmosphere. When the pointer is displaced in a counterclockwise direction from directly upward, the pressure is below the normal undisturbed atmosphere. The arrow below the u indicates the magnitude and direction of the molecule flow in the sound wave at the same point. The direction of the arrow indicates the direction of air flow. The length of the arrow indicates the magnitude of the air flow. The graph depicts the pressure or particle velocity in the sound wave. The gray areas depict the undisturbed atmosphere. The darker gray areas indicate a pressure above the normal undisturbed atmosphere, while the lighter gray areas indicate pressures below the normal undisturbed atmosphere.

termed particle velocity, is depicted by the arrow marked u. It will be seen that the instantaneous value of the pressure corresponds to and is proportional to the particle velocity.

1.4 SOUND GENERATORS

A vibrating body in contact with the atmosphere will produce sound waves. One of the simplest vibrating bodies suitable for the production of sound is the vibrating piston, depicted in Fig. 1.3. The piston is moved back and forth by the crank and connecting-rod arrangement. In Fig. 1.3A, the entire system including the piston and atmosphere is in repose. In Fig. 1.3B, the piston has moved forward, causing a compression of the air in front of the piston. For reasons of simplicity, the conditions existing on the back of the piston are not shown. In Fig. 1.3C, the pressure wave has now advanced one-half cycle. The air pressure next to the piston is the same as that in the free or undisturbed atmosphere. In Fig. 1.3D, the air next to the piston is below that of the undisturbed atmosphere. In Fig. 1.3E, the pressure next to the piston is the same as that in the free atmosphere. Now a cycle has been covered which includes a complete condensation and rarefaction, that is, a pressure above and a pressure below the undisturbed atmosphere. In Fig. 1.3F, two cycles have been completed. In Fig. 1.3G, six cycles have been completed. Examples of this type of sound generator are the soundboards of the piano and harp, the bodies of the instruments of the violin family, the stretched membranes of drums, the surfaces of cymbals, and the diaphragms of loudspeakers.

Another common type of sound generator employs a means for throttling an air stream and thereby converts a steady or constant stream of air into a pulsating one. The throttled air-stream sound generator shown in Fig. 1.4 consists of a valve in the form of a motor-driven eccentric disk covering an aperture and a source of air under pressure. In Fig. 1.4A, the entire system is in repose. In Fig. 1.4B, the aperture is partially open and a pulse of air under pressure is released. In Fig. 1.4C, the aperture is wide open and the pressure is a maximum. In Fig. 1.4D, the aperture is partially closed and the pressure is less than the maximum. In Fig. 1.4E, the aperture is closed and the pressure at the aperture is the same as that in the undisturbed atmosphere and a full cycle has been completed. In Fig. 1.4F, four completed cycles have been emitted. It will be seen that equalization takes place with a result that ultimately the rarefactions are equal to the condensations. Examples of this type of sound generator are the human voice, the trumpet, trombone and other lip-modulated instruments, the clarinet, saxophone, bassoon, organ and other reed-modulated instruments, and sirens.

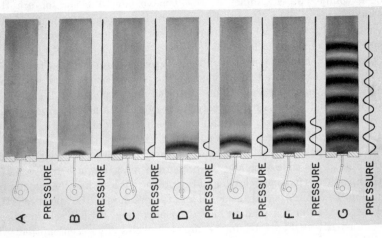

FIG. 1.3. The production of a sound wave by a vibrating or oscillating piston. The piston is moved back and forth by a crank and connecting-rod arrangement. The graph depicts the pressure. For reasons of simplicity, the conditions on the back of the piston are now shown. As in the preceding examples, the gray areas depict the normal undisturbed atmosphere, the dark-gray areas depict condensations or pressures above the normal undisturbed atmosphere, and the light-gray areas depict rarefactions or pressures below the normal undisturbed atmosphere. A. The piston is at rest. B. The crank has made a one-quarter turn. The motion of the piston produces a pressure above that of the normal undisturbed atmosphere. C. The crank has made a half turn, and the pressure at the piston is the same as that in the normal undisturbed atmosphere. D. The crank has made a three-quarter turn, and the pressure at the piston is below that of the normal undisturbed atmosphere. E. The crank has made a complete turn, and a complete cycle has been produced. The sound wave consists of one-half cycle with the pressure above the normal undisturbed atmosphere and one-half cycle with the pressure below the normal undisturbed atmosphere. F. The crank has made two turns, and a sound wave of 2 cycles has been produced. G. The crank has made six turns and a sound wave of 6 cycles has been produced. It will be seen that the amplitude of the sound pressure falls off with the distance from the sound source.

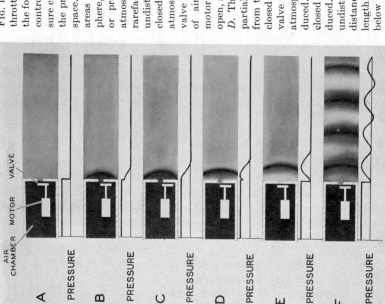

FIG. 1.4. The production of a sound wave by a throttled air stream. The valve mechanism in the form of an eccentric disk driven by a motor controls the passage of air from the pressure chamber to free space. The graph depicts the pressure in the chamber and in the outside space. As in the preceding examples, the gray areas depict the normal undisturbed atmosphere, the dark-gray areas depict condensation or pressures above the normal undisturbed atmosphere, and the light-gray areas depict rarefactions or pressures below the normal undisturbed atmosphere. A. The valve is closed, and no air is released to the outer atmosphere. B. The motor has turned the valve so that it is partially opened, and a pulse of air under pressure is emitted. C. The motor has turned the valve so that it is wide open, and the pressure produced is a maximum. D. The motor has turned the valve so that it is partially closed, and the pressure has decreased from the maximum value. E. The motor has closed the valve, and the pressure next to the valve is the same as the normal undisturbed atmosphere. A complete cycle has been produced. F. The valve has been opened and closed four times, and 4 cycles have been produced. Note that pressures below the normal undisturbed atmosphere are produced when the distance from the valve is greater than a wave-length. Ultimately the pressures above and below the normal undisturbed atmosphere are equal.

1.5   VELOCITY OF PROPAGATION OF A SOUND WAVE

The preceding examples have shown that a sound wave travels with a definite finite velocity.   The velocity of propagation, in centimeters per second, of a sound wave in a gas is given by

$$c = \sqrt{\frac{\gamma p_0}{\rho}} \tag{1.1}$$

where γ = ratio of specific heats for a gas, 1.4 for air

p_0 = static pressure in the gas, in dynes per square centimeter

ρ = density of the gas, in grams per square centimeter

If the pressure is increased, the density is also increased.   Therefore, there is no change in velocity due to a change in pressure.   But this is true only if the temperature remains constant.   Therefore, the velocity can be expressed in terms of the temperature.   The velocity of sound, in centimeters per second, in air is given by

$$c = 33{,}100 \sqrt{1 + 0.00366t} \tag{1.2}$$

where t = the temperature in degrees centigrade.

1.6   FREQUENCY OF A SOUND WAVE

Referring to the Sec. 1.4 on Sound Generators, it will be seen that these generators produce similar recurrent waves.   A complete set of these recurrent waves constitute a cycle.   These recurrent waves are propagated at a definite velocity.   The number of recurrent waves or cycles which pass a certain observation point per second is termed the frequency of the sound wave.

1.7   WAVELENGTH OF A SOUND WAVE

The wavelength of a sound wave is the distance the sound travels to complete one cycle.   The frequency of a sound wave is the number of cycles which pass a certain observation point per second.   Thus it will be seen that the velocity of propagation of a sound wave is the product of the wavelength and the frequency, which may be expressed as follows:

$$c = \lambda f \tag{1.3}$$

where c = velocity of propagation, in centimeters per second

λ = wavelength, in centimeters

f = frequency, in cycles per second

1.8   PRESSURE IN A SOUND WAVE

A sound wave consists of pressures above and below the normal undisturbed pressure in the gas (see Secs. 1.3 and 1.4).

The instantaneous sound pressure at a point is the total instantaneous pressure at that point minus the static pressure, the static pressure being the normal atmospheric pressure in the absence of sound.

The effective sound pressure at a point is the root-mean-square value of the instantaneous sound pressure over a complete cycle at that point. The unit is the dyne per square centimeter. The term "effective sound pressure" is frequently shortened to "sound pressure."

The sound pressure in a spherical sound wave falls off inversely as the distance from the sound source.

1.9 PARTICLE DISPLACEMENT AND PARTICLE VELOCITY IN A SOUND WAVE

The passage of a sound wave produces a displacement of the particles or molecules in the gas from the normal position, that is, the position in the absence of a sound wave. The particle displacement in a normal sound wave in speech and music is a very small fraction of a millimeter. For example, in normal conversational speech at a distance of 10 feet from the speaker, the particle amplitude is of the order of one-millionth of an inch. The particle or molecule in the medium oscillates at the frequency of the sound wave. The velocity of a particle or molecule in the process of being displaced at the frequency of the sound wave is termed the particle velocity.

The relation between sound pressure and particle velocity is given by

$$p = \rho c u \qquad (1.4)$$

where p = sound pressure, in dynes per square centimeter
ρ = density of air, in grams per square centimeter
c = velocity of sound, in centimeters per second
u = particle velocity, in centimeters per second

The amplitude or displacement of the particle from its position in the absence of a sound wave is given by

$$d = \frac{u}{2\pi f} \qquad (1.5)$$

where d = particle amplitude, in centimeters
u = particle velocity, in centimeters per second
f = frequency, in cycles per second

1.10 INTENSITY OR POWER IN A SOUND WAVE

From the foregoing sections it is evident that energy is being transmitted in a sound wave. The sound energy transmitted is termed the intensity of a sound wave.

The intensity of a sound field, in a specified direction at a point, is the sound energy transmitted per unit of time in a specified direction through a unit area normal to this direction at the point. The unit is the erg per second per square centimeter.

The intensity, in ergs per second per square centimeter, of a plane sound wave is

$$I = \frac{p^2}{\rho c} = pu = \rho c u^2 \qquad (1.6)$$

where p = sound pressure, in dynes per square centimeter

     u = particle velocity, in centimeters per second

     c = velocity of propagation of sound, in centimeters per second

     ρ = density of the medium, in grams per cubic centimeter

The intensity level, in decibels, of a sound is ten times the logarithm to the base 10 of the ratio of the intensity of this sound to the reference intensity. Decibels will be described in the section which follows.

TABLE 1.1. THE RELATION BETWEEN DECIBELS
AND POWER AND CURRENT OR VOLTAGE RATIOS

Power ratio	Decibels	Current or voltage ratio	Decibels
1	0	1	0
2	3.0	2	6.0
3	4.8	3	9.5
4	6.0	4	12.0
5	7.0	5	14.0
6	7.8	6	15.6
7	8.5	7	16.9
8	9.0	8	18.1
9	9.5	9	19.1
10	10	10	20
100	20	100	40
1,000	30	1,000	60
10,000	40	10,000	80
100,000	50	100,000	100
1,000,000	60	1,000,000	120

1.11 DECIBELS

In acoustics the ranges of intensities, pressures, and particle velocities are so large that it is convenient to use a condensed scale of smaller numbers termed decibels. The abbreviation db is used for the term decibel. The bel is the fundamental division of a logarithmic scale for expressing the ratio of two amounts of power, the number of bels denoting such a ratio being the logarithm to the base 10 of this ratio. The decibel is one-tenth of a bel. For example, with P_1 and P_2 designating two amounts of power and n the number of decibels denoting their ratio, then

$$n = 10 \log_{10} \frac{P_1}{P_2} \text{ decibels} \qquad (1.7)$$

When the conditions are such that ratios of currents or ratios of voltages (or the analogous quantities such as pressures, volume currents, forces, and particle velocities) are the square roots of the corresponding power ratios, the number of decibels by which the corresponding powers differ is expressed by the following formulas:

$$n = 20 \log_{10} \frac{i_1}{i_2} \text{ decibels} \tag{1.8}$$

$$n = 20 \log_{10} \frac{e_1}{e_2} \text{ decibels} \tag{1.9}$$

where i_1/i_2 and e_1/e_2 are the given current and voltage ratios, respectively.

For relation between decibels and power and current or voltage ratios, see Table 1.1.

1.12 DOPPLER EFFECT IN SOUND WAVES

The Doppler effect is the phenomenon evidenced by the change in the observed frequency of a sound wave in a transmission system caused by a time rate of change in the length of the path of travel between the source and the point of observation. The most common example of the Doppler effect is due to the relative motion of the source and observer. Examples are the change in the frequency of the tones emitted by horns and whistles of passing automobiles or locomotives.

When the source and observer are approaching each other, the frequency observed by the listener is higher than the actual frequency of the sound source. If the source and observer are receding from each other, the frequency is lower.

The frequency at the observation point is

$$f_0 = \frac{v - v_0}{v - v_s} f_s \tag{1.10}$$

where v = velocity of sound in the medium
v_0 = velocity of the observer
v_s = velocity of the source
f_s = frequency of the source

All the velocities must be expressed in the same units.

No account is taken of the effect of wind velocity or motion of the medium in Eq. (1.10). In order to bring in the effect of the wind, the velocity v in the medium must be replaced by $v + w$, where w is the wind velocity in the direction in which the sound is traveling. Making this substitution in Eq. (1.10), the result is

$$f_0 = \frac{v + w - v_0}{v + w - v_s} f_s \tag{1.11}$$

Equation (1.11) shows that the wind does not produce any change in pitch unless there is some relative motion of the sound source and the observer.

1.13 REFRACTION OF A SOUND WAVE

Refraction of a sound wave is the variation in the direction of sound transmission due to a spatial variation in the velocity of sound transmission in the medium. Sound is refracted when the velocity of sound varies over the wavefront. A sound wave traveling along the surface of the earth may be bent upward or downward, depending upon the relative velocities of air in the various strata due to a progressive change in temperature. Refraction phenomena are illustrated in Fig. 1.5. The distance over which sound may be heard is greater when the wave is bent downward than when it is bent upward. The first condition usually prevails during the early morning hours, while the latter condition prevails during the day. The refraction loss is that part of the transmission loss due to refraction resulting from nonuniformity of the medium.

SURFACE OF THE EARTH

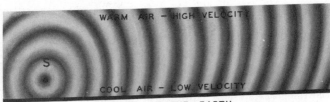

SURFACE OF THE EARTH

FIG. 1.5. Refraction of a sound wave in the atmosphere. S represents a source of sound. A. This figure illustrates a condition of warm air near the surface of the earth and cool air at a distance from the earth. The velocity of sound in the warm air is greater than in the cool air (see Sec. 1.5). Therefore, the sound wave will be bent or refracted upward. B. This figure illustrates cool air near the surface of the earth and warm air at a distance from the earth. The velocity of sound in the warm air is greater than in the cool air. Therefore, the sound wave will be bent or refracted downward. The bending or change in direction of transmission of a sound wave due to a spatial variation in velocity of sound transmission in the medium is termed refraction.

1.14 DIFFRACTION OF A SOUND WAVE

Diffraction is the change in direction of propagation of sound due to the passage of sound around an obstacle. Sound will pass around the corner of a building, over a wall, or through an open window. It is fortunate that the ratio of the wavelength of the sound in the audio-frequency range to the dimension of most obstacles is large; otherwise it would be very inconvenient to conduct our daily tasks. Everyone knows that one does not need to see the source of sound to hear it. In this connection, one also observes that the low frequencies bend round obstacles more easily than do the high frequencies. That is, the larger the ratio of the wavelength to the dimensions of the obstacle, the greater the diffraction. The fundamental characteristics of diffraction phenomena will be illustrated in the examples which follow.

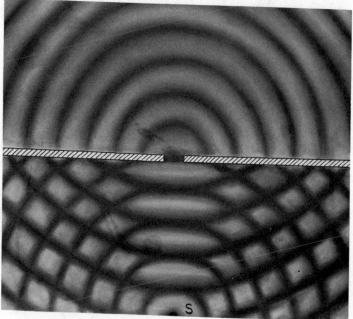

Fig. 1.6. The diffraction of sound. Sound waves emitted by the source S impinge upon the reflecting wall and the small aperture in the wall. The dimensions of the aperture are small compared to the wavelength of the sound. A major portion of the sound wave which impinges upon the wall is reflected. A small portion of the incident sound wave is transmitted through the small aperture. The sound waves which pass through the aperture radiate in all directions in the same way as though the aperture were a source of sound waves. The spreading of sound waves due to the passage through an aperture is termed diffraction.

Figure 1.6 shows the passage of sound through a small aperture in a reflecting wall. The dimensions of the aperture are small compared to the wavelength. It will be seen that the sound which passes through the aperture spreads out in all directions. Of course, the intensity of the diffracted sound is small because the amount of energy transmitted by the aperture is small. Therefore, practically all the incident sound is reflected by the wall.

Figure 1.7 shows the passage of sound through an aperture in a wall in which the dimensions of the aperture are large compared to the wavelength. Here the sound wave, as defined by the aperture, is transmitted through the aperture with no loss in intensity. Furthermore, it will be seen that the transmitted sound beam is geometrically defined by the aperture. The sound waves which impinge upon the wall are reflected.

The diffraction of sound around an obstacle with dimensions which are

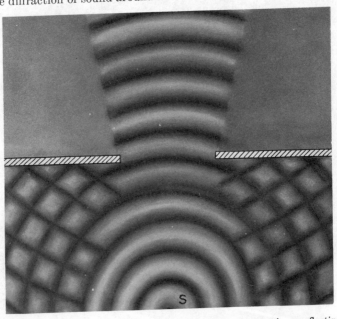

FIG. 1.7. The transmission of sound through a large aperture in a reflecting wall. Sound waves emitted by the source S impinge upon the reflecting wall and the large aperture in the wall. The dimensions of the aperture in the wall are large compared to the wavelength of the sound. Under these conditions, sound is transmitted through the area defined by the aperture with no loss in intensity. Furthermore, the transmitted-sound beam is geometrically defined by the aperture. The sound waves which impinge upon the wall are reflected.

small compared to the wavelength is shown in Fig. 1.8. Under these conditions the sound waves bend around the obstacle, and the sound shadow cast by the obstacle is negligible. Since the dimensions of the obstacle are small compared to the wavelength, the amount of reflected sound is negligible. The diffraction of sound by an obstacle with dimensions large compared to the wavelength is shown in Fig. 1.9. Here the sound intensity back of the obstacle is small. In other words, the obstacle produces an almost perfect sound shadow. Since the dimensions of the obstacle are large compared to the wavelength, practically all the sound which impinges upon the obstacle is reflected.

1.15 REFLECTION OF A SOUND WAVE

If a sound wave in air encounters a large heavy and rigid wall, the sound wave will be reflected backward. The reflected sound wave is the same as that which would be produced by the image of the sound source, as shown in Fig. 1.9. The strength of the image source is the same as that of the original sound source. A sound wave in air will be reflected by any rigid and massive plane object having dimensions which are large com-

Fig. 1.8. The diffraction of sound by a small obstacle. Sound waves emitted by the sound source S impinge upon the small obstacle. The dimensions of the obstacle are small compared to the wavelength of the sound. Under these conditions, the sound waves bend around the obstacle and the sound shadow cast by the obstacle is negligible. The passage of sound around an obstacle is termed diffraction. Since the dimensions of the obstacle are small compared to the wavelength of the sound, the amount of reflected sound is negligible.

pared to the wavelength of the sound, as shown in Fig. 1.9.   If the dimensions of the object are small compared to the wavelength of the impinging sound wave, the reflected sound wave will be negligible, as shown in Fig. 1.8.

The reflection of a sound wave by a plane reflector with the line perpendicular to the plane of the reflector inclined with respect to the line of propagation of the incident sound wave is shown in Fig. 1.10.   The dimensions of the reflector are large compared to the wavelength.   Under these conditions, practically all the incident sound is reflected.   It will be seen that the angle between the direction of propagation of the incident sound and the line perpendicular to the reflector and the angle between the direction of the reflected sound and the line perpendicular to the

FIG. 1.9.   The diffraction of sound by a large obstacle.   The sound waves emitted by sound source S impinge on the large obstacle.   The dimensions of the obstacle are large compared to the wavelength.   Under these conditions, the obstacle produces almost a perfect sound shadow.   Since the dimensions of the obstacle are large compared to the wavelength, practically all the sound which impinges upon the obstacle is reflected.   The wavefront and intensity of the reflected sound wave will be the same as that produced by a sound source of the same strength as the original source located at the image of the original sound source.

reflector are equal. The reflected sound wave is the same as that which would be produced by the image of the sound source.

An echo is a sound which has been reflected or otherwise returned with sufficient magnitude and delay to be perceived in some manner as a sound wave distinct from that directly transmitted. A multiple echo is a succession of separately distinguishable echoes from a single source. A flutter echo is a rapid succession of reflected pulses resulting from a single initial pulse.

1.16 ABSORPTION OF A SOUND WAVE

The absorption of sound is the process by which sound energy is diminished in passing through a medium or in striking a surface. In the

FIG. 1.10. The reflection of a sound wave by a plane reflector inclined at an angle less than 90 degrees with respect to the direction of propagation of the incident sound. The sound waves emitted by the sound source S impinge upon the reflector. The dimensions of the reflector are large compared to the wavelength. Under these conditions, practically all the sound which impinges upon the reflector is reflected. It will be seen that the angle a_I, between the direction of propagation of the incident sound wave and the line perpendicular to the plane of the reflector, is equal to the angle a_R, between the direction of propagation of the reflected sound wave and the line perpendicular to the plane of the reflector. Since the dimensions of the reflector are large compared to the wavelength, the reflector produces practically a perfect sound shadow.

absorption of sound, the mechanism is usually the conversion of sound into heat. In the case of a porous substance, used for the absorption of sound, the transfer of sound energy into heat energy is occasioned by the transmission of sound through narrow passages. The particle velocity or the motion of the air through the narrow passages introduces a slipping of adjacent layers of air. Owing to the viscosity of air, this slippage produces heat. It is a sort of internal friction in the air. Rugs, drapes, and upholstered furniture are good sound absorbers because of the innumerable small passages in the material. In sound-absorbing materials used in the absorption of sound, the dimensions and length of the passages are designed so that the maximum absorption of sound is produced. Absorption loss is that part of the transmission loss due to dissipation or conversion of sound energy into other forms of energy (usually heat) either within the medium or attendant upon a reflection.

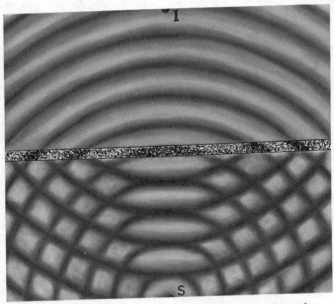

Fig. 1.11. The absorption, reflection, and transmission of a sound wave by a partially transmitting, partially reflecting, and partially absorbing wall. A sound wave emitted by the sound source S impinges upon the wall. A part of the incident sound wave is reflected by the wall. The wavefront of the reflected sound wave is the same as that of a sound source at the image I of the original sound source. A part of the incident sound wave is absorbed by the wall. A part of the incident sound wave is transmitted by the wall.

1.17 TRANSMISSION OF A SOUND WAVE

Sound waves are transmitted in systems or mediums adapted for the transmission of sound.

In practically all reflectors or sound absorbers, some of the sound energy is transmitted through the reflecting or absorbing wall. The three mechanisms, namely, reflection, absorption, and transmission take place in all walls or partitions or boundaries. These three mechanisms are illustrated in Fig. 1.11. In this example, a part of the incident sound energy is reflected, a part is absorbed by the wall, and a part is transmitted by the wall. Transmission loss is a general term used to denote a decrease in power in transmission from one point to another. Transmission loss is usually expressed in decibels.

1.18 STATIONARY SOUND WAVES

Stationary waves result from the interference of two equal wave trains moving in opposite directions and are characterized by the existence in the medium of certain points, lines, or surfaces at which the amplitude of vibration is zero. These points, lines, or surfaces of zero amplitude are termed nodes. Antinodes are the points, lines, or surfaces of a stationary-

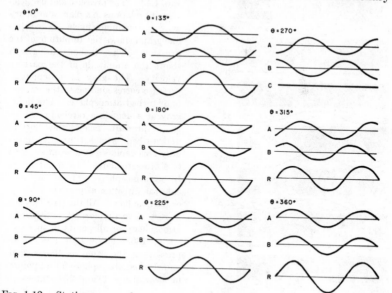

FIG. 1.12. Stationary sound wave R produced by two sound waves A and B of equal amplitude traveling in opposite directions. Sound wave A moves from right to left, and sound wave B moves from left to right. Sound wave R is the resultant stationary sound wave.

wave system which have maximum amplitude. The amplitudes of vibration referred to may be either pressure or particle-velocity amplitudes. A simple stationary-wave system is produced by two equal plane waves moving in opposite directions, as shown in Fig. 1.12. The condensations and rarefactions for the simple standing-wave system are shown in Fig. 1.13. It will be seen that the condensations or rarefactions do not change position in space, which accounts for the term stationary wave. Referring to Fig. 1.13, it will be seen that the maximum amplitudes of the pressure and particle velocity are displaced by an angle of 90 degrees in space and 90 degrees in time. In the case of a plane progressive wave, the phase of the pressure and particle velocity coincide in both time and space (see Sec. 1.3 and Fig. 1.2).

The stationary-wave system described above is of the simplest type. This type of stationary-wave system may be produced by the reflection of a plane wave by a large highly reflecting wall normal to the direction of

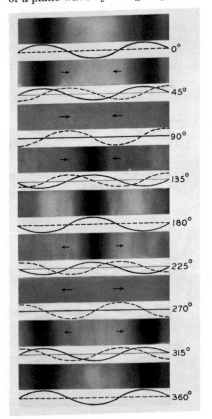

Fig. 1.13. The pressure and particle-velocity relations in a plane stationary wave system of one cycle. The solid-line graph shows the amplitude of the pressure. The dashed-line graph shows the amplitude of the particle velocity. The gray area depicts a pressure corresponding to the normal undisturbed atmosphere. The light-gray area depicts a rarefaction, that is, a pressure below the normal undisturbed atmosphere. A dark-gray area depicts a condensation, that is, a pressure above the normal undisturbed atmosphere. The arrows indicate the direction of particle velocity or particle flow. All the particles are at rest, as in the normal undisturbed atmosphere for all points in space at $t = 0$, 180, and 360 degrees. The pressure is the same as the normal undisturbed atmosphere for all points in space at $t = 90$ and 270 degrees.

propagation. Complex standing-wave systems are produced when a sound source operates in a room with reflecting or partially reflecting walls. The term standing waves is used to designate sound waves having a fixed distribution in space which is the result of interference of progressive sound waves of the same frequency and kind. Such sound waves are characterized by the existence of nodes or partial nodes and antinodes that are fixed in space.

1.19 DISPERSION OF A SOUND WAVE

Acoustical dispersion is the separation of a complex sound wave into its various frequency components, due to the variation with frequency, or its equivalent, of the wave velocity of the medium. A complex sound wave may be separated into its various components by means of an acoustic prism in which the action is similar to that of a light prism.

1.20 EQUATIONS OF SOUND WAVES

A. Plane Sound Waves

The equation for the pressure p, in dynes per square centimeters, in a simple harmonic plane sound wave traveling in the positive x direction is given by

$$p = akc^2\rho \sin k(ct - x) \tag{1.12}$$

where a = amplitude of the particle displacement, in centimeters
$k = 2\pi/\lambda$
λ = wavelength, in centimeters
c = velocity of sound, in centimeters per second
ρ = density of air, in grams per cubic centimeter
t = time, in seconds
x = distance of the observation point along the x axis from $x = 0$, in centimeters

The equation for the particle velocity u, in centimeters per second, in a simple plane sound wave traveling in the positive x direction is given by

$$u = akc \sin k(ct - x) \tag{1.13}$$

Comparing Eqs. (1.12) and (1.13), it will be seen that the pressure and particle velocity in a plane sound wave are in phase (Fig. 1.2).

The ratio of the pressure to the particle velocity in a plane sound wave is

$$\frac{p}{u} = \rho c \tag{1.14}$$

The quantity ρc is the characteristic acoustical resistance of the medium.

B. Spherical Sound Waves

The equation for the pressure p, in dynes per square centimeter, in a spherical sound wave generated by a small sound source is given by

$$p = \frac{S\rho ck}{4\pi r} \sin k(ct - r) \qquad (1.15)$$

where S = strength source, that is, the maximum rate of fluid emission of the small source, in cubic centimeters per second

r = distance from the origin, in centimeters, and all the other quantities as defined in Eq. (1.12)

Equation (1.15) shows that the pressure varies inversely as the distance from the source.

The equation for the particle velocity u, in centimeters per second, in a spherical sound wave generated by a small sound source is given by

$$u = -\frac{Sk}{4\pi r}\left[\frac{1}{kr}\cos k(ct - r) - \sin k(ct - r)\right] \qquad (1.16)$$

Equation (1.16) shows that the particle velocity in a spherical sound wave is an inverse function of r and r^2. The pressure and particle velocity are not in phase in a spherical sound wave save for large distances from the sound source when the spherical wave becomes a plane wave.

C. Stationary Sound Waves

A simple standing-wave system consists of two equal plane waves traveling in opposite directions (Sec. 1.18).

The pressure p, in dynes per square centimeter, in a stationary plane sound-wave system is given by

$$p = 2akc^2\rho[\sin (kct) \cos (kx)] \qquad (1.17)$$

The particle velocity u, in centimeters per second, in a stationary plane sound-wave system is given by

$$u = 2akc\left[\sin\left(kct - \frac{\pi}{2}\right)\cos\left(kx - \frac{\pi}{2}\right)\right] \qquad (1.18)$$

Equations (1.17) and (1.18) show that the maxima of the particle velocity and pressure are separated by a quarter wavelength. The maxima of the pressure and particle velocity differ by 90 degrees in time phase (Figs. 1.12 and 1.13).

Musical Terminology

2.1 INTRODUCTION

Music can be memorized and passed from one person to another by direct sound transmission. Folk, spiritual, and cowboy songs are examples of this form of dissemination of music. However, the direct method is not satisfactory for many obvious reasons. Therefore, recording the music on paper for all time has become the established method. Conventional musical notation employs positioned symbols to denote frequency, duration, quality, intensity, and other factors relating to a tone. It is the purpose of this chapter to outline the fundamentals of musical notation and the correlation between the symbolic notation and the physical characteristics of a tone.

2.2 PITCH NOTATION

Pitch is that attribute of auditory sensation in terms of which sounds may be ordered on a scale extending from low to high, such as a musical scale. Pitch is primarily dependent upon the frequency of the sound stimulus.

The average ear can distinguish 1,400 discrete frequencies. However, in the equally tempered scale covering the hearing range of 16 to 16,000 cycles there are only 120 discrete tones. In other words, musical tones may be said to be quantized, in that only certain discrete frequencies are allowed and others ruled out. One of the principal reasons for the definite and unique frequencies of musical tones is that musical instruments are inherently resonant systems and therefore respond to only certain frequencies. Except in certain instruments, as for example, the trombone and the violin family, the resonant frequencies are fixed and cannot be altered at will. The discrete frequency characteristics of the tones of Western music simplify the means for designating the fundamental frequencies of tones because the number of tones is relatively small.

The staff in music is composed of five lines, as shown in Fig. 2.1. The pitch of a tone is indicated by placing notes ○, ♩, ♪ etc., upon the lines and the spaces between the lines. The definite pitch of a set of lines and spaces is designated by a symbol termed the clef which is placed at the left end of the staff, as shown in Fig. 2.2. The most common clefs are the treble or G clef and the bass or F clef, as shown in Fig. 2.2. The notes which are placed upon the lines and spaces of the clef are designated by letters of the alphabet from A to G. The interval in pitch between two

Fig. 2.1. The musical staff.

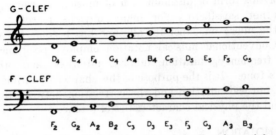

Fig. 2.2. The G, or treble, clef and the F, or bass, clef.

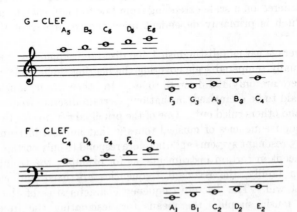

Fig. 2.3. Leger lines—the lines above and below the staff—for the treble and bass clefs.

similar letters is termed an octave. An octave is the interval between two sounds having a basic frequency ratio of 2. By extension, the octave is the interval between any two frequencies having a ratio of 2:1. The pitch interval between adjacent notes or letters on the clef, that is, between a note on a line and a note on the adjacent space, is a half or a whole tone in the equally tempered scale. Pitches above and below the staff are indicated by notes written upon and between short lines termed leger lines, as shown in Fig. 2.3. The number of leger lines may be extended without limit. The sign 8va above the staff indicates that all tones are to be sounded an octave higher than the notes would indicate, as shown in Fig. 2.4. When it is written below the staff, the sign 8va indicates that all the tones are to be sounded an octave lower. The movable, or C, clef is used for two purposes, namely, in music written for certain instruments of extended range, as for example, the cello, viola, and bassoon, and for indicating the soprano, alto, or tenor parts. This latter usage is gradually disappearing. There are three different symbols

used for the C clef, as shown in Fig. 2.5. The C clef is placed on the middle C line. The various positions of the C clef for soprano, alto, and tenor parts are shown in Fig. 2.6. It will be seen that the position of the C clef always corresponds to middle C.

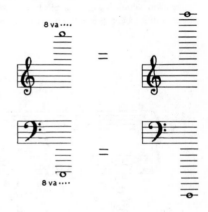

FIG. 2.4. The sign 8va above the staff indicates that all tones are sounded an octave higher. The sign 8va below the staff indicates that all tones are sounded an octave lower.

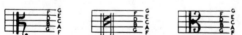

FIG. 2.5. Three symbols used to indicate the C, or movable, clef.

FIG. 2.6. The positions of the C-clef symbol for the soprano, alto, and tenor clefs.

A note may be moved up in pitch a semitone, or half step, by means of a designation termed a sharp, as shown in Fig. 2.7. A note may be moved down a semitone, or half step, by means of a designation termed a flat, as shown in Fig. 2.7. A natural designation nullifies a sharp or flat and restores the note to normal, as shown in Fig. 2.7. A note may be moved up a whole tone, or whole step, by a double sharp. A double sharp is indicated either by ## or ×. A note may be moved down a whole tone,

FIG. 2.7. Sharp designation which raises the pitch of a note by a semitone. Flat designation which lowers the pitch of a note by a semitone. Natural designation which nullifies a sharp or a flat and restores a note to normal.

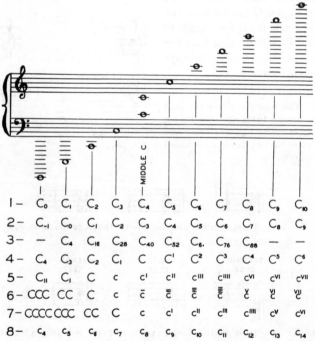

FIG. 2.8. Eight systems used for identifying the tones employed in music.

or whole step, by a double flat.
A double flat is indicated by $\flat\flat$.

Various systems have arisen for identifying the tones employed in music without the aid of the staff. These abbreviations should possess some significance. In addition, they should be easy to use and identify. The suggested and most logical standard,[1] in the opinion of the author, is given in line 1 of Fig. 2.8. The reference frequency C_0 is 16.352 cycles. The lowest pitch audible to the average ear is produced by a frequency in the neighborhood of 16 cycles. It has become the custom to consider C as the point to count complete octaves. Therefore it seems logical to use 16.352 cycles as the reference point and designate this by C_0. Line 2, Fig. 2.8, has been used extensively in the literature in this country. In line 3, Fig. 2.8, the subscript is derived from the corresponding key on the piano. Two digits are used, and it is difficult to identify a tone. Lines 4 and 7, Fig. 2.8, are middle C reference systems. Line 8, Fig. 2.8, is a system of subscripts with C_0 as approximately one cycle per second. Lines 5 and 6, Fig. 2.8, are complex combinations of upper and lower case letters and subscripts. From the foregoing it seems that line 1, Fig. 2.8, is the most logical and the simplest.

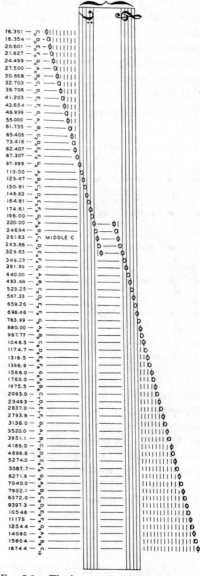

FIG. 2.9.   The frequencies of the notes in the scale of equal temperament in the scale of C from 16 to 16,000 cycles.

[1] Young, Robert W., *Jour. Acoust. Soc. Amer.*, Vol. 11, No. 1, p. 134, 1939.

The frequencies of the notes in the equally tempered scale in the key of C covering the frequency range of 16 to 16,000 cycles are shown in Fig. 2.9. This frequency range corresponds to the frequency range of normal hearing. The terms "key of C" and "equally tempered scale" will be considered in detail in later sections. The notation used to designate the notes is that of line 1 of Fig. 2.8. This notation will be used throughout this book.

FIG. 2.10. Designations for notes of various values.

2.3 DURATION NOTATION

The duration of a tone is the length of time assigned to it in the musical composition or rendition. The durations of tones are indicated by the symbols, termed notes, in Fig. 2.10. The pitch of a tone to be sounded is shown by the position on the note on the staff, as outlined in Sec. 2.2. The kind of note indicates that a certain tone is to be sounded for a certain relative length of time. The magnitude of a particular note is not rigidly fixed and may vary from composition to composition. However, in a particular composition the durations of the tones are proportional to the magnitude of the note.

In musical notation a vertical line or bar is drawn across the staff. The time interval between two vertical bars on a staff is termed a measure. Unfortunately, it has become the custom of some musicians to use bar instead of measure to denote the time interval between two vertical lines on the staff. In general, the time intervals of all measures in a composition are equal. For example, if a whole note constitutes a measure, then a measure, will require two half notes, or four quarter notes, etc., as depicted in Fig. 2.11. A double bar consists of two vertical lines across the staff. The double bar marks the end of a division, movement, or entire composition. A double bar is shown at the end of the staff of Fig. 2.11.

FIG. 2.11. A staff divided into measures.

Under certain conditions, periods of silence may be desired. These are indicated by symbols termed rests, as shown in Fig. 2.12. The duration of a whole rest is equal to a whole note, a half rest is equal to a half note, etc.

A further division of the duration notes and rests is made by the addition of a dot, as shown in Figs. 2.13 and 2.14. The dot increases the duration of a note or rest by one-half.

The duration of a tone represented by a note or rest of a certain denomination does not indicate any absolute time interval but is merely relative and may vary at will. For example, some selections indicate the metronome[2] setting for the quarter note. The setting of the movable bob indicates the number of clicks per minute. This may be a 100, in which case the duration of a quarter note equals $\frac{1}{100}$ of a minute. For other selections, another metronome setting may be indicated.

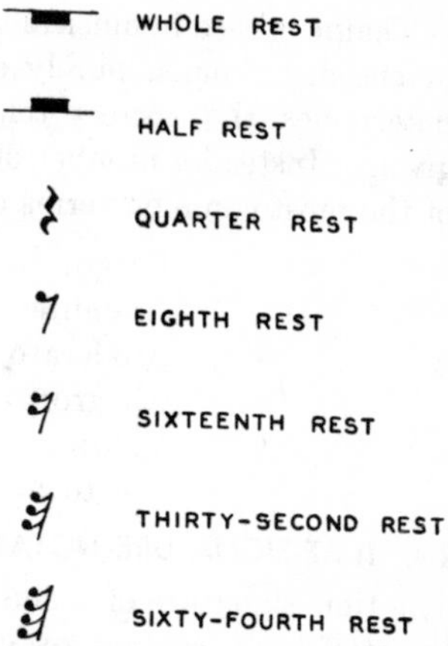

Fig. 2.12. Designations for rests of various values.

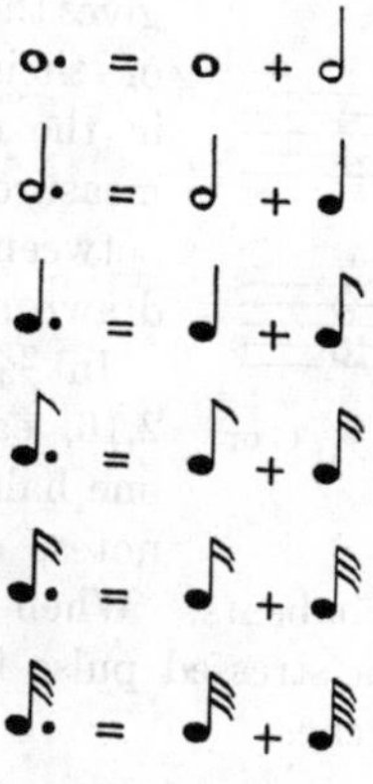

Fig. 2.13. A dot increases the value of a note by 50 per cent.

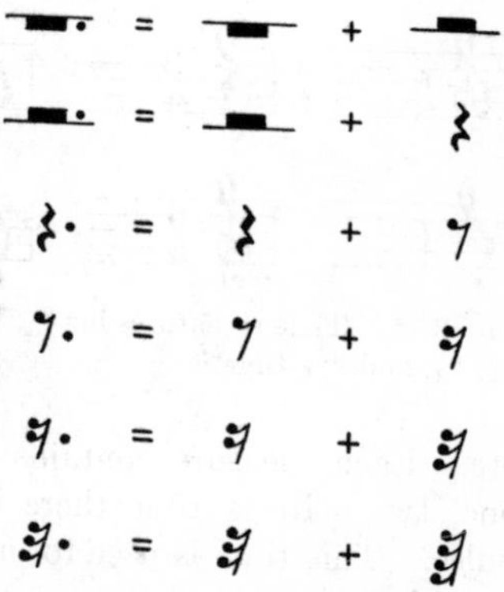

Fig. 2.14. A dot increases the value of a rest by 50 per cent.

[2] A metronome is a mechanical device consisting of a pendulum activated by a clock-type mechanism driven by a spring motor. An audible tick is produced by the pendulum at the extremities of the excursion. The movable bob on the pendulum is used to alter the time interval between ticks. The pendulum is graduated for various time intervals (see Sec. 5.6H and Fig. 5.84). The metronome indication on a musical selection is given by MM followed by a note and a number. The number is the metronome setting for the movable bob. When the pendulum moves, each tick corresponds to a beat.

Tempo is used in musical terminology as meaning the rate of movement or speed. From a purely scientific standpoint the metronome setting determines the correct tempo. The metronome setting is often not given. Instead a number of terms are used to designate tempo. A few of the most common terms describing tempo are as follows:

Largo: Slow tempo.
Andante: Moderately slow tempo.
Moderato: Moderate tempo.
Allegro: Moderately quick tempo.
Vivo: Rapid tempo.
Presto: Very rapid tempo.

2.4 TIME-SIGNATURE NOTATION (RHYTHM)

The time signature of a musical selection is indicated at the beginning of the staff by a fraction, as shown in Fig. 2.15. The most common time signatures are 2/4, 3/4, 4/4, and 6/8. The denominator stands for the unit of measure, that is, the note used to denote a pulse. The numerator gives the number of these units or their equivalent included in the metrical division of a measure, that is, the space between two vertical lines drawn across the staff.

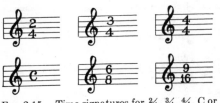

FIG. 2.15. Time signatures for 2/4, 3/4, 4/4, C or 4/4, 6/8, and 9/16 time.

In 2/4 time, shown in Fig. 2.16, each measure contains one half note, or two quarter notes, or four eighth notes, etc. Each measure contains two beats. When playing, the count is one, two. In 2/4 time there is a stressed pulse followed by a relaxed pulse. This time is used for marches.

FIG. 2.16. The notes and the beats per measure for 2/4 time.

FIG. 2.17. The notes and the beats per measure for 3/4 time.

In 3/4 time, shown in Fig. 2.17, each measure is equivalent to three quarter notes. Each measure contains a half note and a quarter note, or three quarter notes, etc. Each measure contains three beats. When playing, the count is one, two, three. In 3/4 time, there is one stressed pulse followed by two relaxed pulses. This time is used for waltzes.

In $\frac{4}{4}$ or common time, shown in Fig. 2.18, each measure is equivalent to four quarter notes. Each measure contains one whole note, or two half notes, or four quarter notes, etc. Each measure contains four beats. The count is one, two, three, four. In $\frac{4}{4}$ time there is one stressed pulse followed by three relaxed pulses. This time is used for dances.

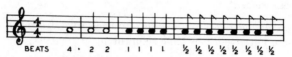

BEATS   4 · 2 2   I I I L   ½ ½ ½ ½ ½ ½ ½ ½

FIG. 2.18. The notes and the beats per measure for common, or $\frac{4}{4}$, time.

In $\frac{6}{8}$ time each measure contains six eighth notes or the equivalent. Each measure contains six beats. In playing, the count is one, two, three, four, five, six. In $\frac{6}{8}$ time the stressed pulses are on one and four.

In listening to music the mind will arrange the regular repetition of sounds into groups of stressed and relaxed pulses. These are termed meters. The meter is given by the numerator of the time signature, and the most common are 2, 3, 4, 6, 9, and 12. A measure contains a certain number of beats or pulses according to the meter. The meters may be classified in terms of the numerator of the time signature as follows:

1. Duple meter. There are two beats to each measure. The first beat is stressed, and the second is relaxed. Examples are $\frac{2}{2}$ and $\frac{2}{4}$ time signatures.

2. Triple meter. There are three beats to each measure. The first one is stressed, and the second and third are relaxed. Examples are $\frac{3}{8}$, $\frac{3}{4}$, and $\frac{3}{2}$ time signatures.

3. Quadruple meter. There are four beats to each measure. The first one is stressed, and the remainder are relaxed. Sometimes there is a secondary stress on the third beat. Examples are $\frac{4}{2}$, $\frac{4}{4}$, and $\frac{4}{8}$ time signatures.

4. Sextuple meter. There are six beats to each measure. The first and fourth are stressed, and the remainder are relaxed. An example is the $\frac{6}{8}$ time signature.

Rhythm is the recurrence of accents in equal intervals of time. The rhythm is determined by the manner in which accents are related to the kind of notes and rests. The numerator of the time signature determines the aspect of the rhythmic pattern. Therefore, it will be seen that the denominator indicates the unit of time measurement. This may be illustrated as follows:

1. For the time signatures, $\frac{2}{2}$, $\frac{3}{2}$, and $\frac{4}{2}$, the half note $\jmath$ is the unit of measurement.

2. For the time signatures $\frac{2}{4}$, $\frac{3}{4}$, and $\frac{4}{4}$, the quarter note $\quarternote$ is the unit of measurement.

3. For the signatures $\frac{3}{8}$, $\frac{4}{8}$, $\frac{6}{8}$, $\frac{9}{8}$, and $\frac{12}{8}$, the eighth note $\eighthnote$ is the unit of measurement.

4. For the time signatures $\frac{9}{16}$ and $\frac{12}{16}$, the sixteenth note $\sixteenthnote$ is the unit of measurement.

The time may change from measure to measure. However, with each change there must be a new time signature.

2.5 KEY-SIGNATURE NOTATION

The keynote is the sound or note with which any given scale begins. The tonic is the keynote of the scale, regardless of whether it is a major or minor scale.

In the case of simple and short pieces, music can be played in a single key. However, in most cases the music begins in one key and shifts temporarily to some other key or may pass into a number of keys one after another. The production of music from one key to another in a smooth manner is known as modulation. The difference in quality between keys is one of the powerful tools for the embellishment of music.

The key signature of a composition is indicated by the number and arrangement of sharps and flats following the clef sign at the beginning of each staff or sometimes only at the beginning of the selection. The most common key signatures for the various major and minor keys are shown in Figs. 2.19 and 2.20. The significance of the various keys will be described in Chap. 3 on Scales.

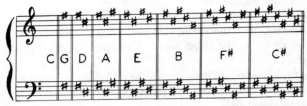

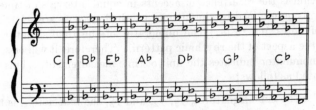

FIG. 2.19.   The key signatures for the various major keys.

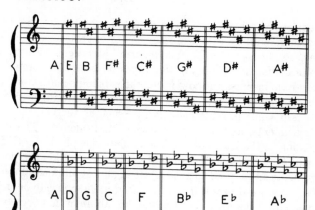

Fig. 2.20. The key signatures for the various minor keys.

2.6 LOUDNESS NOTATION

Loudness is the magnitude of the auditory sensation produced by a sound stimulus. Loudness is primarily dependent upon the intensity of the sound stimulus.

Loudness and the variation of loudness are powerful tools of artistic embellishment. Loudness and the variation of loudness involve the intensity or sound power or variations in intensity or sound power. The absolute value of the intensity of sound can be measured by means of a sound-level meter (Sec. 8.4). The intensity is also given by the volume indicator in recording a phonograph record or a sound motion picture and in the transmission of a broadcast (Secs. 9.7, 9.8, and 9.9). However, the conductor or musician must depend upon his sense of loudness to obtain the proper intensity and range of intensities. The loudness even more than the tempo depends upon the musician or conductor. As in the case of tempo, the expressions referring to loudness are relative and not absolute. Therefore, it is impossible to indicate how loud or soft a selection or a portion of a selection should be.

The common loudness notations and abbreviations are as follows:

Pianisissimo (*ppp*): As soft as possible.
Pianissimo (*pp*): Very soft.
Piano (*p*): Soft.
Mezzo piano (*mp*): Half soft.

Mezzo forte (*mf*): Half loud.
Forte (*f*): Loud.
Fortissimo (*ff*): Very loud.
Fortisissimo (*fff*): As loud as possible.

The variation in loudness is also depicted by signs as well as words. An increase in loudness is indicated by the term crescendo, the abbreviation cres, or the sign $\diagdown\hspace{-0.5em}\diagup$. These terms and signs mean a gradual increase in loudness. There are two types of crescendo, namely, one in which the tone increases in power while it is being sounded and another in which succeeding tones are each sounded louder than the preceding one. The first is possible in some instruments, such as the voice, the bowed-string instruments, and most of the wind instruments. The first kind is not possible in the struck-string instruments, the plucked-string instruments, and the percussion instruments. The second kind is possible in all instruments and is used to good effect in almost all musical renditions. A decrescendo is a decrease in loudness and is the opposite of crescendo. It is indicated by the word decrescendo, the abbreviation decresc, or the sign $\diagup\hspace{-0.5em}\diagdown$. The explanations given for the crescendo apply to the decrescendo in reverse.

2.7 DEFINITIONS OF MUSICAL TERMS

As an aid in the study of this book it appears desirable to provide specific definitions of some musical terms implicitly but not specifically defined in this book.

Tone. A tone is a sound sensation having pitch or a sound wave capable of exciting an auditory sensation having pitch. A tone is a larger successive interval in the major scale.

Note. A note is a conventional sign used to indicate the pitch or the duration or both of a tone sensation. It is also the sensation itself or the vibration causing the sensation. The word serves when no distinction is desired between the symbol, the sensation, and the physical stimulus.

Simple Tone. A simple tone is a sound sensation characterized by its singleness of pitch.

Complex Tone. A complex tone is a sound sensation characterized by more than one pitch.

Fundamental Tone. The fundamental tone is the component tone of the lowest pitch in a complex tone. The fundamental tone is the component of the lowest frequency in a complex tone.

Overtone. An overtone is a component of a complex tone having a pitch higher than the fundamental. An overtone is a physical component of a complex sound having a frequency higher than that of the fundamental tone.

Partial. A partial is a component of a sound sensation which may be distinguished as a simple sound that cannot be further analyzed by the ear and which contributes to the character of the complex tone or complex sound. A partial is a physical component of a complex tone.

Fundamental Frequency. The fundamental frequency is the frequency component of the lowest frequency in a complex sound.

Harmonic. A harmonic is a partial or overtone whose frequency is an integral multiple of the fundamental tone or fundamental frequency.

Subharmonic. A subharmonic is an integral submultiple of the fundamental frequency of the sound to which it is related.

CHAPTER THREE

Musical Scales

3.1 INTRODUCTION

In general, musical sounds are smooth, regular, pleasant, and harmonious. Certain tones when sounded together or immediately following one another produce a pleasing effect, while other combinations of tones lead to an unpleasant sensation. Since musical instruments are capable of producing pleasing sounds, it would appear that these instruments form the basis for musical scales. The vibrations of string instruments give rise to overtones which are multiples of the fundamental. The tones of this series, including the fundamental, are in the ratio 1, 2, 3, 4, 5, 6, 7 . . . . In mathematics it is termed the harmonic series. It can be represented on the musical staff as shown in Fig. 3.1. The notes shown on the musical staff are familiar to all musicians. It will be noted that the series is composed of the following frequency ratios: 2:1, 3:2, 4:3, 5:3, 5:4, 6:5, 8:5, etc. The interval of two frequencies having the ratio of 2:1 is termed an octave. An interval between two sounds is their spacing in pitch or frequency. The frequency interval may be described by the ratio of the frequencies or by the logarithm of this ratio. It has

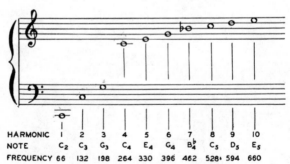

HARMONIC	1	2	3	4	5	6	7	8	9	10
NOTE	C_2	C_3	G_3	C_4	E_4	G_4	$B_4^\flat$	C_5	D_5	E_5
FREQUENCY	66	132	198	264	330	396	462	528·	594	660

Fig. 3.1. Harmonic series based on C_2 as the fundamental. The frequencies shown are taken from the just scale.

been found that the octave produces a pleasant sensation. Another combination of two tones which is next to the octave in pleasantness possesses a ratio of 3:2. These are followed in the order of pleasantness by the ratios 4:3, 5:4, 6:5, 8:5, and 5:3. It is an established fact that the most pleasing combination of two tones is one in which the frequency ratio is expressible by two integers neither of which is large.

From pleasing combinations of tones it is possible to construct a scale. A scale is a series of sounds (symbols, sensation, or stimuli) arranged from low to high frequencies by definite frequency intervals suitable for musical purposes. Intonation is the process of adjusting or selecting the tones of a musical scale with respect to frequency. A scale employing only intervals found in the harmonic series is termed a scale of just intonation. The scale of just intonation presents certain practical difficulties and has been supplanted by the scale of equal temperament in which an octave is divided into twelve equal intervals. It is the purpose of this chapter to consider both the scale of just intonation and the scale of equal temperament.

3.2 SCALE OF JUST INTONATION

A scale of just intonation is a musical scale employing frequency intervals represented by the ratios of the smaller integers of the harmonic series. The complete scale of just intonation is depicted in Table 3.1. The intervals are given in both ratios from the starting point and in cents from the

TABLE 3.1. SCALE OF JUST INTONATION

Interval	Frequency ratio from starting point	Cents from starting point
Unison	1:1	
Semitone	16:15	111.731
Minor tone	10:9	182.404
Major tone	9:8	203.910
Minor third	6:5	315.641
Major third	5:4	386.314
Perfect fourth	4:3	498.045
Augmented fourth	45:32	590.224
Diminished fifth	64:45	609.777
Perfect fifth	3:2	701.955
Minor sixth	8:5	813.687
Major sixth	5:3	884.359
Harmonic minor seventh	7:4	968.826
Grave minor seventh	16:9	996.091
Minor seventh	9:5	1,017.597
Major seventh	15:8	1,088.269
Octave	2:1	1,200.000

starting point. A cent is the interval between two tones whose basic frequency ratio is the twelve-hundredth root of 2. The interval in cents is 1,200 times the logarithm on the base 2 of the frequency ratio. 1,200 cents = 12 semitones = 1 octave.

There are many possible combinations of intervals which constitute a scale of just intonation. However, there are two principal scales of just intonation, namely, the major and the minor. The considerations in this book will be confined to these two because it suffices for purposes of illustrating the properties of just scales.

A major scale of just intonation using C as the tonic constructed from the ratios of Table 3.1 is shown in Table 3.2. An examination of Table 3.2 shows that the ratios of

$$C, E, \text{ and } G$$
$$F, A, \text{ and } C$$
$$G, B, \text{ and } D$$

are all in the ratio $4:5:6$. This combination of tones is termed a major triad. The major scale of just intonation C, D, E, F, G, A, B, with the ratios of Table 3.1 is constructed from these triads.

The intervals between the tones in Table 3.2 are obtained by taking the ratio between adjacent frequency values. The intervals of the major scale are shown in Table 3.3. The major scale is composed of three different intervals, namely, major tones, minor tones, and semitones. The interval $9/8$ is termed a major tone, the interval $10/9$ is termed a minor tone, and the interval $16/15$ is termed a semitone.

The tonic is the keynote of the scale. The key takes its name from the tonic or the starting point. When the key is changed, the notes are related to a new tonic. The intervals in the major scale of just intonation for any keynote are shown in Table 3.4. The keynote in this table is 1.

TABLE 3.2. FREQUENCY RATIOS OF THE TONES IN TERMS OF THE TONIC C IN THE MAJOR SCALE OF JUST INTONATION

C	D	E	F	G	A	B	C
f	$\frac{9}{8}f$	$\frac{5}{4}f$	$\frac{4}{3}f$	$\frac{3}{2}f$	$\frac{5}{3}f$	$\frac{15}{8}f$	$2f$

TABLE 3.3. FREQUENCY RATIOS OF ADJACENT TONES IN THE MAJOR SCALE OF JUST INTONATION WITH THE TONIC C

C		D		E		F		G		A		B		C
	$9/8$		$10/9$		$16/15$		$9/8$		$10/9$		$9/8$		$16/15$	

The major-tone, the minor-tone, and the semitone intervals of the major scale of just intonation for various keynotes are shown in Table 3.5. It will be seen that the pattern for all the keys is the same as that of Table 3.4, with the keynote of Table 3.5 being the numeral 1 of Table 3.4. Furthermore, it will be observed that there is a continuous shift in this pattern from key to key of Table 3.5.

The frequencies of the tones in one octave of the major scale of just

TABLE 3.4. THE INTERVALS BETWEEN THE TONES IN THE MAJOR SCALE OF JUST INTONATION

(The sign ‖ indicates an interval of a major tone. The sign ∥ indicates an interval of a minor tone. The sign | indicates an interval of a semitone. The keynote is 1.)

1 ‖ 2 ∥ 3 | 4 ‖ 5 ∥ 6 ‖ 7 | 8

TABLE 3.5. THE TONES AND INTERVALS BETWEEN THE TONES IN THE MAJOR SCALE OF JUST INTONATION FOR VARIOUS KEYNOTES

(The sign ‖ indicates an interval of a major tone. The sign ∥ indicates an interval of a minor tone. The sign | indicates an interval of a semitone.)

KEY	TONES
C	C ‖ D ∥ E | F ‖ G ∥ A ‖ B | C
D	C♯ | D ‖ E ∥ F♯ | G ‖ A ∥ B ‖ C♯
E	C♯ ‖ D♯ | E ‖ F♯ ∥ G♯ | A ‖ B ∥ C♯
F♯	C♯ ∥ D♯ ‖ E♯ | F♯ ‖ G♯ ∥ A♯ | B ‖ C♯
G	C ‖ D ∥ E ‖ F♯ | G ‖ A ∥ B | C
A	C♯ | D ‖ E ∥ F♯ ‖ G♯ | A ‖ B ∥ C♯
B	C♯ ∥ D♯ | E ‖ F♯ ∥ G♯ ‖ A♯ | B ‖ C♯
C♯	C♯ ‖ D♯ ∥ E♯ | F♯ ‖ G♯ ∥ A♯ ‖ B♯ | C♯
D♭	C | D♭ ‖ E♭ ∥ F | G♭ ‖ A♭ ∥ B♭ ‖ C
E♭	C ‖ D | E♭ ‖ F ∥ G | A♭ ‖ B♭ ∥ C
F	C ‖ D ∥ E | F ‖ G ∥ A | B♭ ‖ C
G♭	C♭ ‖ D♭ ∥ E♭ ‖ F | G♭ ‖ A♭ ∥ B♭ | C♭
A♭	C | D♭ ‖ E♭ ∥ F ‖ G | A♭ ‖ B♭ ∥ C
B♭	C ∥ D | E♭ ‖ F ∥ G ‖ A | B♭ ‖ C
C♭	C♭ ‖ D♭ ∥ E♭ | F♭ ‖ G♭ ∥ A♭ ‖ B♭ | C♭

TABLE 3.6. FREQUENCIES OF THE TONES IN ONE OCTAVE FOR DIFFERENT KEYNOTES IN THE MAJOR SCALE OF JUST INTONATION

NOTE	C	D	E	F#	G	A	B	C#	Db	Eb	F	Gb	Ab	Bb	Cb
Cb4												247.2			247.2
C4	264.0				264.0				260.7	260.7	264.0		260.7	264.0	
C#4		278.4	275.0	278.4		275.0	278.4	278.4							
Db4									278.1			278.1	278.1		278.1
D4	297.0	297.0			297.0	293.3				293.3	293.3			293.3	
D#4			309.4	309.4			309.4	313.2							
Eb4									312.9	312.9		309.0	312.9	312.9	309.0
E4	330.0	334.1	330.0		330.0	330.0	330.0				330.0				
E#4				348.0				348.0							
Fb4															329.6
F4	352.0								347.6	352.0	352.0	347.6	347.6	352.0	
F#4		371.2	371.2	371.2	371.2	366.7	371.2	371.2							
Gb4									370.8			370.8			370.8
G4	396.0	396.0			396.0					391.1	396.0		391.1	391.1	
G#4			412.5	417.6		412.5	412.5	417.6							
Ab4									417.2	417.2		417.2	417.2		412.0
A4	440.0	445.5	440.0		445.5	440.0					440.0			440.0	
A#4				464.1			464.1	464.1							
Bb4									463.5	469.3	469.3	463.5	469.3	469.3	463.5
B4	495.0	495.0	495.0		495.0	495.0	495.0								
B#4								522.1							
Cb5												494.4			494.4
C5	528.0				528.0				521.5	521.5	528.0		521.5	528.0	
C#5		556.9	550.0	556.9		550.0	556.9	556.9							

intonation for different keynotes are shown in Table 3.6. The frequencies were computed from the ratios of Table 3.2. The keynote frequency in each case is f, and the succeeding frequencies of the other tones are $\frac{9}{8} f$, $\frac{5}{4} f$, $\frac{4}{3} f$, $\frac{3}{2} f$, $\frac{5}{3} f$, $\frac{15}{8} f$, and $2f$. The frequencies for the key of C were computed using A_4 as 440 cycles. The frequencies of the keynotes for other keys were obtained from the tones in the key of C or by a second step in the case of keynotes with sharps or flats. Table 3.6 illustrates the practical difficulty of the scale of just intonation, namely, that this scale requires at least 30 discrete frequencies for each octave.

A minor scale of just intonation using A as the tonic constructed from the ratios of Table 3.1 is shown in Table 3.7. An examination of Table 3.7 shows that the ratios of

$$\begin{array}{ccc} A, & C, & \text{and} & E \\ E, & G, & \text{and} & B \\ D, & F, & \text{and} & A \end{array}$$

are all in the ratio $10:12:15$. This combination of tones is termed a minor triad. The minor scale of just intonation, A, B, C, D, E, F, G, with the ratios of Table 3.1 is constructed from these triads.

The interval between the tones in Table 3.7 in the minor scale are shown in Table 3.8. The minor scale is composed of three different intervals, namely, major tones, minor tones, and semitones. The interval $\frac{9}{8}$ is termed a major tone, the interval $\frac{10}{9}$ is termed a minor tone, and the interval $\frac{16}{15}$ is termed a semitone.

The intervals in the minor scale of just intonation for any keynote are shown in Table 3.9. The keynote in this table is 1.

The major-tone, the minor-tone, and the semitone intervals of the minor scale of just intonation for various keynotes are shown in Table

TABLE 3.7. FREQUENCY RATIOS OF THE TONES IN TERMS OF THE TONIC A IN THE MINOR SCALE OF JUST INTONATION

A	B	C	D	E	F	G	A
f	$\frac{9}{8}f$	$\frac{6}{5}f$	$\frac{4}{3}f$	$\frac{3}{2}f$	$\frac{8}{5}f$	$\frac{9}{5}$	$2f$

TABLE 3.8. FREQUENCY RATIOS OF ADJACENT TONES IN THE MINOR SCALE OF JUST INTONATION WITH THE TONIC A

A		B		C		D		E		F		G		A
	$\frac{9}{8}$		$\frac{16}{15}$		$\frac{10}{9}$		$\frac{9}{8}$		$\frac{16}{15}$		$\frac{9}{8}$		$\frac{10}{9}$	

3.10.   It will be seen that the pattern for all the keys is the same as that of Table 3.9, with the keynote of Table 3.10 being the numeral 1 of Table 3.9.   Furthermore, it will be observed that there is a continuous shift in this pattern from key to key of Table 3.10.

The frequencies of the tones in one octave of the minor scale of just intonation for different keynotes are shown in Table 3.11.   The fre-

TABLE 3.9.   THE INTERVALS BETWEEN THE TONES IN THE MINOR SCALE OF JUST INTONATION

(The sign ‖ indicates an interval of a major tone.   The sign ∥ indicates an interval of a minor tone.   The sign | indicates an interval of a semitone.   The keynote is 1.)

1 ‖ 2 | 3 ∥ 4 ∥ 5 | 6 ‖ 7 ∥ 8

TABLE 3.10.   THE TONES AND THE INTERVALS BETWEEN THE TONES IN THE MINOR SCALE OF JUST INTONATION FOR VARIOUS KEYNOTES

(The sign ‖ indicates an interval of a major tone.   The sign ∥ indicates an interval of a minor tone.   The sign | indicates an interval of a semitone.)

KEY	TONES
A	A ‖ B | C ∥ D ‖ E | F ‖ G ‖ A
B	A ‖ B ∥ C# | D ‖ E ‖ F# | G ‖ A
C#	A ‖ B ∥ C# ‖ D# | E ‖ F# ‖ G# | A
D#	A# | B ‖ C# ∥ D# ‖ E# | F# ‖ G# ‖ A#
E	A ‖ B | C ‖ D ‖ E ‖ F# | G ‖ A
F#	A ‖ B ∥ C# | D ‖ E ‖ F# ‖ G# | A
G#	A# | B ‖ C# ∥ D# | E ‖ F# ‖ G# ‖ A#
A#	A# ‖ B# | C# ∥ D# ‖ E# | F# ‖ G# | A#
B♭	A♭ ‖ B♭ | C | D♭ ‖ E♭ ‖ F | G♭ ‖ A♭
C	A♭ ‖ B♭ | C ‖ D | E♭ ‖ F ‖ G | A♭
D	A | B♭ ‖ C | D ‖ E | F ‖ G ‖ A
E♭	A♭ ‖ B♭ | C♭ ‖ D♭ ‖ E♭ ‖ F | G♭ ‖ A♭
F	A♭ ‖ B♭ ‖ C | D♭ ‖ E♭ ‖ F ‖ G | A♭
G	A | B♭ ‖ C ‖ D | E♭ ‖ F ‖ G ‖ A
A♭	A♭ ‖ B♭ | C♭ ‖ D♭ ‖ E♭ | F♭ ‖ G♭ ‖ A♭

TABLE 3.11. FREQUENCIES OF THE TONES IN ONE OCTAVE FOR DIFFERENT KEYNOTES IN THE MINOR SCALE OF JUST INTONATION

NOTE	KEY														
	A	B	C♯	D♯	E	F♯	G♯	A♯	B♭	C	D	E♭	F	G	A♭
A♭3									213.8	211.2		211.2	211.2		211.2
A3	220.0	222.7	222.7		220.0	222.7					220.0			222.7	
A♯3				234.9			234.9	234.9							
B♭3									237.6	237.6	234.7	237.6	234.7	237.6	
B3	247.5	247.5	250.6		247.5	247.5	250.6					253.4			253.4
B♯3								264.3							
C4	264.0				264.0				267.3	264.0	264.0		264.0	264.0	
C♯4		278.4	278.4	281.9		278.4	278.4	281.9							
D♭4									285.1			285.1	281.6		281.6
D4	293.3	297.0			297.0	297.0					293.3			297.0	
D♯4			313.2	313.2			313.2	313.2							
E♭4									316.8	316.8		316.8	316.8	316.8	
E4	330.0	330.0	334.1		330.0	334.1	334.1				330.0				
E♯4				352.4				352.4							
F♭4															337.9
F4	352.0								356.4	352.0	352.0	356.4	352.0	356.4	
F♯4		371.2	371.2	375.9	371.2	371.2	375.9	375.9							
G♭4									380.2			380.2			380.2
G4	396.0				396.0					396.0	391.1		396.0	396.0	
G♯4			417.6	417.6		417.6	417.6	422.8							
A♭4									427.7	422.4		422.4	422.4		422.4
A4	440.0	445.5	445.5		440.0	445.5					440.0			445.5	
A♯4				469.9			469.9	469.9							

quencies were computed from the ratios of Table 3.7, using A_4 as 440 cycles. The keynote frequency in each case is f, and the succeeding frequencies of the other tones are $\frac{9}{8}f, \frac{6}{5}f, \frac{4}{3}f, \frac{3}{2}f, \frac{8}{5}f, \frac{9}{5}f,$ and $2f$. The frequencies for the key of A were computed using A_4 as 440 cycles. The frequencies of the keynotes for other keys were obtained from the tones in the key of A or by a second step in the case of keynotes with sharps or flats. Table 3.11 illustrates the practical difficulty of the scale of just intonation, namely, that this scale requires at least 30 discrete frequencies for each octave.

It is obvious that the number of frequencies in the scale of just intonation is so great that it is impractical to build instruments with fixed tones to play in the just scale. In order to obviate this difficulty, a compromise scale has been evolved, termed the scale of equal temperament.

3.3 SCALE OF EQUAL TEMPERAMENT

Temperament is the process of reducing the number of tones per octave by altering the frequency of the tones from the exact frequencies of just intonation. In the equally tempered scale, the octave is divided into 12 intervals in which the frequency ratios are as follows:

$$1, f, f^2, f^3, f^4, f^5, f^6, f^7, f^8, f^9, f^{10}, f^{11}, f^{12}$$

where $f^{12} = 2$

or $\qquad f = \sqrt[12]{2}$

Thus it will be seen that the scale of equal temperament is a division of the

TABLE 3.12.   SCALE OF EQUAL TEMPERAMENT

Interval	Frequency ratio from starting point	Cents from starting point
Unison................................	1:1	0
Semitone or minor second........	1.059463:1	100
Whole tone or major second......	1.122462:1	200
Minor third....................	1.189207:1	300
Major third....................	1.259921:1	400
Perfect fourth..................	1.334840:1	500
Augmented fourth ⎫ Diminished fifth ⎭	1.414214:1	600
Perfect fifth...................	1.498307:1	700
Minor sixth....................	1.587401:1	800
Major sixth....................	1.681793:1	900
Minor seventh.................	1.781797:1	1,000
Major seventh.................	1.887749:1	1,100
Octave........................	2:1	1,200

octave into 12 equal intervals, called tempered half tones. A semitone, or half tone, in the scale of equal temperament is the frequency ratio between any two tones whose frequency ratio is the twelfth root of 2. The interval, in semitones, between any two tones is twelve times the logarithm on the base 2 of the frequency ratio. There is a further division termed the cent. A cent is the interval between any two tones whose frequency ratio is the twelve-hundredth root of 2. An octave is 1,200 cents. A semitone is 100 cents. Twelve semitones equal one octave. The scale of equal temperament is shown in Table 3.12.

The ratios of the tones in an octave in terms of the tonic in the scale of equal temperament are shown in Table 3.13.

TABLE 3.13. FREQUENCY RATIOS OF THE TONES IN THE SCALE OF EQUAL TEMPERAMENT

(f = frequency of the keynote C)

NOTE	FREQUENCY RATIO
C	1.000000f
C# D♭	1.059463f
D	1.122462f
D# E♭	1.189207f
E	1.259921f
F	1.334840f
F# G♭	1.414214f
G	1.498307f
G# A♭	1.587401f
A	1.681793f
A# B♭	1.781797f
B	1.887749f
C	2.000000f

Using A_4 as 440 cycles, the frequencies of all the tones in 10 octaves, from 16.351 to 15,824.26 cycles, for the scale of equal temperament are shown in Table 3.14.

The intervals in the major scale of equal temperament for any keynote are shown in Table 3.15. The keynote in this table is 1. The major scale is one in which the intervals between the tones three and four and seven and eight are semitones. The intervals between the other tones are whole tones.

The major scale of equal temperament may begin on any of the 12 fre-

TABLE 3.14. FREQUENCIES OF THE TONES IN 10 OCTAVES IN THE SCALE OF EQUAL TEMPERAMENT

NOTE	FREQUENCY	NOTE	FREQUENCY	NOTE	FREQUENCY	NOTE	FREQUENCY	NOTE	FREQUENCY
C_0	16.351	C_1	32.703	C_2	65.406	C_3	130.813	C_4	261.626
$C_0^\#$ D_0^b	17.324	$C_1^\#$ D_1^b	34.648	$C_2^\#$ D_2^b	69.296	$C_3^\#$ D_3^b	138.591	$C_4^\#$ D_4^b	277.183
D_0	18.354	D_1	36.708	D_2	73.416	D_3	146.832	D_4	293.665
$D_0^\#$ E_0^b	19.445	$D_1^\#$ E_1^b	38.891	$D_2^\#$ E_2^b	77.782	$D_3^\#$ E_3^b	155.563	$D_4^\#$ E_4^b	311.127
E_0	20.601	E_1	41.203	E_2	82.407	E_3	164.814	E_4	329.628
F_0	21.827	F_1	43.654	F_2	87.307	F_3	174.614	F_4	349.228
$F_0^\#$ G_0^b	23.124	$F_1^\#$ G_1^b	46.249	$F_2^\#$ G_2^b	92.499	$F_3^\#$ G_3^b	184.997	$F_4^\#$ G_4^b	369.994
G_0	24.499	G_1	48.999	G_2	97.999	G_3	195.998	G_4	391.995
$G_0^\#$ A_0^b	25.956	$G_1^\#$ A_1^b	51.913	$G_2^\#$ A_2^b	103.826	$G_3^\#$ A_3^b	207.652	$G_4^\#$ A_4^b	415.305
A_0	27.500	A_1	55.000	A_2	110.000	A_3	220.000	A_4	440.000
$A_0^\#$ B_0^b	29.135	$A_1^\#$ B_1^b	58.270	$A_2^\#$ B_2^b	116.541	$A_3^\#$ B_3^b	233.082	$A_4^\#$ B_4^b	466.164
B_0	30.868	B_1	61.735	B_2	123.471	B_3	246.942	B_4	493.883

NOTE	FREQUENCY	NOTE	FREQUENCY	NOTE	FREQUENCY	NOTE	FREQUENCY	NOTE	FREQUENCY
C_5	523.251	C_6	1046.502	C_7	2093.005	C_8	4186.009	C_9	8372.02
$C_5^\#$ D_5^b	554.365	$C_6^\#$ D_6^b	1108.731	$C_7^\#$ D_7^b	2217.461	$C_8^\#$ D_8^b	4434.922	$C_9^\#$ D_9^b	8869.84
D_5	587.330	D_6	1174.659	D_7	2349.318	D_8	4698.636	D_9	9397.27
$D_5^\#$ E_5^b	622.254	$D_6^\#$ E_6^b	1244.508	$D_7^\#$ E_7^b	2489.016	$D_8^\#$ E_8^b	4978.032	$D_9^\#$ E_9^b	9956.06
E_5	659.255	E_6	1318.510	E_7	2637.021	E_8	5274.042	E_9	10548.08
F_5	698.456	F_6	1396.913	F_7	2793.826	F_8	5587.652	F_9	11175.30
$F_5^\#$ G_5^b	739.989	$F_6^\#$ G_6^b	1474.978	$F_7^\#$ G_7^b	2959.955	$F_8^\#$ G_8^b	5919.910	$F_9^\#$ G_9^b	11839.82
G_5	783.991	G_6	1567.982	G_7	3135.964	G_8	6271.928	G_9	12543.86
$G_5^\#$ A_5^b	830.609	$G_6^\#$ A_6^b	1661.219	$G_7^\#$ A_7^b	3322.438	$G_8^\#$ A_8^b	6644.876	$G_9^\#$ A_9^b	13289.75
A_5	880.000	A_6	1760.000	A_7	3520.000	A_8	7040.000	A_9	14080.00
$A_5^\#$ B_5^b	932.328	$A_6^\#$ B_6^b	1864.655	$A_7^\#$ B_7^b	3729.310	$A_8^\#$ B_8^b	7458.620	$A_9^\#$ B_9^b	14917.24
B_5	987.767	B_6	1975.533	B_7	3951.066	B_8	7902.132	B_9	15804.26

TABLE 3.15.   THE INTERVALS BETWEEN THE TONES IN THE MAJOR SCALE OF EQUAL TEMPERAMENT

(The sign ‖ indicates an interval of a whole tone.   The sign | indicates an interval of a semitone.   The keynote is 1.)

1 ‖ 2 ‖ 3 | 4 ‖ 5 ‖ 6 ‖ 7 | 8

TABLE 3.16.   THE TONES AND THE INTERVALS BETWEEN THE TONES IN THE MAJOR SCALE OF EQUAL TEMPERAMENT FOR VARIOUS KEYNOTES

(The sign ‖ indicates an interval of a whole tone.   The sign | indicates an interval of a semitone.)

KEY	TONES
C	C ‖ D ‖ E | F ‖ G ‖ A ‖ B | C
D	C# | D ‖ E ‖ F# | G ‖ A ‖ B | C#
E	C# ‖ D# | E ‖ F# ‖ G# | A ‖ B ‖ C#
F#	C# ‖ D# ‖ E# | F# ‖ G# ‖ A# | B ‖ C#
G	C ‖ D ‖ E ‖ F# | G ‖ A ‖ B | C
A	C# | D ‖ E ‖ F# ‖ G# | A ‖ B ‖ C#
B	C# ‖ D# | E ‖ F# ‖ G# ‖ A# | B ‖ C#
C#	C# ‖ D# ‖ E# | F# ‖ G# ‖ A# ‖ B# | C#
Db	C | Db ‖ Eb ‖ F | Gb ‖ Ab ‖ Bb ‖ C
Eb	C ‖ D | Eb ‖ F ‖ G | Ab ‖ Bb ‖ C
F	C ‖ D ‖ E | F ‖ G ‖ A | Bb ‖ C
Gb	Cb ‖ Db ‖ Eb ‖ F | Gb ‖ Ab ‖ Bb | Cb
Ab	C | Db ‖ Eb ‖ F ‖ G | Ab ‖ Bb ‖ C
Bb	C ‖ D | Eb ‖ F ‖ G ‖ A | Bb ‖ C
Cb	Cb ‖ Db ‖ Eb | Fb ‖ Gb ‖ Ab ‖ Bb | Cb

quencies in an octave as follows: C, C♯ or D♭; D, D♯ or E♭; E, F, F♯ or G♭; G, G♯ or A♭; A, A♯ or B♭; and B.   The whole-tone and semitone intervals of the major scale of equal temperament for 15 key signatures as follows: C, D, E, F♯, G, A, B, C♯, D♭, E♭, F, G♭, A♭, B♭, and C♭ are shown in Table 3.16.   These notes are termed keynotes.   It will be noted that the pattern of intervals is the same for all keynotes, the only difference is a shift in the position of the pattern.   In other words, there is only one major scale, but it may be written in many different frequency ranges.

The notes and keynotes of the treble clef of the major scale for the 15 keynotes are shown in Fig. 3.2.

The intervals in the conventional minor scale of equal temperament for any keynote are shown in Table 3.17. The keynote in this table is 1. The conventional minor scale is one in which the interval between the tones two and three and five and six are semitones. The intervals between the other tones are whole tones.

The minor scale of equal temperament may begin on any of the 12 frequencies in an octave as follows: A, A♯ or B♭; B, C, C♯ or D♭; D, D♯ or E♭; E, F, F♯ or G♭; G, G♯ or A♭. The whole-tone and semitone intervals of the minor scale of equal temperament for 15 key signatures as follows: A, B, C♯, D♯, E, F♯, G♯, A♯, B♭, C, D, E♭, F, G, A♭ are shown in Table 3.18. It will be noted that the pattern of the intervals is the same for all keynotes, the only difference is a shift in the position of the pattern. In

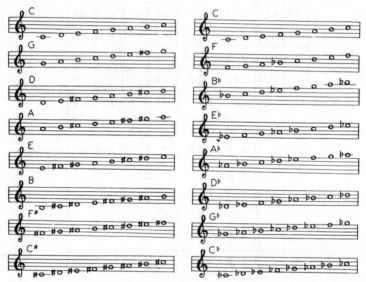

Fig. 3.2.   The notes of the major scale of equal temperament for all the key signatures.

TABLE 3.17.   THE INTERVALS BETWEEN THE TONES IN THE MINOR SCALE OF EQUAL TEMPERAMENT

(The sign ‖ indicates an interval of a whole tone.   The sign | indicates an interval of a semitone.   The keynote is 1.)

1 ‖ 2 | 3 ‖ 4 ‖ 5 | 6 ‖ 7 ‖ 8

TABLE 3.18.   THE TONES AND THE INTERVALS BETWEEN THE TONES IN THE MINOR
SCALE OF EQUAL TEMPERAMENT FOR VARIOUS KEYNOTES
(The sign ‖ indicates an interval of a whole tone.   The sign | indicates an interval
of a semitone.)

KEY	TONES
A	A ‖ B | C ‖ D ‖ E | F ‖ G ‖ A
B	A ‖ B ‖ C# | D ‖ E ‖ F# | G ‖ A
C#	A ‖ B ‖ C# ‖ D# | E ‖ F# ‖ G# | A
D#	A# | B ‖ C# ‖ D# ‖ E# | F# ‖ G# ‖ A#
E	A ‖ B | C ‖ D ‖ E ‖ F# | G ‖ A
F#	A ‖ B ‖ C# | D ‖ E ‖ F# ‖ G# | A
G#	A# | B ‖ C# ‖ D# | E ‖ F# ‖ G# ‖ A#
A#	A# ‖ B# | C# ‖ D# ‖ E# | F# ‖ G# ‖ A#
Bb	Ab ‖ Bb ‖ C | Db ‖ Eb ‖ F | Gb ‖ Ab
C	Ab ‖ Bb ‖ C ‖ D | Eb ‖ F ‖ G | Ab
D	A | Bb ‖ C ‖ D ‖ E | F ‖ G ‖ A
Eb	Ab ‖ Bb | Cb ‖ Db ‖ Eb ‖ F | Gb ‖ Ab
F	Ab ‖ Bb ‖ C | Db ‖ Eb ‖ F ‖ G | Ab
G	A | Bb ‖ C ‖ D | Eb ‖ F ‖ G ‖ A
Ab	Ab ‖ Bb | Cb ‖ Db ‖ Eb | Fb ‖ Gb ‖ Ab

other words, there is only one minor scale, but it may be written in many
different frequency ranges.

The tones and the keynotes of the treble clef of the conventional minor
scale of equal temperament for the 15 keynotes are shown in Fig. 3.3.

There are several forms of the minor scale.   For example, the harmonic
minor scale is another form of the minor scale of equal temperament.
The intervals in the harmonic minor scale of equal temperament are
shown in Table 3.19.   The keynote in this table is 1.   The minor harmonic scale is one in which the intervals between two and three, five and
six, and seven and eight are semitones and between one and two, three
and four, and four and five are whole tones, and between six and seven a
whole tone plus a semitone.   The notes and the keynotes of the treble of
the harmonic minor scale of equal temperament for the following key-

notes: A, B, C♯, D♯, E, F♯, G♯, A♯, B♭, C, D, E♭, F, G, and A♭ are shown in Fig. 3.4.

The intervals between the tones in the melodic minor scale of equal temperament are shown in Table 3.20. It will be noted that the intervals between the tones for the ascending tones differ from the intervals between the tones in the descending scale. The notes and keynotes of the treble clef of the melodic minor scale of equal temperament for the following keynotes: A, B, C♯, D♯, E, F♯, G♯, A♯, B♭, C, D, E♭, F, G, and A♭ are shown in Fig. 3.5.

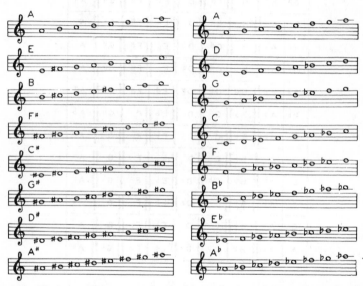

Fɪɢ. 3.3. The notes of the conventional minor scale of equal temperament for all the key signatures.

ᴀʙʟᴇ 3.19. Tʜᴇ Iɴᴛᴇʀᴠᴀʟs ʙᴇᴛᴡᴇᴇɴ ʜᴇ Tᴏɴᴇs ɪɴ ᴛʜᴇ Hᴀʀᴍᴏɴɪᴄ Mɪɴᴏʀ Sᴄᴀʟᴇ ᴏғ Eǫᴜᴀʟ Tᴇᴍᴘᴇʀᴀᴍᴇɴᴛ

The sign ‖ indicates an interval of a whole one. The sign ‖| indicates an interval of a whole tone plus a semitone, or three semitones. The sign | indicates an interval of a semitone. The keynote is 1.)

Tᴀʙʟᴇ 3.20. Tʜᴇ Iɴᴛᴇʀᴠᴀʟs ʙᴇᴛᴡᴇᴇɴ ᴛʜᴇ Tᴏɴᴇs ɪɴ ᴛʜᴇ Asᴄᴇɴᴅɪɴɢ ᴀɴᴅ Dᴇ-sᴄᴇɴᴅɪɴɢ ᴍᴇʟᴏᴅɪᴄ Mɪɴᴏʀ Sᴄᴀʟᴇ ᴏғ Eǫᴜᴀʟ Tᴇᴍᴘᴇʀᴀᴍᴇɴᴛ

(The sign ‖ indicates an interval of a whole tone. The sign | indicates an interval of a semitone. The keynote is 1.)

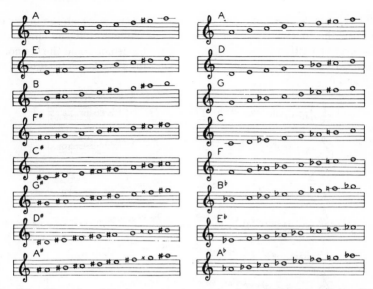

FIG. 3.4. The notes of the harmonic minor scale of equal temperament for all the key signatures.

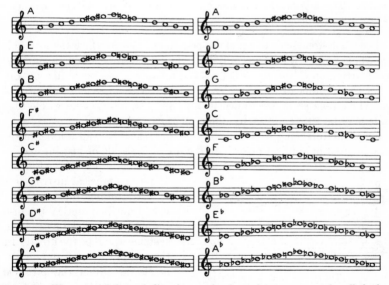

FIG. 3.5. The notes of the melodic minor scale of equal temperament for all the key signatures.

The chromatic scale of equal temperament is one in which the intervals between the tones are all semitones. It is obvious that the chromatic scale may be written in many different ways. The notes and the keynotes in the chromatic scale of equal temperament for a few keynotes are shown in Fig. 3.6.

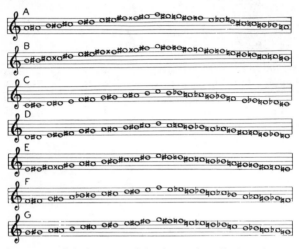

Fig. 3.6. The notes and the keynotes of the chromatic scale of equal temperament for various key signatures.

The whole-tone scale of equal temperament is one in which the intervals are whole tones. There are only two whole-tone scales possible, namely, one starting on C and the other on D♭, because scales beginning on any other tones duplicate these scales in sound. The notes and keynotes of the whole-tone scale are shown in Fig. 3.7. Of course, the whole-tone scale may be written in many different ways.

From the foregoing it is obvious that the scale of equal temperament may be used in applications where the use of the scale of just intonation

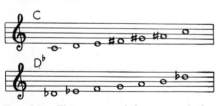

Fig. 3.7. The notes and keynotes of the whole-tone scales of equal temperament for the keynotes C and D♭.

would be impossible. For example, in musical instruments of fixed tuning the just scale would not be practical because the number of fixed resonating systems would be too great. In the case of the voice and instruments in which the pitch is not fixed, either the just or equally tempered scales may be used. Some musicians claim that the equally tempered scale is not so pleasing as the just scale. Comparing the frequencies of an octave of the equally tempered scale with the frequencies in the major and minor just scales, it will be seen that the maximum deviation of the scale of equal temperament from the scale of just temperament is ±2 per cent.

The notes on the piano keyboard, including the sharps and flats, are shown in Fig. 3.8. The piano employs the scale of equal temperament. The interval between adjacent keys which includes both white and black keys is an equally tempered semitone, or 100 cents. Referring to the tables and figures, it will be seen that the piano can be played in all the keys in all the major, minor, chromatic, and whole-tone scales of equal temperament.

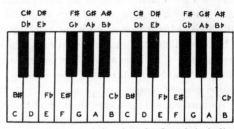

Fig. 3.8. Notes of the piano keyboard, including the sharps and flats.

The interval between two sounds is the spacing between them in pitch or frequency. The frequency interval is expressed by the ratio of the frequencies or by a logarithm of this ratio. Intervals in the scale of just intonation and of equal temperament are given in Tables 3.1 and 3.12, respectively.

When two tones are sounded at the same time, the result is termed a harmonic interval. If the tones are sounded one after the other, the result is termed a melodic interval.

Intervals are designated by means of a number, a second, third, fourth, etc. The number comes from the order of letters in the scale. The interval C to C is a unison, C to D is a second, C to E is a third, C to F is a fourth, C to G is a fifth, C to A is a sixth, C to B is a seventh. The same number names apply regardless of whether one or both letters of the interval are transformed by a sharp or a flat.

Examples of the intervals of Table 3.12 are shown in Fig. 3.9. Actual intervals are expressed by the combination of a name and a number. The

Fig. 3.9. Perfect, major, minor, augmented, and diminished intervals.

interval of the perfect unison is zero.   The interval of the minor second is one semitone.   The interval of the major second is two semitones.   The intervals of the minor third and the augmented second are three semitones. The intervals of the major third and the diminished fourth are four semitones.   The intervals of the perfect fourth and the augmented third are five semitones.   The intervals of the augmented fourth and the diminished fifth are six semitones.   The interval of the perfect fifth is seven semitones.   The intervals of the minor sixth and augmented fifth are eight semitones.   The interval of the major sixth is nine semitones.   The intervals of the minor seventh and the augmented sixth are ten semitones. The intervals of the major seventh and the diminished octave are eleven semitones.   The intervals of the perfect octave and the augmented seventh are twelve semitones.   The interval of the augmented octave is thirteen semitones.

A chord consists of a combination of three, four, five, six, or seven tones which are sounded simultaneously.   Therefore, when written, a chord consists of a series of three, four, five, six, or seven notes written or arranged one above the other and bearing a harmonic relation to one another (Fig. 3.10).   The lowest note, or fundamental, of a chord is termed the root.   The notes above the root or fundamental of a chord are designated as third, fifth, seventh, ninth, eleventh, and thirteenth.

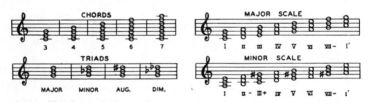

FIG. 3.10.   Chords consisting of three, four, five, six, or seven notes.   Major, minor, augmented, and diminished triads.   Triads in the major scale.   Triads in the minor scale.

TABLE 3.21.   THE INTERVALS BETWEEN THE TONES OF THE FOUR TRIADS
(The sign ||| indicates an interval of a whole tone plus a semitone, or three semitones. The sign |||| indicates an interval of two whole tones, or four semitones.   The numbers 1, 3, and 5 indicate, respectively, the root, the third, and the fifth above it.)

The simplest and most common chord, termed a triad, consists of three tones, the root, the third, and the fifth above it.   Triads are of four kinds, namely, major, minor, augmented, and diminished (Fig. 3.10).   The intervals between the three tones of these four triads are shown in Table 3.21.   A triad may be formed with each of the seven tones of the major scale as a root, as shown in Fig. 3.10.   Roman numerals are used to show the tone in the scale upon which the triad is based.   The different triads are designated as follows: a major triad by a large Roman numeral, a minor triad by a small Roman numeral, an augmented triad by a Roman numeral followed by a plus sign, and a diminished triad by a Roman numeral followed by a minus sign.   The triads in the minor scale are also shown in Fig. 3.10.

3.4   COMPOSITION

In this and the preceding chapter the following subjects have been considered: the principles of notation, such as staffs, clefs, notes, rests, sharps, flats, and naturals; the just and equally tempered scale; the major, minor, melodic, harmonic, chromatic, and whole-tone scales; the different keys; intervals; chords; and meters.   These subjects form the basis for the mechanics of the theory of music.

The musical notation is translated into sound by the musician employing a musical instrument or his own voice or both.   The subject of musical instruments and the physical and psychological characteristics of music will be considered in subsequent chapters.

With the establishment of musical notation and scales, the next step is the musical composition.   A musical composition consists of a series of pleasing, expressive, or intelligible tones of definite structure and significance according to the laws of melody, harmony, and rhythm.   The role of the composer is to invent and write a musical composition.   Reduced to its simplest form, a musical composition consists of a melody and the accompaniment.

A melody consists of a succession of tones with rhythmic and tonal organization.   The melody is the essence of the composition.   The arrangement of the tones, that is, the melodic shape, can produce many and varied subjective effects.   Some of the subjective effects which melodies are designed to produce are as follows: pleasantness, expressiveness, intelligence, gentleness, tenderness, sweetness, humor, calmness, relaxation, tension, excitation, agitation, exhilaration, fury, animation, passion, valor, patriotism, joyousness, and depression.   Save for the conventional mechanics of composition, there are no set rules for the assembly of notes to form a melody which will accomplish a certain objective.

The simplest accompaniment is a series of chords which are fitted to

the melody.   A chord consists of a series of three, four, five, six, or seven
tones, arranged one above the other and bearing a harmonic relation to
each other.   The lowest note of a chord is numbered with what is termed a
root as the reference point.   The smallest chord is a triad consisting of
three tones.   Triads may be classified as follows: major, minor, diminished, and augmented.

A triad may be built on any scale tone as, for example, on C: the major
triad on C is C-E-G, the minor triad on C is C-E♭-G, the diminished
triad on C is C-E♭-G♭, and the augmented triad on C is C-E-G♯; or on G:
the major triad on G is G-B-D, the minor triad on G is G-B♭-D, the diminished triad on G is G-B♭-D♭, and the augmented triad on G is G-B-D♯.

Harmony is a factor which must be considered in the process of composition.   Harmony refers to tones sounding simultaneously, that is,
chords, as contrasted to tones sounding consecutively as in the melody.
Harmony is the process of fitting suitable chords to the melody employing chord constructions and combinations.   A few of the chords have
been given above.   Harmony is a subject dealing with tones as related to
consonance and disonance (Sec. 7.4H).   The laws of harmony have been
derived by empirical and subjective means.   Each era evolves new laws
of harmony.

Counterpoint is the art of making two parts of a musical selection move
together with such independence that each one appears to have a design of
its own.   It may be said to be the process of adding one or more parts or
melodies to a given melody.

In the opinion of the author, it is beyond the scope of musical engineering at this time to outline the theory of music as related to the process of
composition.   It appears that the process of musical composition is art,
rather than a science.   It is an art in which the composer is a gifted person
who does not obtain his creative ability to compose from a set of rules or
studies.   However, this does not mean that studies in the different aspects
of composition are not an aid to those persons endowed with the creative
ability to compose.   The ultimate objective of a composer is the production of a composition which, when rendered, will be acceptable to a certain
cross section of the population.

CHAPTER FOUR

Resonators and Radiators

4.1 INTRODUCTION

The musical tones used in Western music consist of certain definite and discrete fundamental frequencies. Therefore, musical instruments must be capable of producing these tones and discriminate against other tones. Highly resonant vibrating systems are excited by driving frequencies which correspond to the resonant frequencies of the system. There are a large number of vibrating systems which exhibit resonant characteristics that are suitable for use in musical instruments. These systems include strings, bars, membranes, plates, pipes, horns, and electronic generators. In some of these systems, the resonant frequencies are fixed, and in others the resonant frequencies are variable. The resonant systems are provided with radiating means for producing sound waves in air which correspond to the mechanical vibrations. It is the purpose of this chapter to describe resonant vibrating systems used in musical instruments to produce musical sound waves in the surrounding air.

As an aid in describing the action of vibrating systems in musical instruments, it is useful to employ analogies between electrical circuits and these vibrating systems. Therefore, in this book dynamical analogies will be employed to describe the action of the vibrating systems of musical instruments. As a consequence, one of the first considerations in developing the theory and action of vibrating systems will be a brief development of dynamical analogies between electrical, mechanical, and acoustical systems.

4.2 DYNAMICAL ANALOGIES[1,2]

Analogies are useful when it is desired to compare an unfamiliar system with one that is better known. The relations and actions are more easily visualized, the mathematics more readily applied, and the analytical

[1] Olson, H. F., *Dynamical Analogies*, D. Van Nostrand Company, Inc., Princeton 1958.

[2] Olson, H. F., *Acoustical Engineering*, D. Van Nostrand Company, Inc., Princeton, 1957.

solutions more readily obtained in the familiar system. Analogies make it possible to extend the line of reasoning into unexplored fields.

All the analysis in musical engineering is concerned with vibrating systems. Although not generally so considered, the electrical circuit is the most common example and the most widely exploited vibrating system. The electrical circuit is a vibrating system in which the kinetic energy, potential energy, and dissipation may be expressed by general dynamic equations. This immediately suggests analogies between electrical circuits and other dynamical systems as, for example, mechanical and acoustical vibrating systems.

In view of the tremendous amount of study which has been directed toward the solution of circuits and the engineer's familiarity with electrical circuits, it seems logical to apply this knowledge to the solution of vibrating problems in the field of musical engineering by the same theory as that used in the solution of electrical circuits.

It is the purpose of this section to develop the analogies between elements in electrical, mechanical, and acoustical systems.

A. Definitions

A few of the terms used in dynamical analogies will be defined here.

Abvolt. An abvolt is the unit of electromotive force.

Instantaneous Electromotive Force. The instantaneous electromotive force between two points is the total instantaneous electromotive force. The unit is the abvolt.

Effective Electromotive Force. The effective electromotive force is the root mean square of the instantaneous electromotive force over a complete cycle between two points. The unit is the abvolt.

Maximum Electromotive Force. The maximum electromotive force for any given cycle is the maximum absolute value of the instantaneous electromotive force during that cycle. The unit is the abvolt.

Peak Electromotive Force. The peak electromotive force for any specified time interval is the maximum absolute value of the instantaneous electromotive force during that interval. The unit is the abvolt.

Dyne. A dyne is the unit of force or mechanomotive force.

Instantaneous Force (Instantaneous Mechanomotive Force). The instantaneous force at a point is the total instantaneous force. The unit is the dyne.

Effective Force (Effective Mechanomotive Force). The effective force is the root mean square of the instantaneous force over a complete cycle. The unit is the dyne.

Maximum Force (Maximum Mechanomotive Force). The maximum force for any given cycle is the maximum absolute value of the instantaneous force during that interval. The unit is the dyne.

Peak Force (Peak Mechanomotive Force). The peak force for any specified time interval is the maximum absolute value of the instantaneous force during that interval. The unit is the dyne.

Dyne per Square Centimeter. A dyne per square centimeter is the unit of sound pressure.

Static Pressure. The static pressure is the pressure that would exist in a medium with no sound waves present. The unit is the dyne per square centimeter.

Instantaneous Sound Pressure (Instantaneous Acoustomotive Force). The instantaneous sound pressure at a point is the total instantaneous pressure at the point minus the static pressure. The unit is the dyne per square centimeter.

Effective Sound Pressure (Effective Acoustomotive Force). The effective sound pressure at a point is the root-mean-square value of the instantaneous sound pressure over a complete cycle at the point. The unit is the dyne per square centimeter.

Maximum Sound Pressure (Maximum Acoustomotive Force). The maximum sound pressure for any given cycle is the maximum absolute value of the instantaneous sound pressure during that cycle. The unit is the dyne per square centimeter.

Peak Sound Pressure (Maximum Acoustomotive Force). The peak sound pressure for any specified time interval is the maximum absolute value of the instantaneous sound pressure in that interval. The unit is the dyne per square centimeter.

Abampere. An abampere is the unit of current.

Instantaneous Current. The instantaneous current at a point is the total instantaneous current at that point. The unit is the abampere.

Effective Current. The effective current at a point is the root-mean-square value of the instantaneous current over a complete cycle at that point. The unit is the abampere.

Maximum Current. The maximum current for any given cycle is the maximum absolute value of the instantaneous current during that cycle. The unit is the abampere.

Peak Current. The peak current for any specified time interval is the maximum absolute value of the instantaneous current in that interval. The unit is the abampere.

Centimeter per Second. A centimeter per second is the unit of velocity.

Instantaneous Velocity. The instantaneous velocity at a point is the total instantaneous velocity at that point. The unit is the centimeter per second.

Effective Velocity. The effective velocity at a point is the root-mean-square value of the instantaneous velocity over a complete cycle at that point. The unit is the centimeter per second.

Maximum Velocity. The maximum velocity for any given cycle is the maximum absolute value of the instantaneous velocity during that cycle. The unit is the centimeter per second.

Peak Velocity. The peak velocity for any specified time interval is the maximum absolute value of the instantaneous velocity in that interval. The unit is the centimeter per second.

Cubic Centimeter per Second. A cubic centimeter per second is the unit of volume current.

Instantaneous Volume Current. The instantaneous volume current at a point is the total instantaneous volume current at that point. The unit is the cubic centimeter per second.

Effective Volume Current. The effective volume current at a point is the root-mean-square value of the instantaneous volume current over a complete cycle at that point. The unit is the cubic centimeter per second.

Maximum Volume Current. The maximum volume current for any given cycle is the maximum absolute value of the instantaneous volume current during that cycle. The unit is the cubic centimeter per second.

Peak Volume Current. The peak volume current for any specified time interval is the maximum absolute value of the instantaneous volume current in that interval. The unit is the cubic centimeter per second.

Electrical Impedance. Electrical impedance is the complex quotient of the alternating electromotive force applied to the system by the resulting current. The unit is the abohm.

Electrical Resistance. Electrical resistance is the real part of the electrical impedance. This is the part responsible for the dissipation of energy. The unit is the abohm.

Electrical Reactance. Electrical reactance is the imaginary part of the electrical impedance. The unit is the abohm.

Inductance. Inductance in an electrical system is that coefficient which, when multiplied by 2π times the frequency, gives the positive imaginary part of the electrical impedance. The unit is the abhenry.

Electrical Capacitance. Electrical capacitance in an electrical system is that coefficient which, when multiplied by 2π times the frequency, is the reciprocal of the negative imaginary part of the electrical impedance. The unit is the abfarad.

Mechanical Impedance. Mechanical impedance is the complex quotient of the alternating force applied to the system by the resulting linear velocity in the direction of the force at its point of application. The unit is the mechanical ohm.

Mechanical Resistance. Mechanical resistance is the real part of the mechanical impedance. This is the part responsible for the dissipation of energy. The unit is the mechanical ohm.

Mechanical Reactance. Mechanical reactance is the imaginary part of the mechanical impedance. The unit is the mechanical ohm.

Mass. Mass in a mechanical system is that coefficient which, when multiplied by 2π times the frequency, gives the positive imaginary part of the mechanical impedance. The unit is the gram.

Compliance. Compliance in a mechanical system is that coefficient which, when multiplied by 2π times the frequency, is the reciprocal of the negative imaginary part of the mechanical impedance. The unit is the centimeter per dyne.

Acoustical Impedance. Acoustical impedance is the complex quotient of the alternating pressure applied to the system by the resulting volume current. The unit is the acoustical ohm.

Acoustical Resistance. Acoustical resistance is the real part of the acoustical impedance. This is the part responsible for the dissipation of energy. The unit is the acoustical ohm.

Acoustical Reactance. Acoustical reactance is the imaginary part of the acoustical impedance. The unit is the acoustical ohm.

Inertance. Inertance in an acoustical system is that coefficient which, when multiplied by 2π times the frequency, gives the positive imaginary part of the acoustical impedance. The unit is the gram per centimeter to the fourth power.

Acoustical Capacitance. Acoustical capacitance in an acoustical system is that coefficient which, when multiplied by 2π times the frequency is the reciprocal negative imaginary part of the acoustical impedance. The unit is the centimeter to the fifth power per dyne.

Element. An element or circuit parameter in an electrical system defines a distinct activity in its part of the circuit. In the same way, an element in a mechanical or acoustical system defines a distinct activity in its part of the system. The elements in an electrical circuit are electrical resistance, inductance, and electrical capacitance. The elements in a mechanical system are mechanical resistance, mass, and compliance. The elements in an acoustical system are acoustical resistance, inertance, and acoustical capacitance.

Electrical System. An electrical system is a system adapted for the transmission of electrical currents consisting of one or all of the following electrical elements: electrical resistance, inductance, and electrical capacitance.

Mechanical System. A mechanical system is a system adapted for the transmission of vibrations consisting of one or all of the following mechanical elements: mechanical resistance, mass, and compliance.

Acoustical System. An acoustical system is a system adapted for the transmission of sound consisting of one or all of the following acoustical elements: acoustical resistance, inertance, and acoustical capacitance.

Electrical Abohm. An electrical resistance, reactance, or impedance is said to have a magnitude of one abohm when an electromotive force of one abvolt produces a current of one abampere.

Mechanical Ohm. A mechanical resistance, reactance, or impedance is said to have a magnitude of one mechanical ohm when a force of one dyne produces a velocity of one centimeter per second.

Acoustical Ohm. An acoustical resistance, reactance, or impedance is said to have a magnitude of one acoustical ohm when a pressure of one dyne per square centimeter produces a volume current of one cubic centimeter per second.

B. Elements

An element or circuit parameter in an electrical system defines a distinct activity in its part of the circuit. In an electrical system these elements are resistance, inductance, and capacitance. As indicated in the introduction to this chapter, the study of mechanical and acoustical systems is facilitated by the introduction of elements analogous to the elements in an electrical circuit. In this procedure, the first step is to develop the elements in these vibrating systems. It is the purpose of the section which follows to describe electrical, mechanical, and acoustical elements.

C. Resistance

1. Electrical Resistance. Electrical energy is changed into heat by the passage of an electrical current through a resistance. Energy is lost by the system when a charge q of electricity is driven through a resistance by a voltage e. Resistance is the circuit element which causes dissipation.

Electrical resistance r_E, in abohms, is defined as

$$r_E = \frac{e}{i} \tag{4.1}$$

where e = voltage across the electrical resistance, in abvolts

i = current through the electrical resistance, in abamperes

Equation (4.1) states that the electromotive force across an electrical resistance is proportional to the electrical resistance and the current.

2. Mechanical Resistance. Mechanical energy is changed into heat by a rectilinear motion which is opposed by linear resistance (friction). In a mechanical system, dissipation is due to friction. Energy is lost by the system when a mechanical resistance is displaced a distance x by a force f_M.

Mechanical resistance r_M, in mechanical ohms, is defined as

$$r_M = \frac{f_M}{u} \tag{4.2}$$

where f_M = applied force, in dynes

u = velocity at the point of application of the force, in centimeters per second

Equation (4.2) states that the driving force applied to a mechanical resistance is proportional to the mechanical resistance and the linear velocity.

3. Acoustical Resistance. In an acoustical system, dissipation may be due to the fluid resistance or radiation resistance. At this point the former type of acoustical resistance will be considered. Acoustical energy is changed into heat by the passage of a fluid through an acoustical resistance. The resistance is due to viscosity. Energy is lost by the system when a volume displacement X of a fluid or gas is driven through an acoustical resistance by a pressure p.

Acoustical resistance r_A, in acoustical ohms, is defined as

$$r_A = \frac{p}{U} \tag{4.3}$$

where p = pressure, in dynes per square centimeter

U = volume current, in cubic centimeters per second

Equation (4.3) states that the driving pressure applied to an acoustical resistance is proportional to the acoustical resistance and the volume current.

D. Inductance, Mass, Inertance

1. Inductance. Electromagnetic energy is associated with inductance. Electromagnetic energy increases as the current in the inductance increases. It decreases when the current decreases. It remains constant when the current in the inductance is a constant. Inductance is the electrical circuit element which opposes a change in current. Inductance L, in abhenrys, is defined as

$$e = L\frac{di}{dt} \tag{4.4}$$

where e = electromotive or driving force, in abvolts

di/dt = rate of change of current, in abamperes per second

Equation (4.4) states that the electromotive force across an inductance is proportional to the inductance and the rate of change of current.

2. Mass. Mechanical inertial energy is associated with mass in the mechanical system. Mechanical energy increases as the linear velocity

of a mass increases, that is, during linear acceleration.   It decreases when
the velocity decreases.   It remains constant when the velocity is constant.   Mass is the mechanical element which opposes a change of
velocity.   Mass m, in grams, is defined as

$$f_M = m \frac{du}{dt} \tag{4.5}$$

where du/dt = acceleration, in centimeters per second per second
         f_M = driving force, in dynes

Equation (4.5) states that the driving force applied to the mass is proportional to the mass and the rate of change of velocity.

3. Inertance.   Acoustical inertial energy is associated with inertance
in the acoustical system.   Acoustical energy increases as the volume current of an inertance increases.   It decreases when the volume current
decreases.   It remains constant when the volume current of the inertance
is a constant.   Inertance is the acoustical element that opposes a change
in volume current.   Inertance M, in grams per (centimeter)^4 is defined as

$$p = M \frac{dU}{dt} \tag{4.6}$$

where M = inertance, in grams per (centimeter)^4
    dU/dt = rate of change of volume current, in cubic centimeters per
                 second per second
         p = driving pressure, in dynes per square centimeter

Equation (4.6) states that the driving pressure applied to an inertance
is proportional to the inertance and the rate of change of volume current.

Inertance may be expressed as

$$M = \frac{m}{S^2} \tag{4.7}$$

where m = mass, in grams
         S = cross-sectional area, in square centimeters, over which the
                 driving pressure acts to drive the mass

The inertance of the circular tube is

$$M = \frac{\rho l}{\pi R^2} \tag{4.8}$$

where R = radius of the tube, in centimeters
         l = effective length of the tube, that is, length plus end correction,
                 in centimeters
         ρ = density of the medium in the tube, in grams per cubic
                 centimeter

E. Electrical Capacitance, Compliance, Acoustical Capacitance

1. Electrical Capacitance. Electrostatic energy is associated with the separation of positive and negative charges, as in the case of the charges on the two plates of an electrical capacitance. Electrostatic energy increases as the charge of opposite polarity are separated. It is constant and stored when the charges remain unchanged. It decreases as the charges are brought together and the electrostatic energy is released. Electrical capacitance is the electrical-circuit element which opposes a change in voltage. Electrical capacitance C_E, in abfarads, is defined as

$$i = C_E \frac{de}{dt} \tag{4.9}$$

Equation (4.9) may be written

$$e = \frac{1}{C_E} \int i \, dt = \frac{q}{C_E} \tag{4.10}$$

where q = charge on electrical capacitance, in abcoulombs

e = electromotive force, in abvolts

Equation (4.10) states that the charge on an electrical capacitance is proportional to the electrical capacitance and the applied electromotive force.

2. Compliance. Mechanical potential energy is associated with the compression of a spring or compliant element. Mechanical energy increases as the spring is compressed. It decreases as the spring is allowed to expand. It is constant and is stored when the spring remains immovably compressed. Compliance is the mechanical element which opposes a change in the applied force. Compliance C_M, in centimeters per dyne, is defined as

$$f_M = \frac{x}{C_M} \tag{4.11}$$

where x = displacement, in centimeters

f_M = applied force, in dynes

Equation (4.11) states that the displacement of a compliance is proportional to the compliance and the applied force.

Stiffness is the reciprocal of compliance.

3. Acoustical Capacitance. Acoustical potential energy is associated with the compression of a fluid or gas. Acoustical energy increases as the gas is compressed. It decreases as the gas is allowed to expand. It is constant and is stored when the gas remains immovably compressed. Acoustical capacitance is the acoustic element which opposes a change in the applied pressure.   The pressure,[3] in dynes per square centimeter, in terms of the condensation, is

$$p = c^2 \rho s \qquad (4.12)$$

where c = velocity, in centimeters per second
ρ = density, in grams per cubic centimeter
s = condensation, defined in Eq. (4.13)

The condensation in a volume V due to a change in volume from V to V' is

$$s = \frac{V - V'}{V} \qquad (4.13)$$

The change in volume $V - V'$, in cubic centimeters, is equal to the volume displacement, in cubic centimeters.

$$V - V' = X \qquad (4.14)$$

where X = volume displacement, in cubic centimeters

From Eqs. (4.12), (4.13), (4.14), the pressure is

$$p = \frac{\rho c^2}{V} X \qquad (4.15)$$

Acoustical capacitance C_A, in (centimeters)[5] per dyne, is defined as

$$p = \frac{X}{C_A} \qquad (4.16)$$

where p = sound pressure, in dynes per square centimeter
X = volume displacement, in cubic centimeters

Equation (4.16) states the volume displacement in an acoustical capacitance is proportional to the pressure and the acoustical capacitance.

From Eqs. (4.15) and (4.16) the acoustical capacitance of a volume is

$$C_A = \frac{V}{\rho c^2} \qquad (4.17)$$

where V = volume, in cubic centimeters

F. Representation of Electrical, Mechanical, and Acoustical Elements

Electrical, mechanical, and acoustical elements have been defined in the preceding sections.   Figure 4.1 illustrates schematically the four elements in each of the four systems.

The electrical elements, electrical resistance, inductance, and electrical capacitance, are represented by the conventional symbols.

[3] Olson, H. F., *Acoustical Engineering*, D. Van Nostrand Company, Inc., Princeton, 1957.

Mechanical resistance is represented by sliding friction which causes dissipation. Acoustical resistance is represented by narrow slits which cause dissipation due to viscosity when fluid is forced through the slits. These elements are analogous to electrical resistance in the electrical system.

Inertia in the mechanical system is represented by a mass. Inertance in the acoustical system is represented as the fluid contained in a tube in which all the particles move with the same phase when actuated by a force due to pressure. These elements are analogous to inductance in the electrical system.

Compliance in the mechanical system is represented as a spring. Acoustical capacitance in the acoustical system is represented as a volume which acts as a stiffness or spring element. These elements are analogous to electrical capacitance in the electrical system.

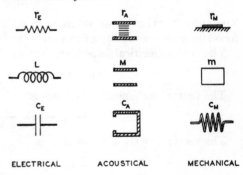

FIG. 4.1. Graphical representation of the three basic elements in electrical, mechanical, and acoustical systems. In the electrical system, r_E = electrical resistance, L = inductance, C_E = electrical capacitance. In the mechanical system, r_M = mechanical resistance, m = mass, C_M = compliance. In the acoustical system, r_A = acoustical resistance, M = inertance, C_A = acoustical capacitance.

ELECTRICAL   ACOUSTICAL   MECHANICAL

4.3 ELECTRICAL, MECHANICAL, AND ACOUSTICAL SYSTEM OF ONE DEGREE OF FREEDOM[4,5]

A resonant system must contain at least one element in which kinetic energy may be stored and another element in which potential energy may be stored. At resonance, energy flows from one type of element to the other, and vice versa. The simplest resonators in electrical, mechanical, and acoustical systems are shown in Fig. 4.2. The element in which kinetic energy may be stored in each of these systems is inductance, mass, and inertance. The element in which potential energy may be stored in each of these systems is electrical capacitance, compliance, and acoustical capacitance. A practical vibrating system without dissipation is impossible of attainment. The element in which dissipation occurs in each of

[4] Olson, H. F., *Dynamical Analogies*, D. Van Nostrand Company, Inc., Princeton, 1958.

[5] Olson, H. F., *Acoustical Engineering*, D. Van Nostrand Company, Inc., Princeton, 1957.

the systems is termed electrical resistance, mechanical resistance, and acoustical resistance.

The current, velocity, and volume current in each of the systems is
Electrical:

$$\dot{q} = i = \frac{E\epsilon^{j\omega t}}{r_E + j\omega L - \dfrac{j}{\omega C_E}} = \frac{e}{z_E} \qquad (4.18)$$

Mechanical:

$$\dot{x} = \frac{F\epsilon^{j\omega t}}{r_M + j\omega m - \dfrac{j}{\omega C_M}} = \frac{f_M}{z_M} \qquad (4.19)$$

Acoustical:

$$\dot{X} = \frac{P\epsilon^{j\omega t}}{r_A + j\omega M - \dfrac{j}{\omega C_A}} = \frac{p}{z_A} \qquad (4.20)$$

The current, velocity, or volume-current response to an electromotive force, force, or pressure as a function of the frequency is shown in Fig. 4.2.

The vector electrical impedance is

$$z_E = r_E + j\omega L - \frac{j}{\omega C_E} \qquad (4.21)$$

The vector mechanical impedance is

$$z_M = r_R + j\omega m - \frac{j}{\omega C_M} \qquad (4.22)$$

The vector acoustical impedance is

$$z_A = r_A + j\omega M - \frac{j}{\omega C_A} \qquad (4.23)$$

For a certain value of L and C_E, m and C_M, and M and C_A there will be a

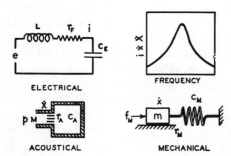

Fig. 4.2. Electrical, mechanical, and acoustical systems of 1 degree of freedom and the current, velocity, and volume-current response characteristics. In the electrical system, e = electromotive force, L = inductance, C_E = electrical capacitance, r_E = electrical resistance, i = current. In the mechanical system, f = force, m = mass, C_M = compliance, r_M = mechanical resistance, $\dot{x}$ = velocity. In the acoustical system, p = pressure, M = inertance, C_A = acoustical capacitance, r_A = acoustical resistance, $\dot{X}$ = volume current.

certain frequency at which the imaginary component of the impedance is zero. This frequency is called the resonant frequency. At this frequency, the ratio of the current to the applied voltage, or the ratio of the velocity to the applied force, or the ratio of the volume current to the applied pressure is a maximum. At the resonant frequency the current and voltage, the velocity and force, and the volume current and pressure are in phase.

The resonant frequency f_r, in the three systems, is
Electrical:

$$f_r = \frac{1}{2\pi\sqrt{LC_E}} \tag{4.24}$$

Mechanical:

$$f_r = \frac{1}{2\pi\sqrt{mC_M}} \tag{4.25}$$

Acoustical:

$$f_r = \frac{1}{2\pi\sqrt{MC_A}} \tag{4.26}$$

From the equations in this section it will be seen that the current, velocity, and volume current increase as the magnitude of the dissipative or resistive element decreases. The sharpness of the resonance also increases as the resistance is decreased. The response characteristics of the resonant systems for different values of the ratio

$$Q_r = \frac{\omega_r L}{r_E} = \frac{\omega_r m}{r_M} = \frac{\omega_r M}{r_A} \tag{4.27}$$

where $\omega_r = 2\pi f_r$
f_r = resonant frequency
and as a function of $f \div f_r$ are shown in Fig. 4.3.

4.4 HELMHOLTZ RESONATOR

The Helmholtz resonator consists of an enclosed volume of air coupled to the outside free air by means of an aperture (Fig. 4.4). The Helmholtz resonator, being a system of one degree of freedom, is the simplest type of resonant system used in musical instruments. The resonant system of the whistle, ocarina, or celesta or the body of the guitar, violin, etc., constitutes a Helmholtz resonator.

Perspective and sectional views and the acoustical circuit of a Helmholtz resonator are shown in Fig. 4.4. Since the dimensions of the resonator and the hole for the fundamental resonant frequency are small compared to the wavelength in the systems in which this type of resonator is effectively employed, the lumped-constant representation may be used

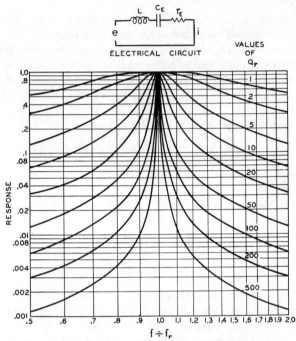

FIG. 4.3. The current-response characteristic of a simple circuit as a function of the ratio $f \div f_r$, where f_r = the resonant frequency and f = the frequency under consideration. The numbers refer to the value Q_r, $Q_r = 2\pi f_r L/r_E$. The above characteristics are applicable to acoustical and mechanical systems by the substitution of the elements and quantities which are analogous to the electrical system. (*After Olson, Acoustical Engineering, D. Van Nostrand Company, Inc., Princeton, 1957.*)

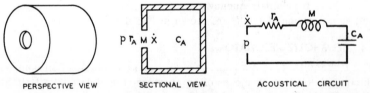

FIG. 4.4. Perspective and sectional views and the acoustical circuit of a Helmholtz resonator. In the acoustical circuit, p = actuating pressure, r_A = acoustical radiation resistance, M = inertance of the aperture, C_A = acoustical capacitance of the volume, $\dot{X}$ = volume current.

to describe the performance of the system. The expressions for the constants of the resonator in terms of the dimensions are as follows:

The acoustical radiation resistance r_A, in acoustical ohms, is given by

$$r_A = \frac{\rho \omega^2}{4\pi c} \tag{4.28}$$

where ρ = density of air, in grams per cubic centimeter

c = velocity of sound, in centimeters per second

$\omega = 2\pi f$

f = frequency, in cycles per second

The inertance M, in grams per (centimeter)^4, for a circular orifice is given by

$$M = \frac{\rho(l + 1.7R)}{\pi R^2} \tag{4.29}$$

where ρ = density of air, in grams per cubic centimeter

l = thickness of the boundary wall of the orifice, in centimeters

R = radius of the orifice, in centimeters

The acoustical capacitance C_A, in (centimeters)^5 per dyne, for the enclosed volume is given by

$$C_A = \frac{V}{\rho c^2} \tag{4.30}$$

where ρ = density of air, in grams per cubic centimeter

c = velocity of sound, in centimeters per second

V = volume of the enclosure, in cubic centimeters

The volume current $\dot{X}$, in cubic centimeters per second, through the aperture is given by

$$\dot{X} = \frac{p}{r_A + j\left(\omega M - \dfrac{1}{\omega C_A}\right)} \tag{4.31}$$

where p = actuating pressure, in dynes per square centimeter

It will be seen that the volume current is a maximum when

$$\omega = \sqrt{\frac{1}{M C_A}} \tag{4.32}$$

This is the resonant frequency of the resonator.

In some of the acoustical systems the orifice is not circular. Under these conditions, Eq. (4.29) does not apply. An approximate expression for the inertance of an orifice of any shape is given by

$$M = \frac{\rho(l + \sqrt{A})}{A} \tag{4.33}$$

where ρ = density of air, in grams per cubic centimeter

l = thickness of the boundary wall of the orifice, in centimeters

A = area of the orifice, in square centimeters

The shapes of the response characteristics of the Helmholtz resonator are the same as those of the system for a single degree of freedom shown in Fig. 4.3.

4.5   STRINGS[6-10]

Strings are used as the resonant system in many musical instruments. In string musical instruments the transverse and not the longitudinal vibrations are used. In the transverse vibrations, each particle of the string vibrates in a plane perpendicular to the line of the string. It is assumed that the mass per unit length is a constant, that it is perfectly flexible (the stiffness being negligible), and that it is connected to massive

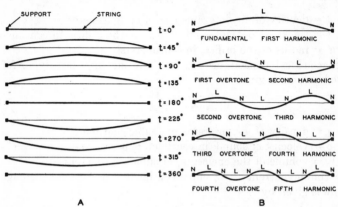

FIG. 4.5.   *A.* The shape of a stretched string vibrating in the fundamental mode for eight equal intervals of a complete cycle.   *B.* The first five modes of vibration of a stretched string.   The nodes and loops are indicated by N and L.

nonyielding supports (Fig. 4.5).   Since the string is fixed at the supports, nodes will occur at these points.   The fundamental resonant frequency of the string is given by

$$f_r = \frac{1}{2l} \sqrt{\frac{T}{m}} \tag{4.34}$$

where T = tension, in dynes

m = mass per unit length, in grams

l = length of the string, in centimeters

The relative positions of a string vibrating in the fundamental mode for a complete cycle are shown in Fig. 4.5*A*.   The shape of the string for all positions during a cycle is sinusoidal.

[6] Rayleigh, *Theory of Sound*, Macmillan & Co., Ltd., London, 1926.

[7] Crandall, *Theory of Vibrating Systems and Sound*, D. Van Nostrand Company, Inc., Princeton, 1926.

[8] Wood, *A Text Book of Sound*, George Bell & Sons, Ltd., London, 1930.

[9] Morse, *Vibration and Sound*, McGraw-Hill Book Company, Inc., New York, 1948.

[10] Lamb, *Dynamical Theory of Sound*, Edward Arnold & Co., London, 1931.

In addition to the fundamental, other modes of vibration may occur, the frequencies being 2, 3, 4, 5, etc., times the fundamental. In other words, the resonant vibration frequencies of a string correspond to the terms in the harmonic series. The first few modes of vibration of a string are shown in Fig. 4.5*B*. The points which are at rest are termed nodes and are marked *N*. The points between the nodes where the amplitude is a maximum are termed antinodes or loops and are marked *L*.

The string can be represented by a lumped-constant analogy shown in Fig. 4.6. The analogy is a multiresonant system. The resonant frequencies are determined by the constants of the system.

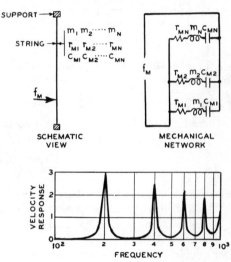

FIG. 4.6. The vibrating system, the mechanical network, and a velocity-response frequency characteristic of a string. In the mechanical system, m_1, m_2, m_3, . . . m_N; C_{M1}, C_{M2}, C_{M3}, . . . C_{MN}; and r_{M1}, r_{M2}, r_{M3}, . . . r_{MN} represent the masses, the compliances, and the mechanical resistances of the different modes of vibration; f_M = driving force. The velocity-response frequency characteristic shown depicts the response as a function of frequency of a typical string.

The response-frequency characteristic of a string is shown in Fig. 4.6. It will be seen that the harmonics are integral multiples of the fundamental.

Strings are used as the resonant system in the violin, viola, violon cello, contrabass, banjo, guitar, ukulele, mandolin, zither, piano, and many other string musical instruments.

4.6 TRANSVERSE VIBRATION OF BARS[11-14]

In the preceding section the perfectly flexible string was considered where the restoring force due to stiffness is negligible compared with that due to

[11] Rayleigh, *Theory of Sound*, Macmillan & Co., Ltd., London, 1926.

[12] Wood, *A Text Book of Sound*, George Bell & Sons, Ltd., London, 1930.

[13] Morse, *Vibration and Sound*, McGraw-Hill Book Company, Inc., New York, 1948.

[14] Lamb, *Dynamical Theory of Sound*, Edward Arnold & Co., London, 1931.

tension.   The bar under no tension is the other limiting case, the restoring force being entirely due to stiffness.   For the cases to be considered it is assumed that the bars are straight, the cross section is uniform and symmetrical about a central plane, and, as in the case of the string, only the transverse vibrations will be considered.

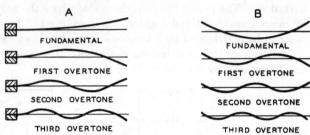

FIG. 4.7.   Modes of transverse vibrations of bars.   *A*. A bar clamped at one end and free at the other.   *B*. A bar free at both ends.

A. Bar Clamped at One End

Consider a bar clamped in a rigid support at one end with the other end free (Fig. 4.7*A*).   The fundamental frequency is given by

$$f_1 = \frac{0.5596}{l^2} \sqrt{\frac{QK^2}{\rho}} \qquad (4.35)$$

where l = length of the bar, in centimeters
     ρ = density, in grams per cubic centimeter
     Q = Young's modulus, in dynes per square centimeter
     K = radius of gyration
For a rectangular cross section, the radius of gyration is

$$K = \frac{a}{\sqrt{12}} \qquad (4.36)$$

where a = thickness of the bar, in centimeters, in the direction of vibration
For a circular cross section,

$$K = \frac{a}{2} \qquad (4.37)$$

where a = radius of the bar, in centimeters
For a hollow circular cross section,

$$K = \frac{\sqrt{a^2 + a_1^2}}{2} \qquad (4.38)$$

where a = outside radius of the pipe, in centimeters
     a_1 = inside radius of the pipe, in centimeters

The modes of vibration of a bar clamped at one end are shown in Fig. 4.7A. Table 4.1 gives the position of the nodes and the frequencies of the overtones. It will be seen that the overtones are not harmonics.

TABLE 4.1

No. of tone	No. of nodes	Distances of nodes from free end in terms of the length of the bar	Frequencies as a ratio of the fundamental
1	0		f_1
2	1	0.2261	$6.267f_1$
3	2	0.1321, 0.4999	$17.55f_1$
4	3	0.0944, 0.3558, 0.6439	$34.39f_1$

The first overtone of a bar or reed has a higher frequency than the sixth harmonic of a string. The reed organ, harmonica, clarinet, oboe, bassoon, saxophone, and tuning fork are examples of instruments employing a bar or reed clamped at one end.

B. Bar Free at Both Ends

Consider a perfectly free bar (Fig. 4.7B). The fundamental frequency is given by

$$f_1 = \frac{1.133\pi}{l^2} \sqrt{\frac{QK^2}{\rho}} \tag{4.39}$$

where l = length of the bar, in centimeters. All the other quantities are the same as in Eq. (4.38).

The modes of vibration of a perfectly free bar are shown in Fig. 4.7B. Table 4.2 gives the position of the nodes and the frequencies of the overtones.

TABLE 4.2

No. of tone	No. of nodes	Distances of nodes from one end in terms of the length of the bar	Frequencies as a ratio of the fundamental
1	2	0.2242, 0.7758	f_1
2	3	0.1321, 0.50, 0.8679	$2.756f_1$
3	4	0.0944, 0.3558, 0.6442, 0.9056	$5.404f_1$
4	5	0.0734, 0.277, 0.5, 0.723, 0.9266	$8.933f_1$

The most common examples of musical instruments employing a free bar are the xylophone, marimba, glockenspiel, and chimes.

4.7 STRETCHED MEMBRANES[15–19]

The ideal membrane is assumed to be flexible and very thin in cross section and stretched in all directions by a force which is not affected by the motion of the membrane. For cases of practical interest the membrane in musical instruments is assumed to be rigidly clamped and stretched by a massive surround. The considerations will be confined to circular membranes, because no other shapes are used in musical instruments.

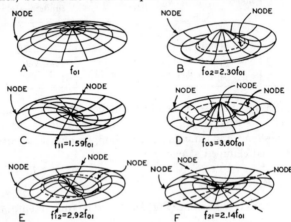

FIG. 4.8. Modes of vibration of a stretched circular membrane. (*After Morse Vibration and Sound, McGraw-Hill Book Company, Inc., New York, 1948.*)

The fundamental frequency of a circular stretched membrane is given by

$$f_{01} = \frac{0.382}{R} \sqrt{\frac{T}{m}} \qquad (4.40)$$

where m = mass, in grams per square centimeter of area

R = radius of the membrane, in centimeters

T = tension, in dynes per centimeter

The fundamental vibration is with the circumference as a node and a maximum displacement at the center of the circle (Fig. 4.8A). The frequencies of the next two overtones with nodal circles are

$$f_{02} = 2.30f_{01}$$
$$f_{03} = 3.60f_{01}$$

[15] Lamb, *Dynamical Theory of Sound*, Edward Arnold & Co., London, 1931.

[16] Rayleigh, *Theory of Sound*, Macmillan & Co., Ltd., London, 1926.

[17] Morse, *Vibration and Sound*, McGraw-Hill Book Company, Inc., New York, 1948.

[18] Wood, *A Text Book of Sound*, George Bell & Sons, Ltd., London, 1930.

[19] Crandall, *Theory of Vibrating Systems and Sound*, D. Van Nostrand Company, Inc., Princeton, 1926.

and are shown in Figs. 4.8B and 4.8D. The frequencies of the first and second overtones with nodal diameters are

$$f_{11} = 1.59f_{01}$$
$$f_{21} = 2.14f_{01}$$

These modes are shown in Figs. 4.8C and 4.8F. Following these simpler forms of vibration are combinations of nodal circles and nodal diameters. The frequency of one nodal circle and one nodal diameter, Fig. 4.8E, is

$$f_{12} = 2.92f_{01}$$

A stretched circular membrane is used in the banjo, tambourine, and all types of drums.   In this case, the air enclosure as well as the characteristics of the membrane controls the modes of vibration.

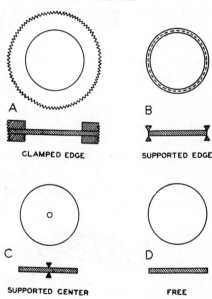

Fig. 4.9. Circular plates. A. A circular plate clamped at the edge. B. A circular plate supported at the edge. C. A circular plate supported at the center. D. A free circular plate. (*After Olson, Acoustical Engineering, D. Van Nostrand Company, Inc., Princeton, 1957.*)

CLAMPED EDGE   SUPPORTED EDGE

SUPPORTED CENTER   FREE

4.8   CIRCULAR PLATES[20-24]

The circular plates shown in Fig. 4.9 are assumed to be of uniform cross section and under no tension.   It is the purpose of this section to consider the vibration of circular plates for the various support means of Fig. 4.9.

[20] Rayleigh, *Theory of Sound*, Macmillan & Co., Ltd., London, 1926.

[21] Morse, *Vibration of Sound*, McGraw-Hill Book Company, Inc., New York, 1948.

[22] Wood, *A Text Book of Sound*, George Bell & Sons, Ltd., London, 1930.

[23] Crandall, *Theory of Vibrating Systems and Sound*, D. Van Nostrand Company, Inc., Princeton, 1926.

[24] Lamb, *Dynamical Theory of Sound*, Edward Arnold & Co., London, 1931.

A. Circular Clamped Plate

Consider a circular clamped plate, as shown in Fig. 4.9A. The fundamental frequency is given by

$$f_{01} = \frac{0.467t}{R^2} \sqrt{\frac{Q}{\rho(1 - \sigma^2)}} \tag{4.41}$$

where t = thickness of the plate, in centimeters

R = radius of the plate up to the clamping boundary, in centimeters

ρ = density, in grams per cubic centimeters

σ = Poisson's ratio

Q = Young's modulus, in dynes per square centimeter

The fundamental frequency is with the circumference as a node and with maximum displacement at the center (Fig. 4.10A). The frequencies of the next two overtones with nodal circles, Figs. 4.10B and 4.10D, are

$$f_{02} = 3.91f_{01}$$
$$f_{03} = 8.75f_{01}$$

The frequencies of the first and second overtones with nodal diameters, Figs. 4.10C and 4.10F, are

$$f_{11} = 2.09f_{01}$$
$$f_{21} = 3.43f_{01}$$

Following these simpler forms of vibration are the combination of nodal circles and nodal diameters. The frequency of the first overtone with one nodal diameter and one nodal circle, Fig. 4.10E, is

$$f_{12} = 5.98f_{01}$$

The clamped circular plate is used in telephone receivers and small condenser microphones.

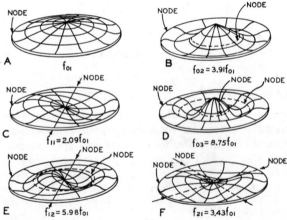

FIG. 4.10. Modes of vibration of a clamped circular plate.

B. Circular Free Plate

Consider a circular plate under no tension, uniform in cross section, and perfectly free (Fig. 4.9D). For a vibration with nodal circle, Fig. 4.11A, the frequency is

$$f = \frac{0.412t}{R^2} \sqrt{\frac{Q}{\rho(1 - \sigma^2)}} \qquad (4.42)$$

where t = thickness of the plate, in centimeters

R = radius of the plate, in centimeters

ρ = density, in grams per cubic centimeter

σ = Poisson's ratio

Q = Young's modulus, in dynes per square centimeter

For a vibration with two nodal diameters, Fig. 4.11B, the frequency is

$$f = \frac{0.193t}{R^2} \sqrt{\frac{Q}{\rho(1 - \sigma^2)}} \qquad (4.43)$$

The gong is an example of the circular free plate.

C. Circular Plate Supported at the Center

Consider a circular plate under no tension, uniform in cross section, edges perfectly free, and supported at the center (Fig. 4.9C). The frequency, for the umbrella mode, shown in Fig. 4.11C, is

$$f = \frac{0.172t}{R^2} \sqrt{\frac{Q}{\rho(1 - \sigma^2)}} \qquad (4.44)$$

The cymbal is an example of a circular plate supported in the center.

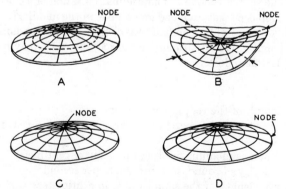

Fig. 4.11. Modes of vibrations of plates. A. Circular free plate with a circular node. B. Circular free plate with two nodal diameters. C. Circular plate supported at the center. D. Circular plate supported at the outside.

D. Circular Plate Supported at the Outside

Consider a plate under no tension, uniform in cross section, edges simply supported at the periphery (Fig. 4.9B). The fundamental frequency for a vibration with an outside edge node, shown in Fig. 4.11D, is

$$f = \frac{0.233t}{R^2} \sqrt{\frac{Q}{\rho(1 - \sigma^2)}} \tag{4.45}$$

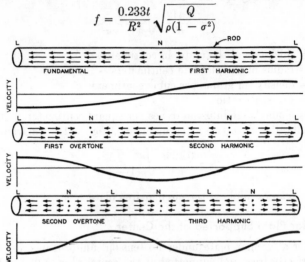

FIG. 4.12. Modes of longitudinal vibrations of a free rod. The nodes and loops are indicated by N and L. The arrows indicate the directions and relative magnitudes of the motion. The graphs depict the relative amplitude of the motion along the rod.

4.9 LONGITUDINAL VIBRATION OF BARS[25-28]

Consider an entirely free rod of homogeneous material and constant cross section. The simplest mode of longitudinal vibration of a free rod is one in which a loop occurs at each end and a node in the middle, that is, when the length of the rod is one-half wavelength. The fundamental frequency of longitudinal vibration of a free rod, Fig. 4.12, may be obtained from Eq. (1.3) as follows:

$$f_1 = \frac{c}{\lambda} = \frac{c}{2l} = \frac{1}{2l} \sqrt{\frac{Q}{\rho}} \tag{4.46}$$

where l = length of the rod, in centimeters

ρ = density of the material, in grams per cubic centimeter (see Table 1.1)

Q = Young's modulus, in dynes per square centimeter

c = velocity of sound, in centimeters per second

λ = wavelength of the sound wave, in centimeters

[25] Rayleigh, *Theory of Sound*, Macmillan & Co., Ltd., London, 1926.
[26] Morse, *Vibration of Sound*, McGraw-Hill Book Company, Inc., New York, 1948.
[27] Wood, *A Text Book of Sound*, George Bell & Sons, Ltd., London, 1930.
[28] Lamb, *Dynamical Theory of Sound*, Edward Arnold & Co., London, 1931.

The overtones of the free rod are harmonics of the fundamental, that is, $f_2 = 2f_1$, $f_3 = 3f_1$, $f_4 = 4f_1$, etc. (Fig. 4.12).

The fundamental resonance frequency occurs when the length of the rod is one-half wavelength. This fact provides a means of computing the velocity of sound when the density, Young's modulus, and the frequency are known, or the frequency of sound when the velocity, density, and Young's modulus are known.

Rods in which the longitudinal waves are excited by striking the ends are used as standards of high-frequency sounds, 5,000 cycles and above, where a tuning fork is not very satisfactory.

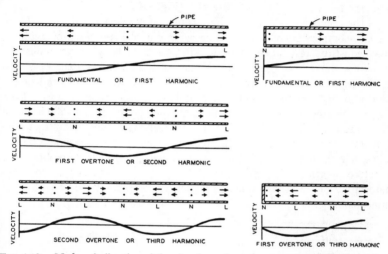

FIG. 4.13. Modes of vibration of the air column in a pipe open at both ends and in a pipe closed at one end and open at the other end. The velocity nodes and loops are indicated by N and L. The arrows indicate the directions and the relative magnitudes of the motion of the air particles in the pipe. The graphs depict the relative amplitudes of the air particles in the pipe.

4.10 OPEN AND CLOSED PIPES[29]

The vibrations of a column of gas or fluid in a cylindrical tube are analogous to the longitudinal vibrations in a solid bar. For the open pipe there must be a loop of displacement at the open ends.

The fundamental resonant frequency of a pipe open at both ends (Fig. 4.13) is

$$f_1 = \frac{c}{\lambda} = \frac{c}{2l} \tag{4.47}$$

[29] Olson, H. F., *Acoustical Engineering*, D. Van Nostrand Company, Inc., Princeton, 1957.

where l = length of the pipe, in centimeters

$\quad$ c = velocity of sound, in centimeters per second

$\quad$ λ = wavelength, in centimeters

The resonant overtones of an open pipe are harmonics of the fundamental, that is, $f_2 = 2f_1$, $f_3 = 3f_1$, $f_4 = 4f_1$, etc. (Fig. 4.13).

The fundamental resonant frequency of a pipe closed at one end and open at the other end (Fig. 4.13) is

$$f_1 = \frac{c}{\lambda} = \frac{c}{4l} \qquad (4.48)$$

The resonant overtones of the pipe closed at one end are the odd harmonics, that is, $f_2 = 3f_1$, $f_3 = 5f_1$, etc. (Fig. 4.13).

In the above considerations there is no end correction applied. There are two components of acoustical impedance at the end of a pipe, namely, the acoustical resistance due to radiation and the positive acoustical reactance due to the quadrature motion of the air beyond the open end of the pipe. The end correction depends upon whether the end of the pipe is flanged or unflanged. The unflanged pipe radiates into a solid angle of 4π steradians, while the flanged pipe radiates into a solid angle of 2π steradians. The end correction of an unflanged pipe is $0.62R$, where R is the radius of the pipe. The end correction of a flanged pipe is $0.82R$. These lengths must be added to the actual length of the pipe to obtain the effective length. The effective length of the pipe is used in accurate determinations of the resonant frequencies of pipes from the physical dimensions.

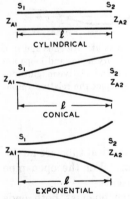

Fig. 4.14. Finite cylindrical, conical, and exponential horns. z_{A1} = input acoustical impedance at the throat; S_1 = cross-sectional area at the throat, in square centimeters; z_{A2} = terminating acoustical impedance at the mouth; S_2 = cross-sectional area at the mouth, in square centimeters; l = length, in centimeters.

The resonant frequencies and the response of a pipe may be determined theoretically by employing the expressions for the acoustical impedance of a pipe. The elements of a pipe are shown in Fig. 4.14. The expression[30] for the acoustical impedance z_{A1}, at the throat, in terms of the length and

[30] Olson, H. F., *Acoustical Engineering*, D. Van Nostrand Company, Inc., Princeton, 1957.

cross-sectional area of the pipe and the acoustical impedance z_{A2}, at the mouth, is

$$z_{A1} = \frac{\rho c}{S_1} \left[\frac{S_1 z_{A2} \cos (kl) + j\rho c \sin (kl)}{j S_1 z_{A2} \sin (kl) + \rho c \cos (kl)} \right] \quad (4.49)$$

where ρ = density of the medium, in grams per cubic centimeter

$k = \dfrac{2\pi}{\lambda}$

λ = wavelength, in centimeters

c = velocity of sound, in centimeters per second

S_1 = cross-sectional area of the pipe, in square centimeters

l = length of the pipe, in centimeters

z_{A2} = acoustical impedance at the mouth, in acoustical ohms

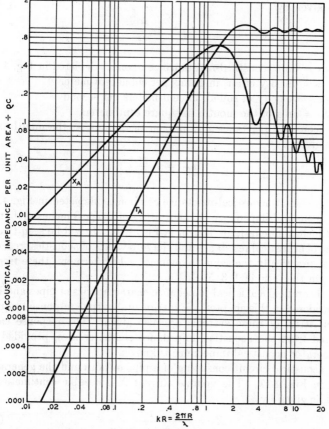

FIG. 4.15. The acoustical-resistance r_A and the acoustical-reactance x_A load per unit area divided by ρc as a function of kR for a piston of radius R set in an infinite baffle.

The acoustical impedance z_{A2}, at the mouth or open end of the flanged pipe, is assumed to be the same as the load upon a piston in an infinite baffle. The acoustical-resistance and acoustical-reactance components of the air load upon a piston are shown in Fig. 4.15. Note that the characteristics depicted in Fig. 4.15 are the acoustical resistance and acoustical reactance per unit area divided by ρc. The acoustical impedance z_{A2}, in acoustical ohms, at the mouth of the horn is given by

$$z_{A2} = r_{A2} + x_{A2} \qquad (4.50)$$

where r_{A2} = acoustical resistance at mouth, in acoustical ohms
$\quad x_{A2}$ = acoustical reactance at the mouth, in acoustical ohms

The acoustical resistance r_{A2}, in acoustical ohms, at the mouth of the horn is given by

$$r_{A2} = r_A \frac{\rho c}{S_2} \qquad (4.51)$$

where r_A = unit acoustical-resistance component depicted in Fig. 4.15
$\quad \rho$ = density of air, in grams per cubic centimeter
$\quad c$ = velocity of sound, in centimeters per second
$\quad S_2$ = area of the mouth, in square centimeters

The acoustical reactance x_{A2}, in acoustical ohms, at the mouth of the horn is given by

$$x_{A2} = x_A \frac{\rho c}{S_2} \qquad (4.52)$$

where x_A = unit acoustical-reactance component depicted in Fig. 4.15.

The acoustical-impedance characteristics of a pipe 3 inches in diameter and 12 inches in length are shown in Fig. 4.16. It will be seen that the maxima of acoustical resistance and the zero values of acoustical reactance occur at practically the same frequencies. Furthermore, these frequencies are in the ratio of 1:3:5, etc.

The theoretical acoustical-impedance characteristics of pipes and horns are considered in this chapter. Experimental output response-frequency characteristics of these systems are also given. The response can be predicted from the acoustical-impedance characteristics, provided that the generator acoustical impedance is known.

The acoustical circuit consisting of a pipe or horn and the generator is shown in Fig. 4.17. The volume current X_1, in cubic centimeters per second, is given by

$$\dot{X}_1 = \frac{p}{z_{AG} + r_{A1} + x_{A1}} \qquad (4.53)$$

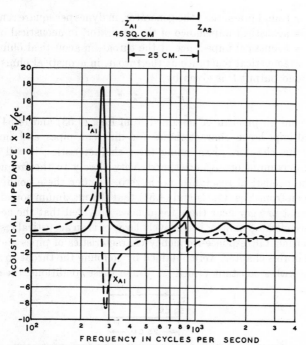

Fig. 4.16. The throat acoustical-resistance and acoustical-reactance frequency characteristics of the finite pipe shown above. r_{A1} = acoustical resistance, x_{A1} = acoustical reactance. Note: The characteristics shown are the throat acoustical resistance and acoustical reactance multiplied by S_1 and divided by ρc. (*After Olson, Acoustical Engineering, D. Van Nostrand Company, Inc., Princeton, 1957.*)

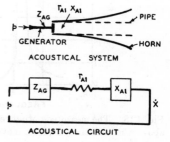

Fig. 4.17. An acoustical generator coupled to a horn or pipe and the acoustical circuit of the system. In the acoustical circuit, p = driving sound pressure, z_{AG} = acoustical impedance of the generator, r_{A1} = acoustical resistance at the throat of the horn or pipe, x_{A1} = acoustical reactance at the throat of the horn or pipe.

where p = sound pressure of the generator, in dynes per square centimeter

z_{AG} = acoustical impedance of the generator, in acoustical ohms

r_{A1} = acoustical impedance at the throat, in acoustical ohms

x_{A1} = acoustical reactance at the throat, in acoustical ohms

The sound output P is given by

$$P = r_{A1}|\dot{X}_1|^2 \qquad (4.54)$$

The sound output can be obtained from Eqs. (4.53) and (4.54). The acoustical-impedance characteristics for different pipes and horns are given in this chapter. Examining these characteristics, it will be seen that the maximum of r_{A1} corresponds very closely to the frequency at which the acoustical reactance x_{A1} is zero. Under these conditions, it will be seen from Eqs. (4.53) and (4.54) that the maximum sound output will occur at or very near these frequencies provided that the acoustical impedance of the generator is relatively large or is an acoustical resistance. The experimental response-frequency characteristics of pipes and horns, depicted in this chapter, were obtained by coupling the throat to a piston vibrating with a constant velocity. Under these conditions, the acousti-

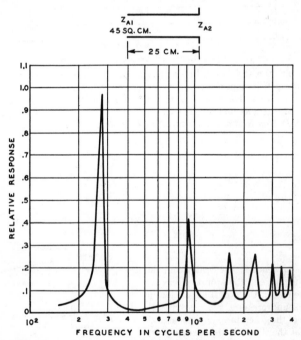

FIG. 4.18. The measured output response-frequency characteristic of the pipe shown above when driven by a piston moving with constant velocity.

cal impedance of the generator is for all practical purposes infinite. The experimental response-frequency characteristic of the pipe of Fig. 4.16 is shown in Fig. 4.18. Comparing Figs. 4.16 and 4.18, it will be seen that resonance occurs when the reactance is zero. It will also be seen that the resonant frequencies are in the ratio 1, 3, 5, etc.

Pipes are used for the resonant system in many musical instruments. In the pipe organ the pipes are fixed in length. Therefore, a pipe is required for each frequency. However, in other instruments, where a single tube is used to cover a range of frequencies, some means must be provided to vary the effective length of the pipe and thereby effect a corresponding change in the resonant frequency. A change in the effective length of the pipe may be accomplished by means of holes in the side of the pipe, which may be open or closed. Examples of musical instruments employing this system are the piccolo, flute, clarinet, saxophone, oboe, and bassoon.

A pipe with a hole in the side and the acoustical network for this system are shown in Fig. 4.19. The acoustical impedance z_{A2} of the pipe length l_2 at the mid-point of the pipe may be obtained in terms of the length l_2 and the acoustical impedance z_{A3} from Eq. (4.49). The acoustical impedance z_{A1} may also be obtained from Eq. (4.49). However, the mouth acoustical impedance z'_{A2} for the length l_1 is now the combination of the hole inertance M_2 in series with the acoustical resistance r_{A2}, in parallel with z_{A2}. The acoustical impedance z'_{A2} is

$$z'_{A2} = \frac{z_{A2}(r_{A2} + j\omega M_2)}{z_{A2} + r_{A2} + j\omega M_2} \tag{4.55}$$

To obtain the acoustical impedance z_{A1}, from Eq. (4.49), the mouth impedance z'_{A2} is used with the length l_1.

FIG. 4.19. A pipe with a hole in the side and the acoustical network of the vibrating system. z_{A1} = acoustical impedance at the throat, r_{A2} and M_2 = acoustical resistance and inertance of the hole, r_{A3} and M_3 = acoustical resistance and inertance of the mouth of the pipe, l_1 and l_2 lengths of the two sections of pipes, z_{A2} = acoustical impedance of the section l_2 considered as a simple pipe.

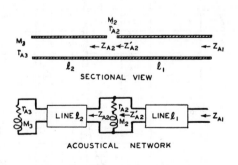

To illustrate the effect of a hole in the side of a pipe, the acoustical impedance and response-frequency characteristics of a pipe with and

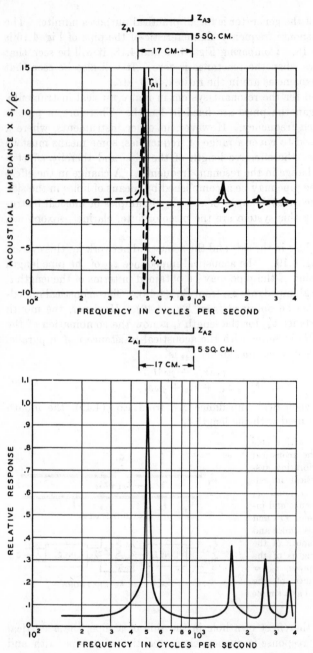

Fig. 4.20. The throat acoustical-resistance and acoustical-reactance frequency characteristic of the finite pipe shown above. r_{A1} = acoustical resistance, x_{A1} = acoustical reactance. Note: The characteristics shown are the throat acoustical resistance and acoustical reactance multiplied by S_1 and divided by ρc.

Fig. 4.21. The measured output response-frequency characteristic of the pipe shown above when driven by a piston moving with constant velocity.

without a hole will now be considered. The acoustical-impedance characteristic and the measured response-frequency characteristic of a pipe 1 inch in diameter and 6 inches in length are shown in Figs. 4.20 and 4.21. The fundamental resonant frequency is 500 cycles, both from the theoretical and experimental characteristics.

The acoustical-impedance characteristic of a pipe with a hole in the side is shown in Fig. 4.22. The measured response-frequency characteristic is shown in Fig. 4.23. The fundamental resonant frequency is 800 cycles. Comparing Figs. 4.20 and 4.21 with Figs. 4.22 and 4.23, it will be seen that the fundamental resonant frequency has been shifted from 500 cycles for a pipe without a hole to 800 cycles for a pipe with a hole. This example illustrates how the resonant frequency of a pipe can be varied by means of a hole in the side of the pipe with provisions for opening or closing the hole.

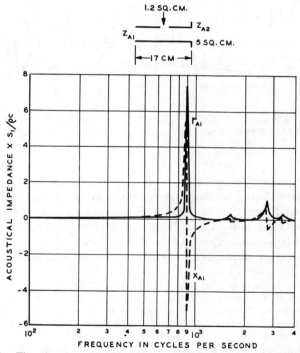

FIG. 4.22. The throat acoustical-resistance and acoustical-reactance frequency characteristics of the finite pipe with a hole as shown above. r_{A1} = acoustical resistance, x_{A1} = acoustical reactance. Note: The characteristics shown are the throat acoustical resistance and acoustical reactance multiplied by S_1 and divided by ρc.

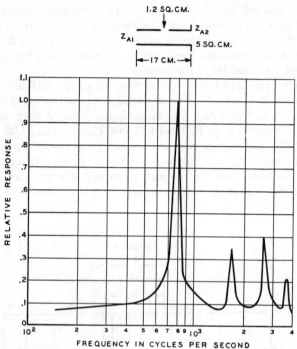

FIG. 4.23. The measured output response-frequency characteristic of the finite pipe with a hole as shown above when driven by a piston moving with a constant velocity.

4.11   HORNS[31-38]

A horn is an acoustical transducer consisting of a tube of varying sectional area.   Horns have been used for centuries for increasing the radiation from a sound source.   The shape or the rate of flare is the distinguishing characteristic between various horns.   The most common rates of flare are parabolic, conical, exponential, and hyperbolic.   In horn-type musical instruments, the horn may be any one of these or some combinations of

[31] Webster, A. G., *Jour. Nat. Acad. Sci.*, Vol. 5, p. 275, 1919.

[32] Stewart, G. W., *Phys. Rev.*, Vol. 16, No. 4, p. 313, 1920.

[33] Goldsmith and Minton, *Proc. Inst. Radio Engrs.*, Vol. 12, No. 4, p. 423, 1924.

[34] Slepian and Hanna, *Jour. Amer. Inst. Elec. Engrs.*, Vol. 43, p. 250, 1924.

[35] Ballantine, G., *Jour. Franklin Inst.*, Vol. 203, No. 1, p. 85, 1927.

[36] Crandall, *Theory of Vibrating Systems and Sound*, D. Van Nostrand Company, Inc., Princeton, 1926.

[37] Stewart and Lindsay, *Acoustics*, D. Van Nostrand Company, Inc., Princeton, 1930.

[38] Olson, H. F., *Acoustical Engineering*, D. Van Nostrand Company, Inc., Princeton, 1957.

these. It is beyond the scope of these considerations to consider all types of flare. A consideration of the acoustical-impedance and response-frequency characteristics of the conical and exponential horns, shown in Fig. 4.14, will be carried out and will illustrate the action of horns.

The equation expressing the cross-sectional area as a function of the distance along the axis in the conical horn is given by

$$S = S_1 x^2 \qquad (4.56)$$

where S = cross-sectional area at any point along the axis

x = distance along the axis

S_1 = area at the throat of the horn or at the point $x = x_1$

The expression[39] for the acoustical impedance z_{A1}, at the throat S_1, of a conical horn in terms of the dimensions and the acoustical impedance z_{A2}, at the mouth, is given by

$$z_{A1} = \frac{\rho c}{S_1} \left[\frac{j z_{A2} \dfrac{\sin k(l - \theta_2)}{\sin k\theta_2} + \dfrac{\rho c}{S_2} \sin kl}{z_{A2} \dfrac{\sin k(l + \theta_1 - \theta_2)}{\sin k\theta_1 \sin k\theta_2} - \dfrac{j\rho c}{S_2} \dfrac{\sin k(l + \theta_1)}{\sin k\theta_1}} \right] \qquad (4.57)$$

where S_1 = area of the throat, in square centimeters

S_2 = area of the mouth, in square centimeters

l = length of the horn, in centimeters

$k\theta_1 = \tan^{-1} kx_1$

$k\theta_2 = \tan^{-1} kx_2$

x_1 = distance from the apex to the throat, in centimeters

x_2 = distance from the apex to the mouth, in centimeters

$k = \dfrac{2\pi}{\lambda}$

The acoustical-impedance characteristics at the throat of a conical horn are shown in Fig. 4.24. The acoustical impedance z_{A2} at the mouth of the horn is assumed to be the same as the load upon a piston in an infinite baffle [see Fig. 4.15 and Eqs. (4.50), (4.51), and (4.52)]. The experimental response-frequency characteristic of the same horn is shown in Fig. 4.25. Comparing Figs. 4.24 and 4.25, it will be seen that the resonant frequencies occur when the acoustical reactance is zero.

The equation expressing the cross-sectional area of an exponential horn as a function of the distance along the axis is given by

$$S = S_1 \epsilon^{mx} \qquad (4.58)$$

where S_1 = cross-sectional area at the throat of the horn

x = distance along the axis

m = flare constant of the horn

S = cross-sectional area at a point x along the axis of the horn

[39] Olson, H. F., *Acoustical Engineering*, D. Van Nostrand Company, Inc., Princeton, 1957.

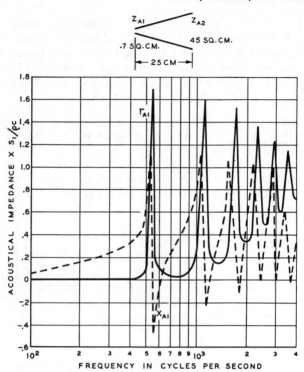

FIG. 4.24.   The throat acoustical-resistance and acoustical-reactance frequency characteristics of the finite conical horn shown above. r_{A1} = acoustical resistance, x_{A1} = acoustical reactance.   Note: The characteristics shown are the throat acoustical resistance and acoustical reactance multiplied by S_1 and divided by ρc.

The expression[40] for the acoustical impedance z_{A1}, at the throat S_1, of an exponential horn in terms of the dimensions, the flare constant, and the acoustical impedance z_{A2}, at the mouth, is given by

$$z_{A1} = \frac{\rho c}{S_1} \left\{ \frac{S_2 z_{A2}[\cos{(bl + \theta)}] + j\rho c[\sin{(bl)}]}{jS_2 z_{A2}[\sin{(bl)}] + \rho c[\cos{(bl - \theta)}]} \right\} \qquad (4.59)$$

where S_1 = area of the throat, in square centimeters
$\quad\quad\ S_2$ = area of the mouth, in square centimeters
$\quad\quad\ \ l$ = length of the horn, in centimeters
$\quad\quad z_{A2}$ = acoustical impedance of the mouth, in acoustical ohms
$\quad\quad\ \theta$ = $\tan^{-1} a/b$
$\quad\quad\ a$ = $m/2$
$\quad\quad\ b$ = $\frac{1}{2} \sqrt{4k^2 - m^2}$

[40] Olson, H. F., *Acoustical Engineering*, D. Van Nostrand Company, Inc., Princeton, 1957.

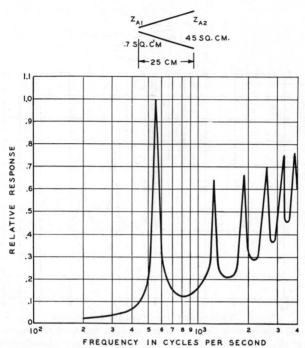

FIG. 4.25.   The measured output response-frequency characteristic of the finite conical horn shown above driven by a piston moving with constant velocity.

The acoustical-impedance characteristics at the throat of an exponential horn are shown in Fig. 4.26.   The acoustical impedance z_{A2} at the mouth of the horn is assumed to be the same as the load upon a piston in an infinite baffle [see Fig. 4.15 and Eqs. (4.50), (4.51), and (4.52)]. The measured response-frequency characteristic of the same horn is shown in Fig. 4.27.   Comparing Figs. 4.26 and 4.27, it will be seen that the resonant frequencies occur when the acoustical reactance is zero.

In musical instruments employing horns, the resonant frequency is varied by means of holes in the side or by varying the length of the horn. The length of the horn is varied in the trumpet, French horn, tuba, and trombone to effect a change in the resonant frequency.   In the first three instruments, the length is varied in discrete steps by introducing additional lengths by means of valves, as shown in Fig. 4.28A.   The use of the three additional side tubes shown in Fig. 4.28A together with valves for introducing or eliminating any of the additional side tubes makes it possible to obtain eight resonant frequencies.   The resonant frequencies correspond to the following lengths, $L, L + A, L + B, L + C, L +$

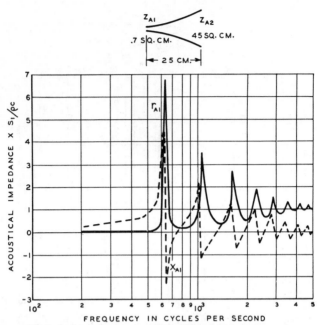

FIG. 4.26. The throat acoustical-resistance and acoustical-reactance frequency characteristics of the finite exponential horn shown above. r_{A1} = acoustical resistance, x_{A1} = acoustical reactance. Note: The characteristics shown are the throat acoustical resistance and acoustical reactance multiplied by S_1 and divided by ρc.

$A + B$, $L + A + C$, $L + B + C$, and $L + A + B + C$, where L is the length of the horn alone and A, B, and C are the lengths of the side tubes. The trombone uses telescopic tubing to vary the length, as shown in Fig. 4.28B. The length can be varied in a uniform and continuous manner. As a consequence, an infinite number of resonant frequencies may be obtained within the range of the instrument.

The acoustical impedance of a horn with an additional length of cylindrical pipe can be determined from Eqs. (4.49) and (4.57) or (4.59). For example, if a pipe is added to the throat of an exponential horn, the acoustical impedance z_{A1} at the opening of the pipe is determined from Eq. (4.49). The acoustical impedance z_{A2} of the pipe is the acoustical impedance at the throat of the exponential horn given by Eq. (4.59).

The acoustical-impedance characteristic of the exponential horn of Fig. 4.26 with an additional cylindrical length of 6 inches attached to the throat is shown in Fig. 4.29. Comparing Figs. 4.26 and 4.29, it will be seen that the impedance characteristic has been shifted downward in

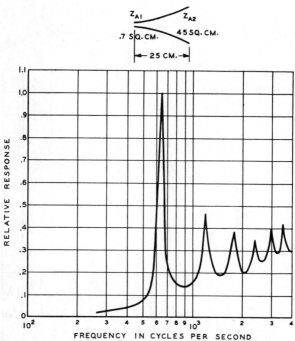

FIG. 4.27. The measured output response-frequency characteristic of the finite exponential horn shown above driven by a piston moving with constant velocity.

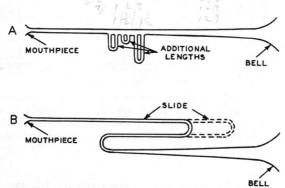

FIG. 4.28. Methods for varying the length of a horn, thereby changing the resonant frequency. In the system of *A* above, the length of the air column may be varied in discrete steps by the addition of the three tube lengths. In the system of *B* above, the length of the air column may be continuously varied by means of the telescopic slide.

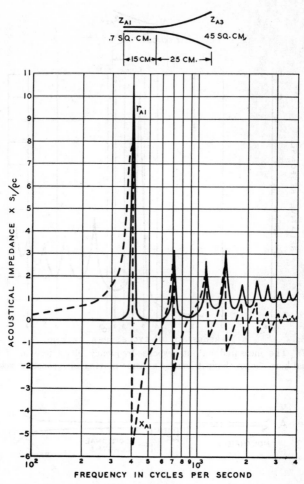

FIG. 4.29. The throat acoustical-resistance and acoustical-reactance frequency characteristics of the finite exponential horn with a 6-inch length of tubing added at the throat as shown above. r_{A1} = acoustical resistance, x_{A1} = acoustical reactance. Note: The characteristics shown are the throat acoustical resistance and acoustical reactance multiplied by S_1 and divided by ρc.

frequency by the additional length. The measured response-frequency characteristic of the exponential horn with the additional cylindrical length is shown in Fig. 4.30. Comparing Figs. 4.27 and 4.30, it will be seen that the fundamental resonant frequency has been lowered from 650

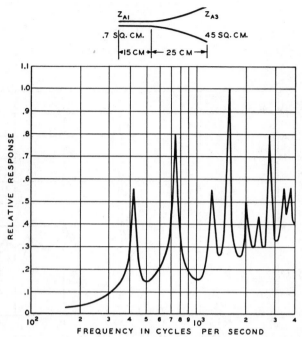

Fɪɢ. 4.30.   The measured output response-frequency characteristic of the finite exponential horn, with a 6-inch length of tubing added at the throat as shown above, driven by a piston moving with constant velocity.

cycles for the exponential horn alone to 420 cycles for the exponential horn with the additional cylindrical length of pipe.

The resonant frequency of a horn may also be varied by means of holes in the side of the horn.   The theory and the action is similar to that of a hole in the side of a cylindrical pipe considered in Sec. 4.10.   The acoustical-impedance frequency characteristics of the exponential horn of Fig. 4.26 with a hole in the side at the mid-point along the axis is shown in Fig. 4.31.   Comparing the impedance frequency characteristics of Figs. 4.26 and 4.31, it will be seen that the acoustical-impedance frequency characteristic has been shifted upward in frequency.   The measured response-frequency characteristic of a horn with a hole in the side is shown in Fig. 4.32.   Comparing Figs. 4.27 and 4.32, it will be seen that the fundamental resonant frequency has been raised from 650 cycles for the exponential horn without a hole to 900 cycles for the exponential horn with a hole in the side.

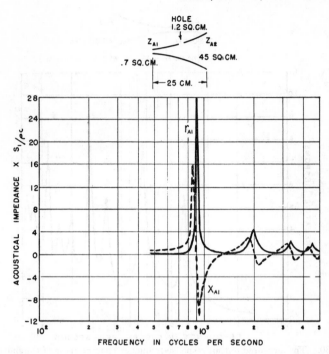

Fig. 4.31. The throat acoustical-resistance and acoustical-reactance frequency characteristics of the finite exponential horn with a hole in the side, as shown above. r_{A1} = acoustical resistance, x_{A1} = acoustical reactance. Note: The characteristics shown are the throat acoustical resistance and acoustical reactance multiplied by S_1 and divided by ρc.

4.12 DIRECTIONAL CHARACTERISTICS OF SOUND SOURCES

The directional characteristic of a sound source is the output as a function of the angle with respect to some axis of the system.   The characteristics are usually plotted as a system of polar characteristics for various frequencies.   The radiating characteristics of sound sources of musical instruments are extremely complex, because, in many instances, each instrument contains many different sound sources.   However, a consideration of some of the simplest sound-radiating systems will give some insight into the directional performance of sound sources.   It is the purpose of this section to consider the directional characteristics of a few of the simplest sound-radiating systems employed in musical instruments.

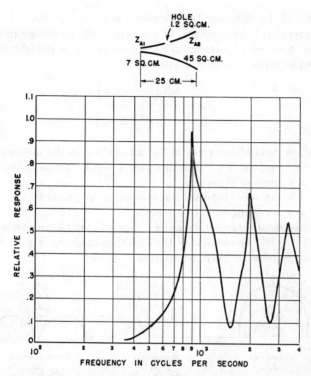

Fig. 4.32. The measured output response-frequency characteristic of the finite exponential horn with a hole in the side as shown above, driven by a piston moving with constant velocity.

A. Series of Point Sources[41–44]

In some musical instruments there are several small sound sources as, for example, in the case of the wind instruments with side holes or a bank of organ pipes. The directional characteristics exhibited by these systems are very complicated because of the unsymmetrical distribution of the sound sources in space and the different amplitude and phase relations between the sound sources. However, a consideration of the directional characteristics of a series of equal point sources located on a line will give

[41] Wolff, I., and Malter, L., *Jour. Acoust. Soc. Amer.*, Vol. 2, No. 2, p. 201, 1930.

[42] Stenzel, H., *Elek. Nachr.-Tech.* Vol. 4, No. 6, p. 239, 1927.

[43] Stenzel, H., *Elek. Nachr.-Tech.* Vol. 6, No. 5, p. 165, 1929.

[44] Olson, H. F., *Acoustical Engineering*, D. Van Nostrand Company, Inc., Princeton, 1957.

some idea of the directional performance of more complicated systems.

The directional characteristic of a source made up of any number of equal point sources, vibrating in phase, located on a straight line, and separated by equal distances is given by

$$R_a = \frac{\sin\left(\dfrac{n\pi d}{\lambda}\sin a\right)}{n\sin\left(\dfrac{\pi d}{\lambda}\sin a\right)} \tag{4.60}$$

where R_a = ratio of the pressure for an angle a to the pressure for an angle $a = 0$.   The direction $a = 0$ is normal to the line

  n = number of sources

  d = distances between the sources, in centimeters

  λ = wavelength, in centimeters

The directional characteristics of a two point sound source are shown in Fig. 4.33.   It will be noted that the directional pattern is quite complex even though there are only two sources.

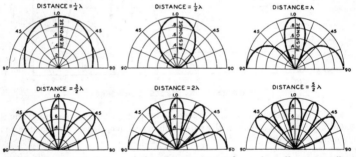

FIG. 4.33.   Directional characteristics of two separated equal small sources vibrating in phase as a function of the distance between the sources and the wavelength.   The polar graph depicts the pressure, at a fixed distance, as a function of the angle.   The pressure for the angle 0 degrees is arbitrarily chosen as unity.   The direction corresponding to the angle 0 degrees is perpendicular to the line joining the two sources.   The directional characteristics in three dimensions are surfaces of revolution about the line joining the two sources as an axis.   (*After Olson, Acoustical Engineering, D. Van Nostrand Company, Inc., Princeton, 1957.*)

B. Plane Circular Surface Source[45-47]

In some musical instruments the radiating surface consists of a circular flat vibrating surface as, for example, drum, banjo, and tambourine.

[45] Stenzel, H., *Elek. Nachr.-Tech.* Vol. 4, No. 6, p. 1, 1927.

[46] Wolff, I., and Malter, L., *Jour. Acoust. Soc. Amer.*, Vol. 2, No. 2, p. 201, 1930.

[47] Olson, H. F., *Acoustical Engineering*, D. Van Nostrand Company, Inc., Princeton, 1957.

The diaphragm of a loudspeaker is also an example of a circular surface source.

The directional characteristics of a circular surface source with all parts of the surface vibrating with the same strength and phase are

$$R_a = \frac{2J_1\left(\dfrac{2\pi R}{\lambda}\sin\,a\right)}{\dfrac{2\pi R}{\lambda}\sin\,a} \tag{4.61}$$

where R_a = ratio of the pressure for an angle a to the pressure for an angle $a = 0$

J_1 = Bessel function of the first order

R = radius of the circle, in centimeters

a = angle between the axis of the circle and the line joining the point of observation and the center of the circle

λ = wavelength, in centimeters

The directional characteristics of a plane circular surface source as a function of the diameter and wavelength are shown in Fig. 4.34. The directional pattern becomes sharper as the ratio of the diameter to the wavelength increases. This is typical of most acoustical radiating systems.

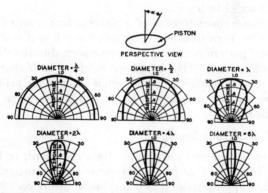

Fig. 4.34. Directional characteristics of a circular piston source as a function of the diameter and wavelength. The polar graph depicts the pressure, at a large fixed distance, as a function of the angle. The pressure for the angle 0 degrees is arbitrarily chosen as unity. The direction corresponding to the angle 0 degrees is the axis. The axis is the center line perpendicular to the plane of the piston. The directional characteristics in three dimensions are surfaces of revolution about the axis.

C. Plane Rectangular Surface Source[48]

In some musical instruments the radiating surface consists of a flat radiating surface, as, for example, the soundboard in the piano and the harp, the flat surfaces of the body of the guitar and ukulele. The directional characteristics of a plane rectangular surface source approximates the performance of these systems.

The directional characteristics of a rectangular surface source with all parts of the surface vibrating with the same strength and phase are

$$R_a = \frac{\sin\left(\dfrac{\pi l_\alpha}{\lambda}\sin\,\alpha\right)}{\dfrac{\pi l_\alpha}{\lambda}\sin\,\alpha} \frac{\sin\left(\dfrac{\pi l_\beta}{\lambda}\sin\,\beta\right)}{\dfrac{\pi l_\beta}{\lambda}\sin\,\beta} \qquad (4.62)$$

where l_α = length of the rectangle

   l_β = width of the rectangle

   α = angle between the normal to the surface source and the projection of the line joining the middle of the surface and the observation point on the plane normal to the surface and parallel to l_α

   β = angle between the normal to the surface source and the projection of the line joining the middle of the surface and the observation point on the plane normal to the surface and parallel to l_β

The directional characteristics of a plane surface source are shown in Fig. 4.35. The characteristics are for either $\alpha = 0$ or $\beta = 0$. If $\beta = 0$, the directional pattern is taken in the plane normal to the vibrating surface and parallel to l_α. The length given above each characteristic is then l_α. If $\alpha = 0$, the directional pattern is taken in the plane normal to the vibrating surface and parallel to l_β. The length given above each characteristic is then l_β.

D. Cylindrical Surface Source[49]

In some of the musical instruments the soundboard, or body of the instrument, is a curved surface. The directional characteristics of this type of surface approximates that of a cylindrical surface. A cylindrical surface radiator is shown in Fig. 4.36. In this radiator it will be assumed that all parts of the surface vibrate, with equal amplitudes and in the same phase, in directions along the radii of the cylinder. The directional characteristics of interest lie in a plane normal to the axis of the cylinder.

[48, 49] Olson, H. F., *Acoustical Engineering*, D. Van Nostrand Company, Inc., Princeton, 1957.

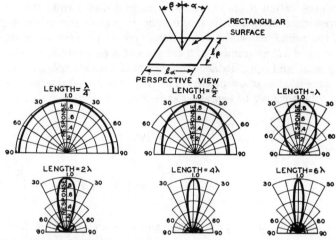

Fig. 4.35.   Directional characteristics of a rectangular surface source as a function of the length and the wavelength.   The polar graph depicts the pressure, at a large fixed distance, as a function of the angle.   The pressure for the angle 0 degrees is arbitrarily chosen as unity.   The direction corresponding to the angle 0 degrees is perpendicular to the surface.

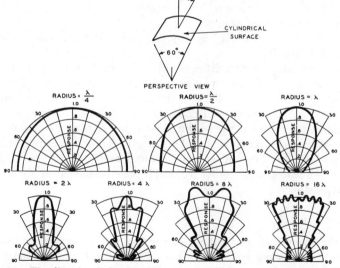

Fig. 4.36.   The directional characteristics of a cylindrical surface source which subtends an angle of 60 degrees.   The polar diagram depicts the pressure, at a large fixed distance, as a function of the angle.   The direction corresponding to the angle 0 degrees is perpendicular to the surface at the mid-point.   The observation points, for the angle a, are located in the plane normal to the axis of the cylindrical surface.   The pressure for the angle 0 degrees is arbitrarily chosen as unity.

For an observation point located at a large distance from the surface, this directional characteristic is the same as that of one of the circular arcs of the cylindrical surface. A circular arc source may be assumed to be made up of a large number of point sources of equal strength separated by very small and equal distances on the arc of a circle, with all the points vibrating with the same phase. The directional characteristic of this type of vibrator may be expressed as follows:

$$R_a = \frac{1}{2m+1} \left| \sum_{k=-m}^{k=m} \cos \left[\frac{2\pi R}{\lambda} \cos (a + k\theta) \right] \right.$$
$$\left. + j \sum_{k=-m}^{k=m} \sin \left[\frac{2\pi R}{\lambda} \cos (a + k\theta) \right] \right| \quad (4.63)$$

where R_a = ratio of the pressure for an angle a to the pressure for an angle $a = 0$

a = angle between the radius drawn through the central point and the line joining the source and the distant observation point

λ = wavelength, in centimeters

R = radius of the arc, in centimeters

$2m + 1$ = number of points

θ = angle subtended by any two points at the center of the arc

k = variable

The directional characteristics of cylindrical surface in a plane normal to the axis of the cylinder for an arc of 60 degrees are shown in Fig. 4.36. The interesting feature of the directional characteristics of a cylindrical surface is that the directional characteristics are very broad for wavelengths large compared to the dimensions, are narrow for wavelengths comparable to the dimensions, and are broad again for wavelengths small compared to the dimensions. The dimensions of the surface must be several wavelengths in order to yield the wedge-shaped directional characteristic.

E. Horn Source[50,51]

In the lip-reed or brass instruments and some of the other wind instruments the radiating system is a horn. The directional characteristics of a horn depend upon the shape, mouth opening, and frequency. It is

[50] Olson, H. F., *RCA Rev.*, Vol. 1, No. 4, p. 68, 1937.
[51] Olson, H. F., *Acoustical Engineering*, D. Van Nostrand Company, Inc., Princeton, 1957.

the purpose of this section to examine and consider some of the factors which influence the directional characteristics of a horn.

The phase and particle velocity of the various incremental areas which may be considered to constitute the mouth determine the directional characteristics of the horn. The particular complexion of the velocities and phase of these areas is governed by the flare and dimensions and shape of the mouth. The mouth of the horn plays a major role in determining the directional characteristics in the range where the wavelength is greater than the mouth diameter. The flare is the major factor in determining the directional characteristics in the range where the wavelength is less than the mouth diameter.

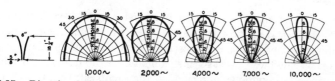

Fig. 4.37. Directional characteristics of a small exponential horn of circular cross section of the dimensions given above. The polar graph depicts the pressure, at a fixed distance, as a function of the angle. The pressure for the angle 0 degrees is arbitrarily chosen as unity. The direction corresponding to 0 degrees is the axis of the horn. The directional characteristics in three dimensions are surfaces of revolution about the horn axis.

The directional characteristics of a small exponential horn are shown in Fig. 4.37. As in the case of other radiating systems the directional characteristic becomes sharper with an increase of the frequency. The directional characteristics shown in Fig. 4.37 are typical of the type of characteristic obtained upon horn musical instruments. For horns of smaller or larger dimensions, the frequency given in Fig. 4.37 should be multiplied by the ratio obtained by dividing the linear dimensions of the horn in Fig. 4.37 by the linear dimensions of the horn in question.

CHAPTER FIVE

Musical Instruments

5.1 INTRODUCTION

A musical instrument is a system for producing one or more pleasing tones. A musical composition consists of a symbolic notation on paper of a combination and sequence of discrete tones which are pleasing, expressive, or intelligible. Musical instruments are used by musicians for translating the symbolic notation of the composition into the corresponding sounds. Musical instruments employ resonant or multiresonant systems for producing the discrete tones of Western music. The preceding chapter showed that there are many types of resonant systems. In addition, there are many different ways of exciting these resonant systems. A musical instrument consists of the combination of one or more resonant systems capable of producing one or more tones and means for exciting these systems, which are under the control of a musician. It is the purpose of this chapter to describe the most common musical instruments in use today.

5.2 TYPES OF MUSICAL INSTRUMENTS

Musical instruments may be divided into the following classes: string, wind, percussion, and electrical instruments. In the case of string instruments, the string may be struck, bowed, or plucked. Wind instruments may be classed as single mechanical reed, double mechanical reed, lip reed, air reed, and vocal-cord reed. Percussion instruments may be classed as definite and indefinite pitch. Musical instruments in these classifications are arranged in the following list.

A. String instruments
   1. Plucked strings

a. Lyre	*d.* Zither	*g.* Mandolin
b. Lute	*e.* Guitar	*h.* Banjo
c. Harp	*f.* Ukulele	*i.* Harpsichord

   2. Bowed strings
     a. Violin   *c.* Violoncello
     b. Viola   *d.* Double bass (contrabass)
   3. Struck strings
     a. Piano   *b.* Dulcimer

B. Wind instruments
   1. Air reed
     a. Whistle   *e.* Ocarina
     b. Flue organ pipe   *f.* Flute
     c. Recorder   *g.* Piccolo
     d. Flageolet   *h.* Fife
   2. Single mechanical reed
     a. Free-reed organ   *e.* Clarinet and bass clarinet
     b. Reed organ pipe   *f.* Saxophone (soprano, alto,
     c. Accordion   tenor, and bass)
     d. Harmonica   *g.* Bagpipe
   3. Double mechanical reed
     a. Oboe   *d.* Bassoon and contra bassoon
     b. English horn   *e.* Sarrusophone
     c. Oboe d'amore
   4. Organ (combination mechanical reed and air reed instrument)
   5. Lip reed
     a. Bugle   *d.* French horn
     b. Trumpet   *e.* Trombone and bass trombone
     c. Cornet   *f.* Tuba
   6. Vocal-cord reed

C. Percussion instruments
   1. Definite pitch
     a. Tuning fork   *d.* Chimes   *g.* Kettledrums (timpani)
     b. Xylophone   *e.* Glockenspiel *h.* Bell
     c. Marimba   *f.* Celesta   *i.* Carillon
   2. Indefinite pitch
     a. Side or snare drum   *e.* Triangle
     b. Military drum   *f.* Cymbals
     c. Bass drum   *g.* Tambourine
     d. Gong   *h.* Castanets

D. Electrical instruments
   1. Siren   5. Electrical guitar
   2. Automobile horn   6. Music box
   3. Electrical organs   7. Electrical carillon
   4. Electrical piano   8. Metronome

5.3   STRING INSTRUMENTS[1]

Vibrating a stretched string is one of the oldest ways of producing a musical tone. A stretched string is capable of producing the full range of overtones, which are harmonics of the fundamental, with the frequencies in the ratio of 1, 2, 3, 4, 5, etc. (see Sec. 4.5). The number and amplitude of the harmonics depend upon how and where the string is excited. String instruments in use today are excited in three different ways: by bowing, by striking, and by plucking. The projected area of a string is quite small, and therefore, a vibrating string alone is not a very efficient sound producer because it merely cuts through the air without causing any appreciable movement of air. For this reason, it is customary to couple a large multiresonant radiating surface to the string to increase the sound output.

A. Plucked-string Instruments

Plucked-string musical instruments are excited by pulling the string to one side and then releasing it. In some of the ancient musical instruments, as, for example, the lyre, lute, and zither, the strings were set into vibration by plucking. Modern examples of plucked-string instruments are the harp, guitar, banjo, and ukulele.

1. Lyre. The lyre is a string musical instrument of ancient Greek origin, consisting of a frame, a finger board, and a hollow body (Fig. 5.1). The lyre is usually considered the forerunner of all modern string instruments since it comprises the essential elements of these instruments. The lyre is played by plucking the strings with the fingers. The length of a string and hence the resonant frequency is varied by pressing the string against the finger board. The sound output of the string is increased by means of the hollow body, or soundboard.

2. Soundboard. All string musical instruments employ a soundboard or the combination of a ported hollow body and a soundboard to couple the string to the air. The vibration of a piston or a board of any shape produces radiation of sound. Referring to Fig. 4.15, it will be seen that the acoustical-radiation resistance increases as the size of the radiator is increased. Therefore, relatively large soundboards are used with all string-type musical instruments. The soundboards exhibit a large acoustical-radiation resistance and, thereby, improve the coupling between the string and the air and, as a result, increase the sound output. The soundboards are multiresonant systems which exhibit complex

[1] The word string rather than stringed is used as the modifier for designating musical instruments employing strings as the resonating system. This is in accordance with other uses of the word string as the modifier as, for example, string galvanometer, and in keeping with the modifiers wind, reed, etc., for other instruments.

modes of vibration similar to the vibrating plates described in Sec. 4.8. In some string instruments such as the lyre, lute, zither, guitar, and the violin family, a hollow body with a hole is coupled to the string. The vibration of the outside of the body produces radiation similar to that of the soundboard described in the above discussion. The hollow body with a hole coupled to the outside air constitutes a Helmholtz resonator (Sec. 4.4). The fundamental frequency of this resonator may be computed from the equations of Sec. 4.4. Since the dimensions of the resonator may be comparable to or greater than the wavelength of sound in air for some of the tones or overtones produced in the instruments, the hollow body together with the hole exhibits other resonant frequencies besides the fundamental resonant frequency. Thus it will be seen that the hollow body is a very complex resonant system.

3. Lute. The lute is a string musical instrument of ancient origin, consisting of a large pear-shaped hollow body, a neck with frets, and a head with pegs for tuning the strings (Fig. 5.2). It was developed about one thousand years ago and is the forerunner of instruments of the guitar class. It comprises all the essential elements of these instruments,

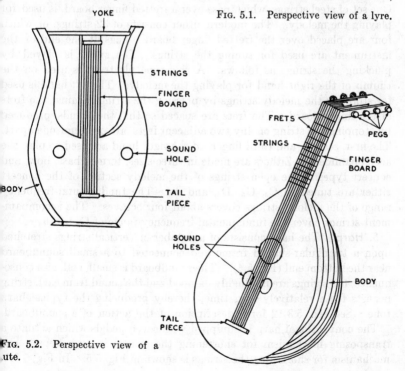

Fig. 5.1. Perspective view of a lyre.

Fig. 5.2. Perspective view of a lute.

namely, a hollow resonant body, turnable pegs located in the head for tuning the strings, and a fretted finger board.

4. Zither. The zither is a string instrument consisting of two sets of strings stretched across a flat hollow body equipped with a large sound hole (Fig. 5.3). One set of gut strings is used for the accompaniment.

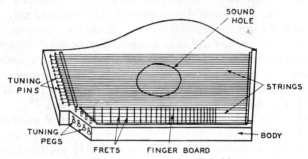

Fig. 5.3. Perspective view of a zither.

One set of steel strings, which pass over a fretted finger board, is used for playing the melody. The modern zither consists of 32 strings, of which four are placed over the fretted finger board. Pegs at one end of the instrument are used for tuning the strings. The zither is played by plucking the strings as follows: A ring-type plectrum is used on the thumb of the right hand for playing the melody. The left hand is used for stopping the melody strings by pressing the strings against the frets of the finger board. The frets are spaced so that the sounds produced by stopping the string on any two adjacent frets are one semitone apart. The first, second, and third fingers of the right hand are used to play the accompaniment. Zithers are made in three sizes, termed bass, bow, and concert types. The open strings of the melody section of the concert zither are tuned to C_3, G_3, D_4, and A_4. The fundamental-frequency range of the melody strings covers about four octaves. The accompaniment strings cover the fundamental-frequency range of C_2 to $A\flat_4$.

5. Harp. The harp consists of a number of vertical strings stretched upon a triangular-shaped frame and connected to a small soundboard near the bottom end (Fig. 5.4). The soundboard is small, and, as a consequence, the strings are not highly damped and the sound from each string persists for a relatively long time, thereby producing the typical harp tone. See Sec. 5.3A2 for a description of the action of a soundboard.

The conventional harp is equipped with seven pedals which actuate a transposing mechanism for shortening the strings in two steps. The mechanism for shortening the strings is shown in Fig. 5.5. In Fig. 5.5A, the string is in the position of maximum length. Depressing the pedal halfway rotates disk 1 and the pins attached to it, so that the string is shortened by an amount corresponding to a semitone, as shown in Fig. 5.5B. Depressing the pedal all the way rotates disk 2 and the pins attached to it, so that the string is shortened by an amount corresponding

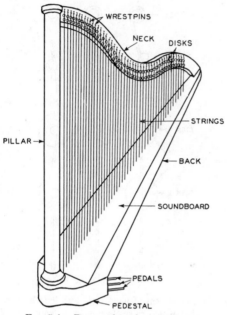

FIG. 5.4. Perspective view of a harp.

to a whole tone, as shown in Fig. 5.5C. Each pedal affects all the strings of the same name. The C pedal controls all the C strings, the D pedal, all the D strings, etc. The harp is usually tuned in the key of C flat. From the foregoing, it will be seen that, by depressing the appropriate pedals by the proper distances, it is possible to obtain all the tones in the major and minor keys.

The column of the harp serves as a structural member for supporting the strains produced by the stretched strings and as a housing for the rods connected to the pedals and tuning mechanism. The curved neck serves as an anchor for the strings and as a housing for the transposing mechanism described above. The strings pass through the soundboard and are anchored to a structural member forming the third side of the triangle. The over-all height of a conventional harp is 5 feet and 8 inches.

The fundamental-frequency range of the harp covers six and one-half octaves from C_1 to G_7.

The harp is played either by plucking with the fingers or by running the fingers up and down the strings in a sliding or gliding fashion termed a glissando.

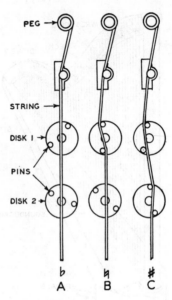

Fig. 5.5. Transposing mechanism of harp. *A*. Full length of string. *B*. String is shortened by turning disk 1 which raises frequency a semitone. *C*. String is shortened by turning disks 1 and 2 which raises frequency by a semitone over B. With this arrangement B is natural, A is flat, and C is sharp.

6. Guitar. The guitar consists of six strings stretched between a combination bridge and tailpiece fastened on the flat top of the body and the end of a fretted finger board (Fig. 5.6*A*). The body is made of two flat parallel boards fastened together along the outside edges. Vibrations of the strings are transmitted from the tailpiece to the top of the body. The bottom of the body is mechanically coupled to the top by means of a post. The cavity of the body, coupled to the outside air by means of a hole, constitutes a resonator (Sec. 4.4). The top and bottom of the body and the cavity and port combination form a multiresonant system for coupling the vibrating strings to the air (Sec. 5.3*A*2). The body of a guitar is made of wood or steel. Since the damping in metal is very small, the sound output of the steel guitar is much greater than that of the wood variety.

The six open strings of the guitar are tuned to E_2, A_2, D_3, G_3, B_3, and E_4.

The guitar can be played by plucking the strings with the fingers. It can also be played by plucking with a pick or plectrum, a flat piece of plastic or metal, held tightly between the thumb and first finger. The length of the string and hence the resonant frequency is varied by pressing the string against the frets. The frets on a guitar are spaced so that the sounds produced by stopping any string on any two adjacent frets are one semitone apart.

In the Hawaiian guitar, the strings are raised by a special steel nut so that the strings cannot be brought in contact with the frets. The lengths of the strings are varied by bringing a relatively heavy bar in contact

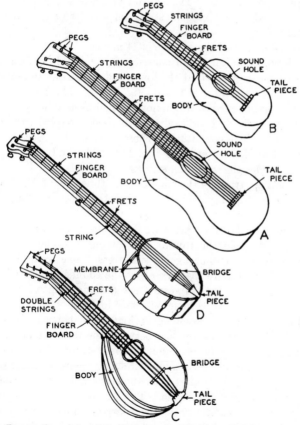

FIG. 5.6. Perspective views of *A*. Guitar. *B*. Ukulele. *C*. Mandolin. *D*. Banjo.

with the strings. Two picks are used, one slips over the finger and the other over the thumb. One of the fascinating effects that can be obtained in the Hawaiian guitar is the frequency glide which is produced by sliding the steel bar along the strings, thereby producing a continuous change in frequency.

The over-all length of the conventional guitar is 36 inches.

7. Ukulele. The ukulele is a small version of the guitar. It consists of four strings stretched between a combination bridge and tailpiece fastened to the flat top of the body and the end of a fretted finger board (Fig. 5.6B).

The four open strings of the ukulele are tuned to D_4, $F\sharp_4$, A_4, and B_4.

The ukulele is played by strumming the strings with the fingers. The length of a string and hence the resonant frequency is varied by pressing the string against the frets. The frets on a ukulele are spaced so that the sounds produced by stopping a string on any two adjacent frets are one semitone apart.

The over-all length of the ukulele is 24 inches.

8. Mandolin. The mandolin consists of four double strings stretched between a tailpiece fastened at the lower end of the flat top of the body and the end of a fretted finger board (Fig. 5.6C). The hollow body consists of a flat top fastened over a hollow hemiellipsoidal body. The combination of the body and the cavity of the body coupled to the outside air by means of a hole constitutes a complex resonator (Sec. 5.3A2). A bridge, placed at the center of the flat top of the body and supporting the strings, couples the vibrating strings to the body.

The four double strings of the mandolin are tuned to G_3, D_4, A_4, and E_5. The two strings of each pair are tuned exactly alike in pitch, and each double string is spoken of simply as one string.

The mandolin is played by plucking the strings with a pick or plectrum which is held between the thumb and forefinger. The length of the strings is varied by pressing against the frets. The frets of the finger board of the mandolin are spaced so that the sounds produced by stopping any string on any two adjacent frets are one semitone apart.

The over-all length of the mandolin is 23 inches.

9. Banjo. The banjo, as shown in Fig. 5.6D, consists of four long strings and a short string stretched between the tailpiece and the finger board and passing over a bridge supported on a stretched-skin membrane. The body is drumlike, in that a skin is stretched over one end of a circular cylinder. The bridge is supported by the stretched skin and couples the vibrating strings to the membrane. The stretched membrane increases the output of the strings by providing a large resonant surface which is coupled to the air (Sec. 5.3A2). The other end of the cylinder is open.

The relatively long neck is equipped with frets. The short string is termed the melody string.

At the present time, the tenor banjo with four metal strings of equal length seems to be the most popular and has supplanted the older classical inst ument with one short string.

The four open strings of the tenor banjo are tuned to C_3, G_3, D_4, and A_4.

The banjo can be played by plucking the strings with the fingers. It is also played by plucking the strings with a plectrum or pick, a flat piece of tortoise shell, held between the thumb and first finger. The length of a string and hence the resonant frequency is varied by pressing the string against the frets. The frets of the finger board of the banjo are spaced so that the sounds produced by stopping any string on any two adjacent frets are one semitone apart.

The over-all length of the banjo is 34 inches.

10. Harpsichord. The harpsichord consists of a large number of steel strings stretched upon a steel frame. The strings are excited by being plucked by a key-actuated mechanism. The keys form a keyboard similar to that of a piano. The shape, the arrangement of the keys, and the strings of the harpsichord are similar to that of a grand piano. The mechanism consists of a key coupled by means of a lever system to a plectrum of leather, fiber, or tortoise shell which plucks the string. The fundamental-frequency range of a typical harpsichord is G_1 to F_6.

11. Electrical Analogy of Plucked-string Instruments. A schematic view and the mechanical network of plucked-string instruments is shown in Fig. 5.7. The string is represented as a multiresonant system contain-

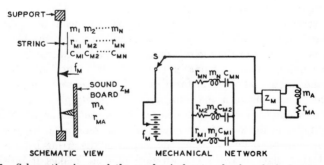

Fig. 5.7. Schematic view and the mechanical network of a plucked-string instrument. In the mechanical network, f_{M1} = driving force; m_1, C_{M1}, r_{M1}, m_2, C_{M2}, r_{M2}, . . . m_n, C_{MN}, r_{MN} = lumped masses, compliances, and mechanical resistances representing the string; m_A and r_{MA} = mass and mechanical resistance of the air load; z_M = quadripole representing the soundboard which couples the string to the air.

ing a combination of elements which will resonate at the fundamental frequency and all the harmonics.   These elements are represented as the masses m_1, m_2, . . . m_N, the mechanical resistances r_{M1}, r_{M2} ,. . . r_{MN}, and the compliances C_{M1}, C_{M2}, . . . C_{MN}.   The string is coupled to the air by means of a soundboard or body which is represented as the quadripole z_M.   The air load upon the soundboard or body is represented as the mass m_A and the mechanical resistance r_{MA}.   The force f_M displaces the string, which is analogous to applying a steady voltage to the resonant system depicting the string.   When the force is removed, the string is released and vibrates in various modes.   This is analogous to switching from the steady voltage to the path of zero resistance across the multiresonant system.

The ratios of the amplitudes of the fundamental and the various harmonics are complex and vary between various fundamental notes, owing to the complex nature of the actuating force, the string, the quadripole coupling the string to the air, and the air load.

B. Bowed-string Instruments

Bowed-string instruments are excited by rubbing the strings with the bow.   There are four modern examples of bowed-string instruments, namely, the violin, viola, violoncello, and double bass.   The essential difference in the four instruments is the physical size and the fundamental-frequency range.

1. Violin.   The violin consists of four strings stretched between the tailpiece and the end of the finger board and passing over a bridge supported on the top of the body (Fig. 5.8).   The body of the violin consists of two soundboards.   The top soundboard, termed the belly, supports the bridge which couples the strings to the belly.   The bottom soundboard, termed the back, is coupled to the belly by means of a post.   The bass bar is a longitudinal rib glued to the underside of the belly.   The cavity of the body in the violin, coupled to the outside air by means of the two holes, forms a resonator (Sec. 5.3A2).   The belly, back, and cavity and hole combination form powerful resonators and serve as a coupling means between the strings and the atmosphere (Sec. 5.3A2). In a good violin the resonant frequencies of these resonators are uniformly distributed over the lower frequency range.   The resonant frequency of a string is controlled by adjusting the length of the string by pressing it against the finger board.   The absence of frets makes it possible to obtain any frequency within the range, as contrasted with discrete frequencies in the case of fretted instruments.

The bow consists of horsehair stretched between the two ends of a thin strip of wood (Fig. 5.8). The movable frog is connected to a screw which makes it possible to vary the tension of the stretched horsehair. The horsehair is rubbed with rosin to increase the friction between the bow and the strings.

The string is kept in a constant state of agitation or vibration by drawing the bow across the string. The action is as follows: when the bow is drawn across the string, it drags the string along until the restoring force becomes too great and the string springs back. Since the friction between moving surfaces is less than between two adhering surfaces, the relative motion continues until the string again moves in the direction of the bow, when it again adheres and the process is repeated. It will be seen that the force produced in this way will have a saw-tooth wave shape. This means that the driving force contains the fundamental and all the harmonics. The motion corresponds to the resonant frequencies of the string. Therefore, the constant agitation by the bow and the waveform thus produced coupled with the multiresonant properties of the strings make the resultant sound output rich in harmonics. The

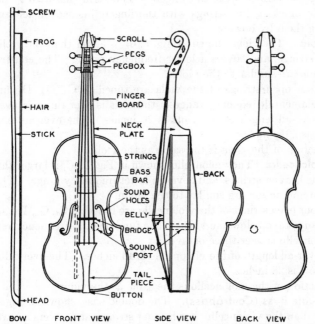

BOW   FRONT VIEW   SIDE VIEW   BACK VIEW

Fig. 5.8.   Front, side, and back views of a violin and a side view of the bow.

amplitude of the string and the resultant sound output are proportional to the pressure on the bow.

The four open strings of the violin are tuned to G_3, D_4, A_4, and E_5. The length of the string and hence the resonant frequency is varied by pressing the string against the finger board. This action is termed stopping. The fundamental-frequency range of the violin is over four octaves.

A mute is a forked clip which can be attached to the bridge of the violin. The added mass reactance of this load reduces the total sound output of the instrument and decreases the amplitudes of the overtones with relation to the fundamentals. This can be deduced from the electrical analogy of Sec. 5.3B5.

The over-all length of the violin is $23\frac{1}{2}$ inches. The over-all length of the bow is $29\frac{1}{2}$ inches.

The four members of the viol family, namely, the violin, viola, violoncello, and double bass, are shown in approximately relative sizes in Fig. 5.9.

The violin and other members of the viol family may also be played by plucking the strings with the fingers. The term pizzicato is used to designate plucking the strings with the fingers instead of actuating or sounding them by bowing.

2. Viola. The viola, shown in Fig. 5.9B, is larger than the violin, has heavier strings, and covers a lower frequency range. The construction and action are similar to the violin.

The four open strings of the viola are tuned to C_3, G_3, D_4, and A_4. The fundamental-frequency range of the viola is over three octaves.

The over-all length of the viola is 26 inches. The over-all length of the bow is $29\frac{1}{2}$ inches.

The action of the viola is the same as the violin.

3. Violoncello. The violoncello, shown in Fig. 5.9C, is larger than the viola, has heavier strings, and covers a lower frequency range. The construction and action are similar to the violin.

The four open strings of the violoncello are tuned to C_2, G_2, D_3, and A_3. This is an octave below the viola. The fundamental-frequency range of the violoncello is over three octaves.

The over-all length of the violoncello is 46 inches. The over-all length of the bow is 28 inches.

The action of the violoncello is the same as the violin.

4. Double Bass (Contrabass). The double bass, shown in Fig. 5.9D, is larger than the violoncello, has heavier strings, and covers a lower frequency range. The construction and action are similar to the violin.

The four open strings of the double bass are tuned to E_1, A_1, D_2, and G_2. The fundamental-frequency range of the double bass is about three octaves.

The over-all length of the double bass is 78 inches. The over-all length of the bow is 26 inches.

The action of the double bass is the same as the violin.

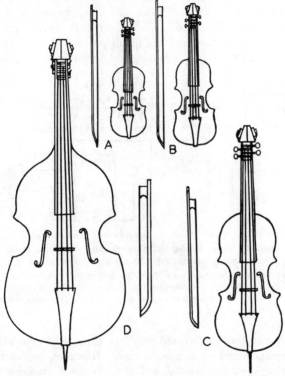

FIG. 5.9. A. Violin and bow. B. Viola and bow. C. Violoncello and bow. D. Double bass and bow.

The double bass is also played by plucking the strings with the fingers. Rhythmic plucking of the strings of the double bass provides a powerful bass accompaniment to the other instruments in a dance band.

5. Electrical Analogy of Bowed-string Instruments. A schematic view and the mechanical network of the vibrating system of bowed-string instruments are shown in Fig. 5.10. The wave shape of the driving force f_M in bowed instruments is approximately of a saw-tooth form. Therefore, the driving force contains the fundamental and all the harmonics (see Sec. 6.3A). The string is represented as a multiresonant system containing individual combination of elements which will resonate at the fundamental frequency and all the harmonics. These elements are represented as the masses m_1, m_2, . . . m_N, the mechanical resistances r_{M1}, r_{M2}, . . . r_{MN}, and the compliances C_{M1}, C_{M2}, . . . C_{MN}. The string is coupled to the air by means of the soundboard or body which is represented by the quadripole z_M. The air load is represented as the mass m_A and the mechanical resistance r_{MA}. The air load, mechanical reactance and mechanical resistance, and the quadripole affect the output of the string by accentuating some harmonics and discriminating against others.

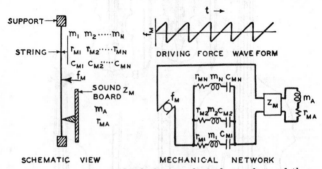

Fig. 5.10. Schematic view, mechanical network, and waveform of the equivalent generator of a bowed-string instrument. In the mechanical network, f_M = driving force; m_1, C_{M1}, r_{M1}, m_2, C_{M2}, r_{M2}, . . . m_N, C_{MN}, r_{MN} = lumped masses, compliances, and mechanical resistances representing the string; m_A and r_{MA} = mass and the mechanical resistance of the air load; z_M = quadripole representing the soundboard which couples the string to the air.

The vibrating system as depicted would be expected to emit the fundamental frequency and all the harmonics. However, the ratios of the amplitudes of the fundamental and the various harmonics would be expected to be complex and to vary between various fundamental notes, owing to the complex nature of the driving force, the string, the quadripole coupling the string to the air, and the air load. Some of these factors are controlled by the operator or musician, while others depend upon the individual instruments. The factors which are not controlled by the musician determine the musical merit of the instrument.

C. Struck-string Instruments

Struck-string instruments are excited by striking the string with a hammer. Examples of struck-string musical instruments are the piano, dulcimer, and clavichord. The modern example of a struck-string instrument is the piano.

1. Piano. The piano consists of a large number of steel strings stretched on a steel frame. The strings are coupled through a bridge to a

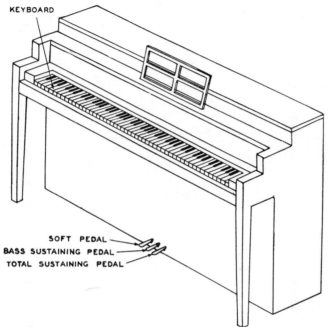

KEYBOARD

SOFT PEDAL
BASS SUSTAINING PEDAL
TOTAL SUSTAINING PEDAL

FIG. 5.11. Perspective view of an upright or spinet piano.

large soundboard. The function of a soundboard is described in Sec. 5.3A2. The strings are excited by being struck by hammers. The hammers are connected to keys forming a keyboard. Depressing a key actuates the hammer, which in turn strikes the string.

The piano is the most common of all musical instruments. It also has the distinction of covering a wide frequency range of 27.5 to 4,186 cycles, or A_0 to C_8—a range of over seven octaves.

The conventional piano is equipped with 88 keys. There is a string or a group of strings for each key. There are three plain steel strings each tuned to the same tone for each note or key in the five upper octaves, that is, from C_8 to C_3. There are two wrapped strings for each note from B_2 to G_1. There is one wrapped string for the remainder of the low-frequency range from $F\sharp_1$ to A_0.   The strings above C_3 are plain steel wires, while the strings below C_3 are wrapped with wire to add mass and to reduce the stiffness.

The piano is made in two forms, namely, the upright or spinet and the grand, shown in Figs. 5.11 and 5.12.   The strings are vertical in the

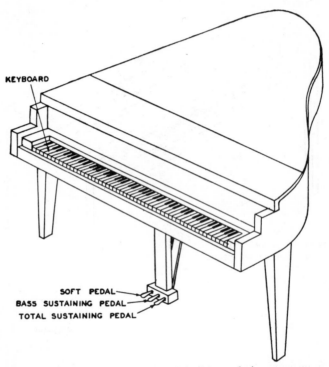

KEYBOARD

SOFT PEDAL
BASS SUSTAINING PEDAL
TOTAL SUSTAINING PEDAL

Fig. 5.12.   Perspective view of a grand piano.

upright and horizontal in the grand.   The width of the keyboard in a standard 88-note piano is 48 inches.   The over-all width of a spinet piano is about 57 inches; the over-all depth is about 25 inches; the over-all height varies from 37 to 50 inches.   The over-all width of a grand piano is about 60 inches; the height is 40 inches; the over-all depth varies depending on the type.   The over-all depth of the baby grand is 56 inches.   The over-all depth of the grand is 60 to 72 inches.   The over-all depth of the concert grand is 108 inches.

The mechanism of the upright, or spinet, piano is shown in Fig. 5.13.

The wires are stretched between pins on the steel frame. The vibrating part of the string is limited by two bridges, one set of bridges being on the steel frame and the other set on the soundboard. In this way, the sound output of the vibrating string is increased by coupling to the large sound-

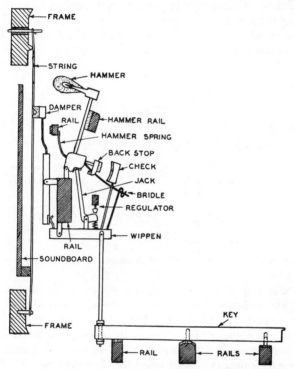

FIG. 5.13. A schematic view of the mechanism of an upright piano.

board (Sec. 5.3A2). The strings are struck at fixed points by felt-covered hammers actuated by means of keys. The action of the mechanism may be deduced from Fig. 5.13. When the key is depressed, the hammer is driven toward the string at relatively high speed. Examination of the jack and regulator mechanism shows that the hammer rebounds after it has struck the string and remains about one-half inch from the string if the key is held down. This position is determined by the jack and regulator. When the key is released, the hammer returns to its normal position against the hammer rail and the jack returns to its cocked position. While the key is depressed, the damping pad does not engage the string. The function of the damping pad is to increase the decay of the sounding note of a string after the key has been released. In general, the blow of the hammer lasts for a time comparable to the period of the string. The struck point moves in advance of more remote parts of the string. Under the above conditions of excitation, it can be shown that it is possible to suppress certain harmonics, depending upon the striking point along the string. For example, the seventh and higher harmonics will be suppressed if the string is struck at a point one-seventh of the length from one end.

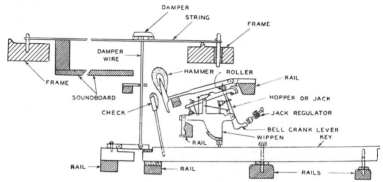

Fig. 5.14. A schematic view of the mechanism of a grand piano.

The mechanism of the grand piano is shown in Fig. 5.14. It will be seen that the strings and soundboard lie in a horizontal plane. In the upright or spinet piano, as shown in Fig. 5.13, the strings and soundboard lie in a vertical plane. This is one of the principal differences between the two instruments. It appears that a faster acting and smoother operating mechanism can be designed for the grand piano. The action of the mechanism may be deduced from Fig. 5.14. When the key is depressed, the wippen moves upward. The lever system of the wippen is coupled to the jack. As a result, the jack also moves upward. The jack, in turn, drives the hammer toward the string at relatively high speed. The hammer rebounds after it has struck the string and remains about one-half inch from the string if the key is held down. This is determined by the jack and jack regulator. While the key is depressed, the damping pad does not engage the string. The function of the pad is to increase the decay of the sounding note after the key has been released. In general, the blow of the hammer lasts for a time comparable to the period of the string. The struck point moves in advance of more remote parts of the string.

There are three pedals on a conventional piano. The right pedal is termed the sustaining pedal and removes all the dampers from all the strings so that the strings are damped only by the soundboard and end supports. The center pedal is termed the bass sustaining pedal and removes the dampers from all the bass strings. The left pedal is termed the soft pedal and reduces the sound output by reducing the length of stroke of the hammers, or by shifting the hammers so that fewer strings are struck, or by allowing the dampers to act.

2. Dulcimer. The dulcimer consists of a large number of horizontal strings stretched upon a frame mounted in an oblong box. The strings pass over bridges which are coupled to a soundboard (Sec. 5.3A2). The oblong box is equipped with legs. The instrument resembles a square piano without keys and is said to be a forerunner of the piano. The instrument is played by striking the strings with hammers—one held in each hand. The strings are equipped with dampers controlled by a foot pedal. The range of the dulcimer is from D_2 to E_6.

3. Electrical Analogy of Struck-string Instruments. A schematic view and the mechanical network of the vibrating system of struck-string instruments are shown in Fig. 5.15. The string is represented as a multi-resonant system containing elements which will resonate at the fundamental frequency and all the harmonics. These elements are represented as the masses m_1, m_2, . . . m_N, the mechanical resistances r_{M1}, r_{M2}, . . . r_{MN}, and the compliances C_{M1}, C_{M2}, . . . C_{MN}. The string is coupled to the air by means of a soundboard which is represented by the quadripole z_M. The air load is represented as the mass m_A and the

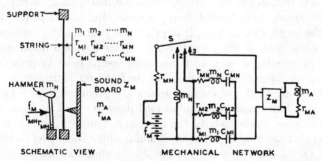

Fig. 5.15. Schematic view and the mechanical network of a struck-string instrument. In the mechanical network: f_M = driving force; r_{MH} = mechanical resistance of the generator; m_H = mass of the hammer; m_1, C_{M1}, r_{M1}, m_2, C_{M2}, r_{M2}, . . . m_N, C_{MN}, r_{MN} = lumped masses, compliances, and mechanical resistances representing the string; m_A and r_{MA} = mass and mechanical resistance of the air load; z_M = quadripole representing the soundboard which couples the string to the air.

mechanical resistance r_{MA}. The hammer is represented as the mass m_H. The force which imparts a velocity $\dot{x}$ to the hammer is represented as a battery of variable voltage. The force f_M causes the hammer to move with a velocity $\dot{x}$. This corresponds to closing the switch on contact 1. The kinetic energy of the hammer is transferred to the string, which causes it to vibrate. This corresponds to moving the switch to contact 3 so that contacts 1 and 3 are joined. The hammer then rebounds. This corresponds to moving the switch to contact 2 so that contacts 2 and 3 are joined. The resonant system transfers its energy to the surrounding air through the soundboard. Certain harmonics are suppressed, depending upon the striking point along the string. In the mechanical network, this is controlled by the mechanical resistance in series with the mass and compliance. From the mechanical network it will be seen that the ratio of the amplitudes of the fundamental and the various harmonics would be expected to be quite complex, owing to the complex nature of the striking force, the string, the quadripole coupling the string to the air, and the air load. The only factor which the musician can control to effect a change in the quality is the velocity with which the hammer strikes the string. This is quite apparent from the mechanical network of the system.

5.4   WIND INSTRUMENTS

A wind musical instrument consists of a resonator coupled to a means for interrupting a steady air stream at audio frequencies. The air stream may be interrupted in the following ways: by the air reed in the whistle, flue organ pipe, recorder, flageolet, ocarina, fife, flute, or piccolo; by the mechanical reed in the reed organ, accordion, harmonica, clarinet, saxophone, bagpipe, oboe, English horn, bassoon, and sarrusophone; by the lips in the bugle, French horn, trumpet, cornet, tuba, and trombone; and by the vocal cords in the human voice. Since the instrument must be capable of covering a range of tones, some means must be provided for obtaining the different frequencies. In the free-reed instruments, a number of reeds are used, one for each tone to be produced (Sec. 5.4B). In the organ, a large number of flue or reed pipes are used, at least one for each tone to be produced. In the lip-reed instruments, means are provided so that the length of the air column, which is coupled to the lips, may be varied, thereby making it possible to obtain a range of resonant frequencies (Sec. 4.11). In the mechanical-reed instruments, the pipe or horn, which is coupled to the reed, is provided with holes along the length, which may be opened or closed to obtain a series of resonant frequencies (Secs. 4.10 and 4.11). In the human voice, the resonant frequencies of the vocal cords is reinforced by the resonant properties of the vocal cavities. The resonant frequencies of the vocal cavities may be varied by altering the shape and dimensions of the cavities, the apertures between the cavities, and the mouth opening.

A. Air-reed Instruments

Air-reed instruments are actuated by a steady stream of air which flips in and out of a pipe or cavity at the resonant frequency of the system, thereby converting a steady air stream into an alternating one. Because of the nonlinear character of the exciting force, a number of the resonant elements in the system are excited, thereby producing a series of overtones in addition to the fundamental.

Modern examples of air-reed instruments are the whistle, flue organ pipe, recorder, flageolet, ocarina, flute, piccolo, and fife.

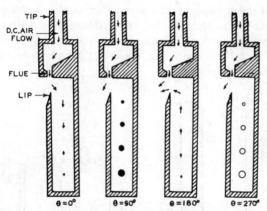

FIG. 5.16. The pressures and particle velocities in a flue organ pipe for a complete cycle. The pressures are shown as circles. The diameter of the circle is proportional to the magnitude of the pressure. A dark circle indicates a pressure above the atmospheric pressure, while a light circle indicates a pressure below the atmospheric. The particle velocities are indicated as arrows. The length of the arrow is proportional to the magnitude of the particle velocity. The direction of the arrow indicates the direction of particle-velocity flow.

1. Whistle and Flue Organ Pipe. The whistle and flue organ pipe consist of a cavity, a closed or an open pipe coupled to an air reed actuated by a steady air stream. The action of a whistle or flue organ pipe is depicted in Fig. 5.16. The pipe is actuated by the air stream which teeters from the inside to the outside of the pipe. When the air stream enters the pipe, it compresses the air in front of it, as shown in Fig. 5.16, $\theta = 0$ degrees. Ultimately the pressure in the pipe is built up to the equilibrium point and no more air will enter the pipe, as shown in Fig. 5.16, $\theta = 90$ degrees. The excess pressure in the pipe forces the air stream out, as shown in Fig. 5.16, $\theta = 180$ degrees. This causes the excess pressure to be relieved and finally causes a rarefaction or decreased pressure due to the inertia of the outrushing air, as shown in Fig. 5.16, $\theta = 270$ degrees. Then the air is again pulled in by the decreased pressure, as shown in Fig. 5.16, $\theta = 0$ degrees, and the cycle of four events is repeated at the resonant frequency of the system. The frequency of the complete cycle of events occurs at the resonant frequency of the closed pipe. The odd harmonics are also produced because the closed pipe resonates at these frequencies as well as the fundamental and, therefore, controls the teetering air stream at these frequencies (Sec. 4.10). The action is the same as for the fundamental frequency described above. The action of the open pipe is the same as for the closed pipe save that all the harmonics, both even and odd, are produced (Sec. 4.10).

Perspective and sectional views of two types of flue organ pipes are shown in Fig. 5.17, namely, an open metal pipe and a closed wood pipe with a movable piston.

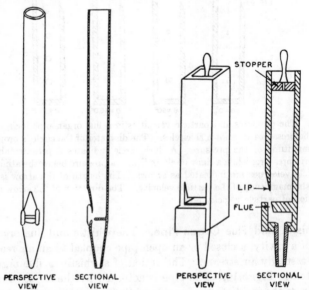

FIG. 5.17. Perspective and sectional views of an open-flue pipe of metal and a stopped-flue pipe of wood.

A whistle blown by the breath and used by policemen, athletic officials, etc., is shown in Fig. 5.18*A*. The action is the same as that of the flue organ pipe. The construction differs in that the resonating chamber is not a cylindrical pipe, but a Helmholtz resonator, that is, a chamber with a narrow neck. The resonant frequency of the whistle can be computed from the volume of the cavity and the area of the sound-radiating hole coupling the chamber to the outside air, as outlined in Sec. 4.4. The sound emitted by a whistle is almost a pure tone corresponding to the fundamental resonant f equency of the resonator, because the dimensions of the hollow body are small compared to the wavelength at the fundamental frequency. Therefore, the overtones which are due to resonances within the small body when the dimensions are comparable to the wavelength occur at a relatively high frequency. As a result, the overtones are highly suppressed by the inertance of the sound-radiating hole, and this accounts for the pure tone of the whistle.

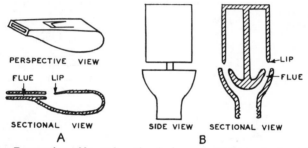

Fig. 5.18. Perspective, side, and sectional views of whistles. *A*. A small whistle blown by the breath. *B*. A large whistle blown by steam or air.

A whistle, blown by steam, air, or some other gas and used as a signal device in locomotives, steam tractors, power plants, etc., and in the musical instrument termed the calliope is shown in Fig. 5.18*B*. The action is the same as the flue organ pipe. The construction differs slightly in that the lip extends around the entire periphery.

A calliope is a musical instrument consisting of a group of whistles with frequencies corresponding to the notes of a musical scale. Each whistle is equipped with a valve connected to a key. The series of keys form a keyboard similar to that of a piano or organ. Steam or compressed air is used to actuate the whistles.

2. Recorder. A recorder is an instrument of the whistle class consisting of a mouthpiece, a fipple hole, and a cylindrical resonating pipe

(Fig. 5.19A). It is blown by the breath. The action is similar to the flue organ pipe. The pipe is equipped with eight finger holes for changing the resonant frequency of the air column (Sec. 4.10). Recorders of all sizes ranging in length from 1 to 10 feet have been built. These have been termed soprano, alto, tenor, and bass recorders.

3. Flageolet. The flageolet is an instrument of the whistle class consisting of a mouthpiece, a fipple hole, and a resonating pipe (Fig. 5.19B). The action of the system is similar to the flue organ pipe (Sec. 5.4A1). It is blown by the breath. The pipe is equipped with six finger holes for changing the resonant frequency (Sec. 4.10).

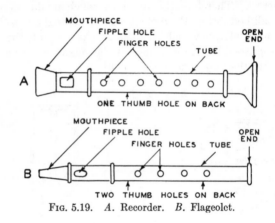

Fig. 5.19. *A*. Recorder. *B*. Flageolet.

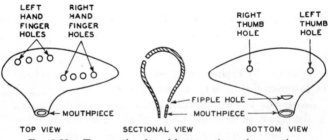

Fig. 5.20. Top, sectional, and bottom views of an ocarina.

4. Ocarina. The ocarina is an instrument of the whistle class consisting of a mouthpiece, a fipple hole, and a hollow air chamber equipped with eight finger holes and two thumb holes (Fig. 5.20). Various resonant frequencies may be obtained by covering the different finger holes in various combinations. The action of the system is similar to that of the flue organ pipe or whistle (Sec. 5.4A1).   The construction differs in that the resonating system is not a pipe, but a cavity-and-hole combination. The resonator consists of a cavity coupled to the atmosphere by several holes including the fipple hole.   The resonant frequency can be computed as outlined in Sec. 4.4.   The open holes are considered to be in parallel. Thus it will be seen that the resonant frequency is increased as the number of holes is increased, because the inertance is decreased as the number of holes is increased.   The ocarina covers about one and one-half octaves. Ocarinas are made of ceramic, plastic, or metal.

5. Flute.   The flute consists of two cylindrical sections joined by a conical section with one end open and the other end closed, as shown in Fig. 5.21A.   The embouchure, or blowhole, is placed a short distance from the closed end.   The effective length of the resonating air column is controlled by a number of holes which may be opened or closed by the fingers either directly or by means of keys (Sec. 4.10).   A change in the effective length of the resonant air column produces a change in the resonant frequency.   In this manner, a series of resonant frequencies corresponding to the musical scale can be obtained.

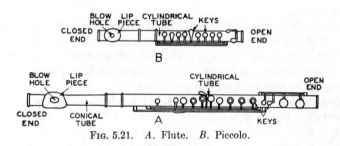

FIG. 5.21.   A. Flute.   B. Piccolo.

The action of the flute is shown in Fig. 5.22.   The air stream produced by the lips impinges upon the embouchure of the flute.   The resonant frequencies are excited by the air stream flipping in and out of the hole. The air stream flipping in and out of the blowhole for a complete cycle of events is shown in Fig. 5.22.   In Fig. 5.22, $\theta = 0$ degrees, the stream of air enters the blowhole.   In Fig. 5.22, $\theta = 90$ degrees, the air has stopped entering the blowhole, because a pressure has been built up within the flute by the incoming air.   In Fig. 5.22, $\theta = 180$ degrees, the excess pressure within the flute forces air out through the blowhole.   In Fig. 5.22, $\theta = 270$ degrees, the air has stopped leaving the blowhole.   Following this event, air is pushed in again, as shown in Fig. 5.22, $\theta = 0$ degrees, and the cycle of four events is repeated.

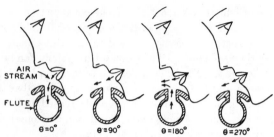

Fig. 5.22. The particle velocities at the embouchure of the flute for a complete cycle. The arrows indicate the direction of the air flow.

The fundamental frequency range of the flute is about three octaves, from C_4 to C_7.

The over-all length of the flute is $26\frac{1}{2}$ inches.

6. Finger-hole Mechanisms. Musical instruments of the single resonating-tube type employ either holes along the side, which may be opened or closed to change the effective length, or a means for changing the actual length for altering the resonant frequency. It is the purpose of this section to describe finger-hole mechanisms used in the wind instruments.

A great step forward in the use of finger holes was effected by the use of valves in the form of covers for finger holes and by the use of key, lever, and shaft systems for remote operation of a finger hole. These systems made it possible to open or close finger holes which lie too far apart or are too large to be stopped by the finger alone.

A valve which opens or closes a hole in the side of the tube of a musical instrument is shown in several views in Fig. 5.23. This valve may be operated either directly by pressing the finger against the cap, or indirectly by pressing a key which is connected to the valve by means of a lever or shaft system. A spring is used to keep the valve in the normal or unactuated position. The normal position of the valve may be either open or closed.

A direct-actuated valve system is depicted in Fig. 5.24A. This valve is normally open. The valve is closed by pressing the cap.

A simple key-actuated valve is shown in Fig. 5.24B. This valve is normally closed. The valve is opened by pressing the key.

A key-actuated remote valve with shaft coupling is shown in Fig. 5.24C. This valve is normally open. The valve is closed by pressing the key.

A key-actuated remote valve with lever coupling is shown in Fig. 5.24D. This valve is normally open. The valve is closed by pressing the key.

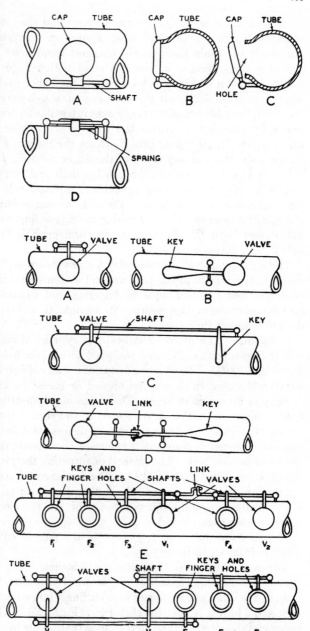

Fig. 5.23. A valve which opens or closes a hole in the side of a tube of a musical instrument. *A.* Top view. *B.* Sectional view with the valve closed. *C.* Sectional view with the valve open. *D.* Side view showing the spring which holds the valve either opened or closed in its normal or unactuated position.

Fig. 5.24. Valve, key, and finger-hole mechanisms. *A.* Direct-actuated valve which is closed by pressing the cap. *B.* Key-actuated valve in which the valve is opened by pressing the key. *C.* Key-actuated remote valve and shaft coupling in which the valve is closed by pressing the key. *D.* Key-actuated remote valve which is closed by pressing the key. *E.* Combination finger-hole and valve system with interconnected shaft and link mechanism. *F.* Combination finger-hole and valve system with interconnected shafts and lever mechanisms.

In some instances it is necessary to open and close a finger hole and operate a valve at the same time, because in many instruments it is impossible to obtain the required resonant frequencies by opening and closing a single hole at a time. A combination finger-hole and valve system is shown in Fig. 5.24E. Pressing the finger over the ring of the finger hole closes the hole in the same way as though the ring were absent. However, the addition of the ring makes it possible to perform other operations by coupling valves to the ring. For example, in Fig. 5.24E, pressing the ring or finger hole F_4 closes the hole at F_4 and valve V_2. Pressing the rings of any or all of the finger holes F_1, F_2, and F_3 closes valve V_1 as well as valve V_2 through the shaft and link system.

Another system employing a combination finger-hole and valve combination is shown in Fig. 5.24F. Pressing the ring or finger hole F_2 closes the hole at F_2 and valve V_2. Pressing the ring or finger hole F_3 closes the hole F_3 and valve V_2. Pressing the ring or finger hole F_1 closes the finger hole F_1 and valves V_1 and V_2.

There are many variations of the systems shown in Fig. 5.24. These systems, together with the fundamental systems described above, have revolutionized the technique of fingering and improved the tone of musical instruments by making it possible to obtain the exact and desired resonant frequency.

7. Piccolo. The piccolo consists of a cylindrical tube closed at one end and open at the other and equipped with a blowhole near the closed end (Fig. 5.21B). The effective length of the air column is controlled by a number of holes which may be opened or closed by the fingers either directly or by means of keys. The action of a cylindrical pipe with side holes is described in Sec. 4.10. Opening and closing the side holes alter the effective length and thereby change the resonant frequency. In this manner, a series of resonant frequencies corresponding to the musical scale can be obtained. The method of actuating the piccolo is similar to the flute, described in Sec. 5.4A5. The fundamental action and acoustical performance are similar to the flute. The essential difference being that the fundamental-frequency range of the piccolo is displaced an octave higher than the flute.

The fundamental-frequency range of the piccolo covers about three octaves, from D_5 to B_7.

The over-all length of the piccolo is 13 inches.

8. Fife. The fife consists of a cylindrical tube closed at one end and open at the other and equipped with six finger holes distributed along the tube and a blowhole near the closed end (Fig. 5.25). The fife is made of either metal or wood. The method of actuating the instrument by blowing with the breath is similar to that of the flute and piccolo. The action of the fife is similar to that of the flute, described in Sec. 5.4A5.

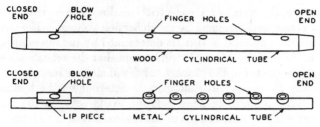

CLOSED END   BLOW HOLE   FINGER HOLES   OPEN END

WOOD   CYLINDRICAL   TUBE

CLOSED END   BLOW HOLE   FINGER HOLES   OPEN END

LIP PIECE   METAL   CYLINDRICAL   TUBE

Fig. 5.25.   Fifes of wood and metal.

9. Electrical Analogy of Air-reed Instruments.

A sectional view, an electrical analogy, and the acoustical network of the vibrating system of a whistle or flue organ pipe are shown in Fig. 5.26.   The air stream which flips in and out of the pipe is analogous to the electron stream in a deflection-type electron tube.   The action of the air stream flipping in and out of the pipe and the resonant system was described in Sec. 5.4A1.   In the flue organ pipe, the flow in and out of the pipe is determined by the side pressure on the air stream.   The direction of the side pressure on the air stream is determined by the air flow in and out of the pipe.   The frequency of the air flow in and out of the pipe is determined by the resonant

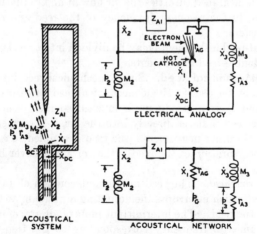

FIG. 5.26.   Sectional view, electrical analogy, and acoustical network of the flue organ pipe.   In the electrical analogy and acoustical network, p_{DC} = pressure in the direct-current air supply; $\dot{X}_{DC}$ = the direct volume current flow; r_{AG} = acoustical resistance of the equivalent acoustical generator; p_G = pressure of the equivalent acoustical generator; z_{A1} = acoustical impedance of the pipe; M_2 = inertance of the pipe opening; M_3 and r_{A3} = inertance and acoustical resistance of the air load; p_2 = pressure drop across M_2; X_1, $\dot{X}_2$, and X_3 = alternating-volume currents. p_3 = pressure drop across r_{A3}, or the output pressure.

frequencies of the pipe. In the electrical analogy, the current flow in the resonant circuit is determined by the deflection voltage. The frequency of the deflection voltage, in turn, is determined by the resonant frequencies of the resonant circuits. It will be seen that the acoustical elements of the pipe resonator or the elements of the analogous resonant electrical circuits play the most important part in determining the resonant frequencies. An appropriate phase relation exists between the deflection pressure and the volume current in the resonant system so that a steady volume current is converted into an alternating volume current. In the acoustical network the elements are the same as those in the electrical analogy, except that the electrical generator has been replaced by the acoustical generator p_G, having an internal acoustical resistance r_{AG}. The waveform of the acoustical generator is the same as that of the electrical generator. The output contains the fundamental and all the harmonics. Again it will be seen that the elements of the resonating system play the important part in determining the nature of the sound output.

B. Mechanical-reed Instruments

Mechanical-reed instruments are actuated by a steady air stream in which a mechanical reed throttles the air flow at an alternating frequency corresponding to the resonant frequency of the reed and an associated acoustical system.

Mechanical-reed instruments may be divided into two classes, namely, single- and double-reed instruments.

1. Single Mechanical Reed. Single mechanical-reed instruments are characterized by the use of a single mechanical reed for throttling a steady air stream and thereby producing a musical sound. When the sound output of the reed radiates directly into the air, it is termed a free-reed instrument. In other instruments, the reed is coupled to a resonant air column. The vibration of the reed is that of a cantilever bar described in Sec. 4.6A.

a. Reed Instrument. The reed instrument consists of an air-actuated vibrating reed which interrupts the actuating air stream at the vibration frequency of the reed. The intermittent puffs of air are, of course, sound with a fundamental frequency corresponding to the frequency of the pulsing air. The action of the reed is shown in Fig. 5.27. With the reed in a normal position, Fig. 5.27, $\theta = 0$ degrees, air is forced through the opening. The high velocity of the air through the opening reduces the pressure on the flow side of the reed in accordance with Bernoulli's theorem. Because of this reduced pressure, the reed is forced from its normal position so that the aperture through which the air moves is reduced, as shown in Fig. 5.27, θ = 90 degrees. When the air flow is reduced by the restricted passage, the pressure on the flow side of the reed increases and the reed springs back to the original position shown in Fig. 5.27, θ = 180 degrees. Then it moves beyond its original position, owing to inertial energy, as shown in Fig. 5.27, θ = 270 degrees. Now the opening is very large and the air flow is correspondingly large. As a result, the pressure on the flow side of the reed is again small and it returns to the normal position of Fig. 5.27, θ = 0 degrees, and the cycle of events is repeated at the resonant frequency of the system. The throttling action of the reed converts the steady air stream into a pulsating one of the saw-tooth type which contains the fundamental and all the harmonics (Sec. 6.3A).

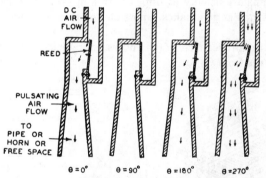

Fig. 5.27. The position of the reed and the particle velocities in a mechanical-reed instrument for a complete cycle. The arrows indicate the direction and the magnitude of the air flow.

The reed is used in two ways: in one form the output from the reed radiates directly into the air; in the other the output of the reed is coupled to a resonator. Modern examples of musical instruments in which the reed feeds directly into the air are the free-reed organ, accordion, and harmonica. Modern examples of musical instruments in which the reed is coupled to a resonator are the reed organ pipe, clarinet, saxophone, and bagpipe.

b. Free-reed Organ (Harmonium). The free-reed organ consists of a series of air-actuated free reeds tuned to the notes of the tempered scale and controlled by means of a keyboard (Fig. 5.28). The vibration of the reed is that of a cantilever bar, described in Sec. 4.6A. The resonant frequency of the reed may be determined from the dimensions of the reed and Eq. (4.35). The air supply which actuates the reeds is obtained from two pedal-operated bellows.

The arrangement of the key, valve, reed, and wind chest is shown in Fig. 5.28. The bellows, operated by the pedals, are connected to the wind chest. The reeds are mounted on the top of the wind chest. Each key actuates a valve. The valve controls the air supply to the reed. There is one reed for each key. Each reed is mounted upon a small brass frame, as shown in Fig. 5.28. The action of the reed has been described in Sec. 5.4B1a.

The air supply is obtained by means of foot pedals connected to a bellows, as shown in Fig. 5.28. It will be seen that the reed organ shown in Fig. 5.28 operates on a vacuum. Practically all free-reed organs built in the United States have been vacuum-operated. Some free-reed organs built in Europe have been pressure-operated.

Reed organs are usually equipped with stops for connecting banks of reeds. The stop action makes it possible to operate several reeds from a single key.

A typical small free-reed organ covers a fundamental frequency range of four octaves from C_2 to B_5.

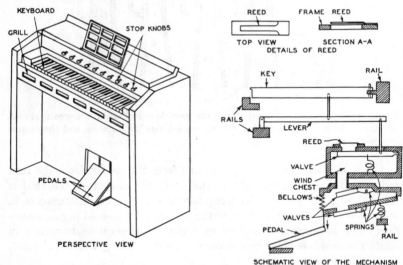

FIG. 5.28. A perspective view and a schematic view of the mechanism of a free-reed organ.

c. Accordion. The accordion consists of a series of air-actuated free reeds tuned to notes of a musical scale and controlled by means of a keyboard (Fig. 5.29). The air for actuating the reeds is supplied by a bellows worked by the player's arms. The air supply is either a pressure or a vacuum, depending upon whether the bellows are being compressed or expanded. Each key is connected to a valve which controls two different reeds. One reed operates upon a pressure and the other reed upon a vacuum In some instruments, two differently tuned reeds are used with each key so that one tone is produced when the bellows are expanded and another tone when they are compressed.

The vibration of the reed is that of a cantilever bar, described in Sec. 4.6A. The resonant frequency of a reed may be determined from the dimensions of the reed and Eq. (4.35). The reeds used in the accordion are similar to those employed in the free-reed organ, shown in Fig. 5.28. The action of the reed has been described in Sec. 5.4B1a.

Accordions are usually equipped with stops for connecting individual reeds to form banks of reeds. The stop action makes it possible to operate several reeds from a single key.

In playing the accordion, the left hand is used to expand and compress the bellows and to play the bass parts and the accompaniment. The right hand is used to play the melody.

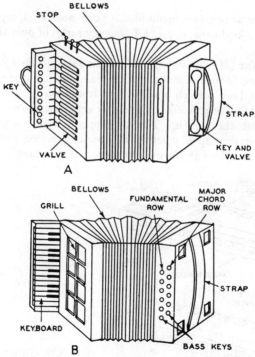

Fig. 5.29. Perspective views of accordions. A. Button keyboard. B. Piano keyboard.

The simple accordion, shown in Fig. 5.29A, is provided with 10 melody buttons or keys and two bass keys. In this accordion, each key produces two different tones, one tone is produced when the bellows are expanded and another when the bellows are compressed.

The piano accordion, shown in Fig. 5.29B, is provided with a two-octave piano-type melody keyboard and 12 bass and chord buttons. The white and black keys of the melody keyboard are analogous to those of a piano. The same tone is produced on both compression and expansion. The size of accordions has not been standardized. For example, piano accordions with melody keyboards up to four octaves have been built. The instrument shown in Fig. 5.29B is provided with 12 bass buttons arranged in two rows. One row produces bass tones, and the other row produces major chords. In some of the larger accordions, 120 bass buttons are used and arranged in six rows. The first and second rows produce bass notes. The third, fourth, fifth, and sixth rows produce the major, minor, dominant-seventh, and diminished-seventh chords, respectively.

The melody section of a medium-size piano accordion, represented by the piano-type keyboard, covers a frequency range of over three octaves from F_3 to A_6.

d. Harmonica (Mouth Organ). The harmonica consists of a series of free reeds tuned to the notes of a musical scale and mounted upon a wood, plastic, or metal box with channels leading from the reeds to orifices along one side of the box (Fig. 5.30). The air for actuating the reeds is supplied by breath from the mouth. The air supply is either a pressure or a vacuum, depending upon whether the person playing the instrument inhales or exhales through the instrument. Each channel is connected to

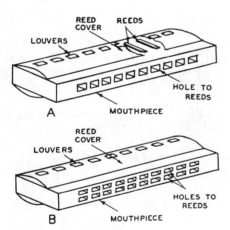

Fig. 5.30. Perspective views of harmonicas. *A*. Simple harmonica. *B*. Concert harmonica.

two differently tuned reeds. One reed operates upon pressure and the other reed upon vacuum. In this way each channel of the harmonica is capable of producing two tones of different frequencies—one frequency by a pressure and another frequency by a vacuum. In order to play a melody, the mouth encircles one or more holes.

There are three principal harmonicas, namely, the simple, the concert, and the chromatic. The simple harmonica, shown in Fig. 5.30A, consists of 10 holes, each of which can produce two different notes. The concert harmonica, shown in Fig. 5.30B, has two rows of holes. The reeds in the upper holes are tuned one octave higher than the reeds in the lower holes. The concert harmonica is louder and somewhat more pleasing. Both of these harmonicas are made in different keys. The chromatic harmonica, similar in appearance to the concert type, consists of two separate harmonicas placed one above the other, with the lower instrument tuned one semitone higher than the upper one. Inside the harmonica is a slide which is pushed by a knob on the right. When the slide is out, the upper holes are open and the lower holes are closed. When the slide is pushed in, the upper holes are closed and the lower holes are open. The chromatic harmonica makes it possible to play in different keys.

The reeds used in the harmonica, shown in the cutaway of the reed cover of Fig. 5.30, are similar to those employed in the free-reed organ, shown in Fig. 5.28. The vibration of the reed is that of the cantilever bar, described in Sec. 4.6A. The action of the reed has been described in Sec. 5.4B1a.

A typical 10-hole harmonica covers a fundamental frequency range of three octaves from C_3 to C_6.

e. Reed Organ Pipe. The reed organ pipe consists of an air-actuated reed coupled to a conical pipe (Fig. 5.31). In some reed organ pipes a combination conical and cylindrical pipe is used.

In general, the fundamental resonant frequency of the reed is made approximately the same as that of the pipe. The vibration of the reed is that of a cantilever bar, described in Sec. 4.6A. The fundamental resonant frequency of the reed may be determined from Eq. (4.35). The resonant frequencies of the pipe may be determined from the equations of Secs. 4.10 and 4.11. Since the reed and pipe are closely coupled, the resonant frequency is determined from the combination of the reed and the pipe. A change in the resonant frequency of the combination can be obtained by changing the resonant frequency of either the reed or the pipe. A reed organ pipe is usually tuned by changing the effective stiffness of the reed by altering the position of the wire which is in contact with the reed (Fig. 5.31). The system may also be tuned by changing the length of the pipe by the metal roll back in the side of the pipe near the open end (Fig. 5.31). The shape of the reed and the shape of the pipe determine the timbre or overtone structure. The process of adjusting and selecting these elements to produce the proper timbre is termed voicing. Reed organ pipes are usually classified as chorus and orchestral reeds. The reed organ pipes are usually designed to imitate the instruments of the orchestra, as, for example, the trumpet, tuba, oboe, and clarinet, as well as human voice.

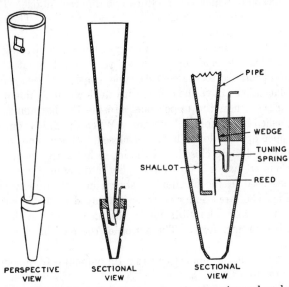

FIG. 5.31. Perspective and sectional views of reed organ pipe and reed mechanism.

f. Clarinet. The clarinet consists of a single mechanical reed coupled to a cylindrical tube with a flared mouth (Fig. 5.32). The effective length of the resonating air column is controlled by a number of holes which may be opened or closed by the fingers either directly or by keys (Sec. 4.10). A change in the effective length of the resonant air column produces a change in the resonant frequency.

With the reed in a normal position, Fig. 5.33, $\theta = 0$ degrees, air is forced through the opening between the reed and the mouthpiece. The

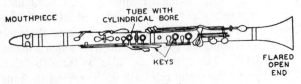

FIG. 5.32. Clarinet.

high velocity of the air through the opening reduces the pressure on the flow side of the reed in accordance with Bernoulli's theorem. Because of this reduced pressure, the reed is forced from its normal position so that the aperture through which the air moves is reduced, thereby reducing the air flow, Fig. 5.33, $\theta = 90$ degrees. When the air stream is reduced by the constricted passage, the internal pressure on the reed becomes greater and the reed springs back to the original position, as shown in Fig. 5.33, $\theta = 180$ degrees. Because of the inertial energy of the reed, it moves beyond its original position to the position in Fig. 5.33, $\theta = 270$ degrees. Now the opening is very large and the air flow is correspondingly large. As a result, the pressure on the flow side of the reed is small, and, as a result, it returns to the normal position of Fig. 5.33, $\theta = 0$ degrees, and the cycle of events is repeated. The fundamental variation in the air stream occurs at the fundamental frequency of vibration of the reed. The throttling action of the reed

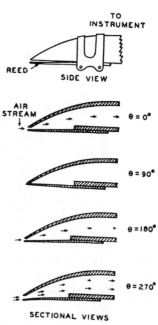

converts a steady air stream into a pulsating one of the saw-tooth type which contains the fundamental and all the harmonics (Sec. 6.3A). The reed when sounded alone produces a sound rich in harmonics. The quality is changed and improved when the reed is coupled to the cylindrical air column. The resonant air column which is coupled to the reed is for all practical purposes a pipe or horn closed at the reed end, because the acoustical impedance is very high at the closed end (Sec. 4.10). When the reed is attached to the pipe or horn, the motion of the reed is essentially controlled by the closely coupled resonant air column. The frequency of the emitted tone is varied by changing the effective length of the air column in the pipe by the openings in the side (Sec. 4.10). A change in the effective length of the air column produces a change in the resonant frequency.

The fundamental frequency range of the clarinet is over three octaves, from D_3 to F_6.

The over-all length of the clarinet is 25 inches.

Fig. 5.33. The shape of the reed and the particle velocities in a single mechanical-reed instrument for a complete cycle. The arrows indicate the direction and magnitude of the air flow.

g. Bass Clarinet. The bass clarinet covers a lower frequency range than the clarinet, described in the preceding section. In order to obtain this lower frequency range in a portable and easily handled instrument, the tube is doubled back on itself (Fig. 5.34). The sound output is considerably greater than that of the clarinet.

The bass clarinet covers a fundamental frequency range of three octaves, from D_2 to F_5.

The over-all length of the bass clarinet is 37 inches.

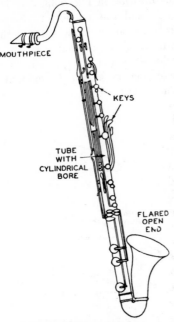

FIG. 5.34. Bass clarinet.

h. Saxophone. The saxophone consists of a single mechanical reed coupled to a conical tube with a slightly flared mouth. The effective length of the resonating tube is varied by a number of holes which are opened or closed by cover valves operated by keys (Figs. 5.35 and 5.36, Secs. 4.11 and 5.4A6).

The action of the saxophone is similar to the clarinet. A constructional difference is that the diameter of the tube at the reed end is relatively large. This means that the acoustical impedance will be relatively small. Therefore, the coupling between the pipe and the reed is not so intimate as in the case of the clarinet. The result is that the build-up of the vibration of the reed is very fast, which gives the sharp attack characteristic of the saxophone.

The conventional saxophone is made in many sizes which include soprano, alto, tenor, baritone, and bass. The soprano saxophone consists of a straight tube with a slightly flared mouth (Fig. 5.35). The others employ a curved mouth pipe and an upturned bell (Fig. 5.36).

Each type of saxophone covers a fundamental range of about two and

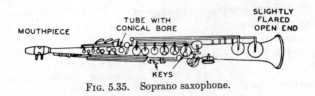

FIG. 5.35. Soprano saxophone.

one-half octaves, as follows: bass, Ab_1 to Db_4; baritone, Db_2 to Ab_4; tenor, Ab_2 to Eb_5; alto, Db_3 to Ab_5; and soprano, Ab_3 to Eb_6.

The over-all length of some of the saxophones is as follows: soprano, $15\frac{3}{4}$ inches; alto, $15\frac{3}{4}$ inches; tenor, $27\frac{1}{2}$ inches; baritone, 33 inches; and bass, 39 inches.

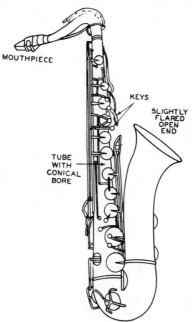

MOUTHPIECE

KEYS

SLIGHTLY FLARED OPEN END

TUBE WITH CONICAL BORE

FIG. 5.36.   Alto saxophone.

i. Bagpipe.   The bagpipe consists of one or more combinations of an air-actuated reed coupled to a resonating pipe (Fig. 5.37).   The air supply which actuates the reeds is obtained from a leather bag which serves as a reservoir.   Air is supplied to the bag by blowing with the breath.   Since the reeds are supplied by a steady air stream, the reeds sound continuously without interruption.   The bagpipe differs from other breath-blown musical instruments on account of the characteristic steady sound output.   There are usually two or three fixed-frequency reed-pipe combinations termed drones, and there is a reed-pipe combination of variable pitch termed the chanter.   The pipe of the chanter, as shown in Fig. 5.37, is provided with eight finger holes so that the discrete frequencies covering an octave can be obtained.   The chanter produces the melody, and the drones produce a harmonious steady tone.

The reed and pipe combination is similar to that of the reed-organ pipe shown in Fig. 5.31. In some instruments the reeds in both the chanter and the drones are of the single mechanical type, while in other instruments the reeds are a form of the double mechanical type as described in Sec. 5.4*B*2. In general, in the modern instrument the reeds in the drones are of the single mechanical type, while the reed in the chanter is a form of the double mechanical type.

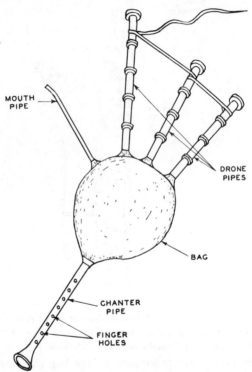

MOUTH
PIPE

DRONE
PIPES

BAG

CHANTER
PIPE

FINGER
HOLES

FIG. 5.37.   Perspective view of a bagpipe.

j. Electrical Analogy of the Single-reed Instrument. A schematic view, an electrical analogy, and the acoustical network of the vibrating system of a single-reed musical instrument is shown in Fig. 5.38. At the resonant frequency of the reed, the system may be represented by the lumped inertance M_1 and the lumped acoustical capacitance C_{A1}. In the case of the free reed, the air load z_{AP} is an acoustical resistance and a positive acoustical reactance. The mutual coupling between the reed and the aperture and air load is represented as M. The throttled air stream produced by the free reed is of the saw-tooth type which contains the fundamental and all the harmonics (Sec. 6.3A). The fundamental resonant frequency is determined by the resonant frequency of the reed (Sec. 4.6A). In the case of a reed coupled to a resonant pipe, the pipe and reed are usually tuned to the same frequency. The pipe load, under these conditions, is represented as z_{AP}. The mutual coupling between the reed and the aperture and the pipe load is represented as M. It will be seen that the electrical analogy is a conventional electronic oscillator. The acoustical network, under these conditions, is depicted in Fig. 5.38. The resonant frequencies are determined by all the elements of the vibrating system. However, in the case of a pipe- or horn-loaded reed, the pipe plays an important part in determining the resonant frequencies. In the acoustical network the elements are the same as those in the electrical analogy, except that the electrical generator has been replaced by the acoustical generator p_G, having an internal acoustical resistance r_{AG}. The waveform of the acoustical generator is the same as that of the electrical generator. The output contains the fundamental and all the harmonics. Again it will be seen that elements of the resonating system play an important part in determining the sound

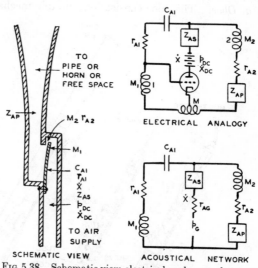

Fig. 5.38. Schematic view, electrical analogy, and acoustical network of a mechanical-reed instrument. In the electrical analogy and acoustical network, M_1, C_{A1}, and r_{A1} = inertance, acoustical capacitance, and acoustical resistance of the reed; M_2 and r_{A2} = inertance and acoustical resistance of the throttling aperture; z_{AP} = acoustical impedance of the pipe or horn or free space; z_{AS} = acoustical impedance of the supply system; X = alternating volume current; p_{DC} = pressure of the air-supply system; X_{DC} = direct volume current flow; r_{AG} = acoustical resistance of the equivalent acoustical generator; p_G = pressure of the equivalent acoustical generator; M = mutual coupling between branches 1 and 2.

output. The vibrating system of reed organ pipes is equipped with means for changing the stiffness of the reed and the length of the pipe (Sec. 5.4$B1e$). In the clarinet and the saxophone the resonant frequency of the reed and tube combination is determined primarily by the resonant frequency of the pipe or horn of the system (Secs. 4.10 and 4.11). The cavities of the mouth and the loading of the lips also influence the resonant frequency.

2. Double Mechanical Reed. Double mechanical-reed instruments are characterized by the use of two mechanical reeds for throttling a steady air stream and thereby producing a musical sound. All double-reed instruments combine a resonant air column with the vibrating reeds. Examples of double mechanical-reed instruments are the oboe, English horn, oboe d'amore, bassoon, and sarrusophone.

a. Oboe. The oboe consists of a double mechanical reed coupled to a

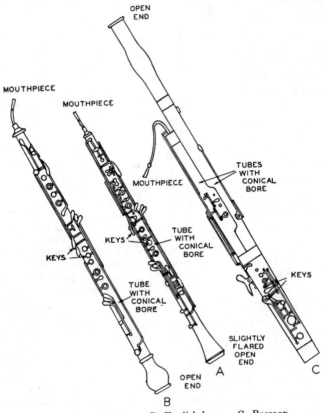

Fig. 5.39.   *A.* Oboe.   *B.* English horn.   *C.* Bassoon.

conical tube with a slightly flared mouth (Fig. 5.39A). The effective length of the resonating air column is determined by the number of holes which may be opened or closed by the fingers, either directly or by keys (Sec. 5.4A6). A change in the effective length of the resonant air column produces a change in the resonant frequency (Sec. 4.11). In this manner a series of resonant frequencies corresponding to the musical scale can be obtained.

The sound generator in the oboe, shown in Fig. 5.40, is a double reed. The action of the double reed in interrupting the air stream is depicted in Fig. 5.40. With the reeds in the normal position, Fig. 5.40, $\theta = 0$ degrees, air is forced through the opening between the reeds. The high velocity of the air through the opening reduces the pressure between the reeds in accordance with Bernoulli's theorem. Because of this reduced pressure, the reeds are forced closer together so that the aperture through which the air moves is reduced, thereby reducing the air flow, Fig. 5.40, $\theta = 90$ degrees. When the air flow is reduced by the constricted passage, the internal pressure on the two reeds increases and the reeds spring back to the original position, as shown in Fig. 5.40, $\theta = 180$ degrees. Owing to the inertial energy of the reeds, they move beyond the original position to the position shown in Fig. 5.40, $\theta = 270$ degrees. Now the opening is very large and the air flow is correspondingly large. Under these conditions, the internal pressures on the reeds are small, and, as a result, they return to the normal position of Fig. 5.40, $\theta = 0$ degrees, and the cycle of events is repeated at the resonant frequency of the system. The variation of flow in the air stream occurs at the frequency of vibration of the reed. The throttling action of reeds converts a steady air stream into a pulsating one of the saw-tooth type which contains the fundamental and all the harmonics (Sec. 6.3A). These variations are sound waves which are altered by the resonances in the tube of the oboe. The tube in the oboe is conical with a slight flare at the mouth. The acoustical length of the air column is controlled by finger holes

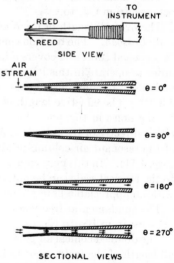

FIG. 5.40. The positions of the reeds and the air-particle velocities in a double mechanical-reed instrument for a complete cycle. The arrows indicate the direction and magnitude of the air flow.

and keys which operate valves on the holes. These openings alter the resonant frequencies of the air column, as described in Secs. 4.11 and 5.4A6.

The fundamental frequency range of the oboe is about three octaves, from B♭$_3$ to G$_6$.

The over-all length of the oboe is 24½ inches.

b. English Horn. The English horn consists of a double mechanical reed coupled to a conical tube terminating in a hollow spherical bulb with a relatively small mouth opening (Fig. 5.39B). This type of termination plays an important role in the unique timbre of this instrument. The key and fingering system and the action of the English horn are the same as the oboe.

The fundamental frequency range of the English horn is less than three octaves from E$_3$ to B♭$_5$.

The over-all length of the English horn is 35½ inches.

c. Oboe d'Amore. The oboe d'amore consists of a double mechanical reed coupled to a conical tube terminating in a hollow spherical bulb with a relatively small mouth opening. In appearance and in action it resembles the English horn. The instrument is keyed and played like the English horn.

The fundamental frequency range of the oboe d'amore is about three octaves, from G♯$_3$ to C♯$_6$.

The over-all length of the oboe d'amore is 28 inches.

d. Bassoon. The bassoon consists of a double mechanical reed coupled to a conical tube. It covers a lower frequency range than the oboe. In order to operate in this lower frequency range and still retain a portable and easily handled instrument, the tube is doubled back on itself (Fig. 5.39C). The effective length of the resonating air column is determined by the holes in the side which may be opened or closed by the fingers either directly or by keys (Sec. 5.4A6). A change in the effective length of the resonant air column produces a change in the resonant frequency (Sec. 4.11). In this manner a series of resonant frequencies corresponding to the musical scale can be obtained. The tube in the bassoon is conical with no appreciable flare at the mouth.

The fundamental frequency range of the bassoon covers about three octaves, from B♭$_1$ to E♭$_5$.

The doubled conical air column is about 93 inches in length. The over all length of the instrument is 48½ inches.

e. Contra Bassoon. The contra bassoon consists of a double mechanical reed coupled to a conical tube (Fig. 5.41). It covers a lower range than the bassoon. The total length of the air column is about 190 inches. The tube is folded several times so that the over-all length of the instruments is 50 inches. The effective length of the resonating air column is determined by the number of holes which may be opened or closed by the fingers either directly or by keys (Sec. 5.4A6). A change in the effective length of the resonant air column produces a change in the resonant frequency (Sec. 4.11). The tube in the contra bassoon is conical with a slight flare at the mouth.

The fundamental frequency range of the contra bassoon is B_0 to F_3.

f. Sarrusophone. The sarrusophone consists of a double mechanical reed coupled to a brass tube of conical bore with a flare at the open end (Fig. 5.42). The effective length of the resonating tube is varied by a number of holes which are opened or closed by cover valves operated by keys (Secs. 4.11 and 5.4A6). In this manner a series of resonant frequencies corresponding to the musical scale can be obtained.

The sarrusophone is made in several different sizes covering different

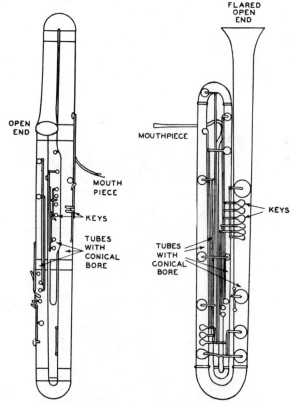

FIG. 5.41. Contra bassoon.   FIG. 5.42. Sarrusophone.

fundamental frequency ranges. The most common instrument is the contrabass type shown in Fig. 5.42.

The fundamental frequency range of the contrabass sarrusophone is Db_1 to B_3.

g. Electrical Analogy of the Double-reed Instruments. The electrical analogy of the double-reed instrument is the same as that of the single-reed instrument.

C. Organ (Combination Mechanical-reed and Air-reed Instrument)

The modern organ consists of a large number of flue and reed-type pipes controlled directly by manual and pedal keyboards and indirectly by stops, couplers, and pistons.

An organ console consisting of three manuals and a pedal keyboard, stops, tablet couplers, thumb and toe pistons, and swell pedals is shown in Fig. 5.43. A schematic view of the elements of a three-manual organ is depicted in Fig. 5.44. This diagram will be useful in the description of the elements of the organ which follows.

The organ may be built with one or more manuals and with or without a pedal keyboard. In general, the modern organ consists of two or more manuals and a pedal keyboard. The functions of the manuals for any organ up to five manuals are as follows:

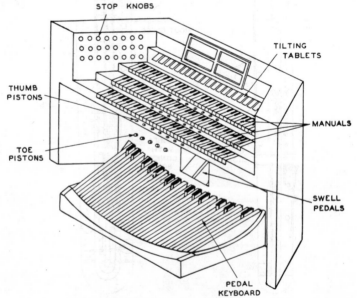

FIG. 5.43. Perspective view of an organ console.

The lowest or first manual controls the pipes in the choir organ. The second manual controls the pipes in the great organ. The third manual controls the pipes in the swell organ. The fourth manual controls the pipes in the solo organ. The fifth manual controls the pipes in the echo organ. The pedals control the pipes in the bass organ. In some cases, other instruments besides pipes have been added, as, for example, celesta, chimes, glockenspiel, drums, triangle, cymbals, and gongs. Additional manuals, as many as a total of seven, are used in some special organs. However, the conventional pipe organ employs two to five manual keyboards and a pedal keyboard.

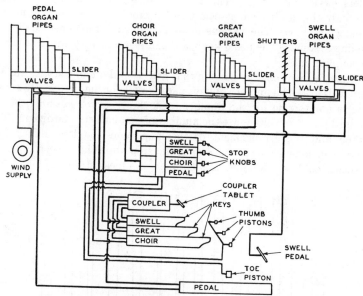

FIG. 5.44. Schematic diagram of the elements of an organ.

From the above, it will be apparent that the organ in reality is a combination of several organs. These organs are composed of different kinds of pipes. The flue organ pipe has been described in Sec. 5.4A1 and the reed organ pipe in Sec. 5.4B1e. The most common flue and reed organ pipes are shown in Fig. 5.45. The basic groups of pipes in the organ are the diapasons, flutes, strings, and reeds.

As the organ was evolved, it was found that the tone could be improved by grouping a set of pipes which were sounded simultaneously by depressing a single key. Thus, for each key many pipes are represented. Each combination of these pipes is termed a stop. The term stop is also applied to the knob along the sides of the console (Fig. 5.43). Each stop knob controls a certain group of pipes.

The pedal organ contains the largest flue and reed stops. The great organ, the choir organ, and the swell organ contain decreasing steps of flue and reed stops.

The swell organ is placed in an enclosed space. The space is equipped with shutters in the wall between the organ and the listener. The swell pedals on the organ control the opening and closing of these shutters. If there is a solo organ, it is also encased in a swell box.

A row of tilting tablets is located above the top manual. These tablets serve as couplers. The coupling arrangement makes it possible to actuate the mechanism associated with more than one key by merely pressing one key. The various manuals can be interconnected by the coupling system. The pedal keyboard can be hitched to any manual by the coupling means.

It may be mentioned in passing that organs and organ consoles are not standardized. The number of keys per manual is usually 61. However, there is no set rule for the number as is the case, for example, in the standard piano. Neither are the number of manuals standardized. The method and the arrangement of the stops and couplers differ. For example, in some of the newer organs the stop knobs have been replaced by tilting tablets. Finally, the number of pipes, stops, and couplers

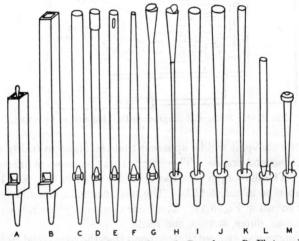

FIG. 5.45. Perspective views of organ pipes. *A.* Bourdon. *B.* Flute and diapason. *C.* Open diapason. *D.* Salicional diapason. *E.* Dulciana. *F.* Viola celeste. *G.* Bell gamba. *H.* Oboe. *I.* Trumpet. *J.* Tuba. *K.* Cornopean. *L.* Clarinet. *M.* Vox humana.

varies from organ to organ. The action also varies, depending upon the time the organ was built and the builder.

The organ action has undergone many changes through the years. In the very early instruments the keys were connected directly to the valves which controlled the air supply to a pipe or pipes. The objection in this action was the large force required to press a key.

Later, the action was improved by the use of a pneumatic system, in which the key operated a small pneumatic valve requiring only a small force, which in turn operated a valve connected with a pipe or pipes. The principal disadvantage was the slow action due to the slow rate of propagation of the air impulses from the key to the valve at the organ pipes. As a consequence, the console had to be located very close to the pipes. Except for small organs the pneumatic action has been replaced by the electropneumatic action or the electric action. These two systems will be described.

A schematic view of an electropneumatic action is shown in Fig. 5.46. The actual construction varies among the different builders. However, the general principle is the same in all electropneumatic actions. Referring to Fig. 5.46, it will be seen that when the key is pressed the electrical circuit is closed, which energizes the electromagnet. The electromagnet attracts valve A which opens chamber 1 to the atmosphere. Since there is now a pressure on the bottom of the diaphragm, the diaphragm moves upward, opening valve B and closing valve C. The air supply under pressure in air chamber 2 enters chamber 3 of the motor. The motor expands and opens valve D, which allows the air supply under pressure to enter chamber 5 and actuate the pipes. When the key is released, the electromagnet is no longer energized and valve A moves upward. The pressure is now the same on the two sides of the diaphragm, and the diaphragm

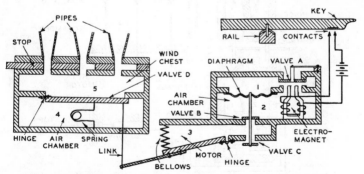

FIG. 5.46. Schematic view of an electropneumatic organ action.

returns to its normal or unstrained position owing to the inherent restoring force, which closes valve B and opens valve C.   The air under pressure in chamber 3 escapes and allows the spring to collapse the motor which closes valve D.   The supply of air to the pipes ceases, and the pipes are no longer actuated.

The advantages of the combination electric and pneumatic system are as follows: a very small force is required to press the key, the action is very fast, keys can be interconnected by merely throwing a switch, the console can be located at any distance from the organ, and the console can be moved about, because all the wires connecting the console to the different organs can be enclosed in a relatively small flexible cable.

A schematic view of an electropneumatic stop action for moving the slider which controls the pipes which a key can actuate is shown in Fig. 5.47.   The actual construction varies among the different builders. However, the general principle is the same for all electropneumatic stop actions.   Referring to Fig. 5.47, it will be seen that when the stop knob is pulled out the electrical circuit is closed, which energizes the electromagnet.   The electromagnet attracts valve A which opens chamber 1 to the atmosphere.   Since there is air under a pressure in chamber 2 on the left side of the diaphragm, the diaphragm moves to the right.   This moves valves B and C so that air under pressure from chamber 2 is

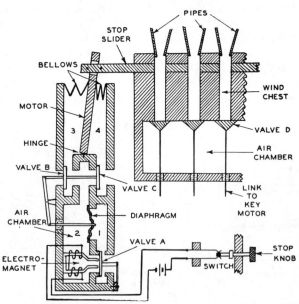

Fig. 5.47.   Schematic view of an electropneumatic organ-stop action.

admitted to chamber 3 and valve C so that the atmosphere is connected to chamber 4. This causes the bellows of chamber 3 of the motor to expand and the bellows of chamber 4 of the motor to contract. As a result, the slider moves to the right. In this position, all the pipes may be actuated when the keys are depressed, thereby actuating valve D, which is also valve D of Fig. 5.46. When the stop knob is pushed in, the electrical circuit is opened and valve A closes the chamber to the atmosphere. Since the pressure on the two sides of the diaphragm is the same, the diaphragm moves to the left to its unstrained position, because of to the inherent restoring force. This moves valves B and C so that air under pressure from chamber 2 is admitted to chamber 4 and the atmosphere is connected to chamber 3. This causes chamber 3 of the motor to contract and chamber 4 of the motor to expand. As a result, the slider moves to the left. In this position the pipes in this stop cannot be actuated when the key is depressed, thereby opening valve D.

In another arrangement, there is an electropneumatic valve for each pipe. Under these conditions, the stop action is electrically interconnected with the electrical key action. This makes it possible to eliminate the slider type of wind chest. Three different types of individual electropneumatic actions are shown in Fig. 5.48. Each action controls a single pipe. A simple pneumatic valve is shown in Fig. 5.48A. The disk valve is actuated by a bellows motor. Several pipes may be connected to a

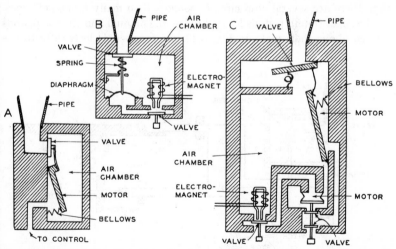

FIG. 5.48. Schematic views of organ-valve actions for single pipes. A. Pneumatic with disk valve and bellows motor. B. Electropneumatic with disk valve and diaphragm motor. C. Electropneumatic with double motor and valve actions.

single control, thereby constituting a stop.   In Fig. 5.48*B* a disk valve is actuated by a diaphragm.   Groups of pipes are connected together to form a stop by interconnecting the electrical system.   A compound motor and valve system is shown in Fig. 5.48*C*.   This is a high-speed mechanism because the electropneumatic valve controls a relatively small motor which requires a very small amount of air.   This motor, in turn, is used to actuate a much larger valve than the electrical valve.   Therefore, the action of the pipe valve will be faster, owing to the larger actuating valve and a correspondingly larger flow of air to the pipe valve.   This system is also used to actuate large pipes requiring large amounts of air.

A schematic arrangement of an all-electrical action is shown in Fig. 5.49.   The detailed construction varies among the different builders. However, the principle is the same in all electrical actions.   In the all-electric system, there is a valve for each pipe.   When the stop knob is pulled out as shown, closing switch S_1, the stop relay R is actuated, which closes the switch S_2.   Under these conditions, if the key is pressed, the switch S_3 will be closed, which energizes the electromagnet M_2.   This causes the armature A to move downward, opening valve V.   When the valve is open, air from the wind chest is admitted to the pipe.   When the stop knob is pushed in, the stop relay R is not actuated and the switch S_2 is open.   Under these conditions, the key circuit is open and M_2 will not be actuated when the key is pressed.   In an actual system, each key has several contacts, so that it is connected to several stop circuits.   In addition, there are several stop circuits connected to many groups of pipes.

The small thumb pistons located below the manual and the toe pistons located above the pedals, in Fig. 5.43, are connected electrically to the

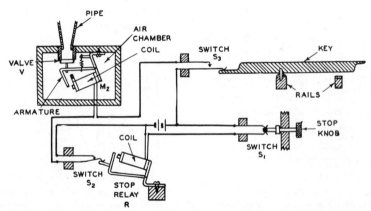

Fig. 5.49.   Schematic view of an electrical-organ action and electrical-stop action.

stop system so that a combination of stops can be brought in and out of action without reaching for a stop knob or stop tablet.

The swell pedals are used to obtain expression. The organ pipes are usually located in an enclosed space. This space is equipped with shutters in the portion between the listening area and the organ loft. The swell pedals control the openings of the shutters.

The fundamental-frequency range of an organ depends on the size. For example, in some small organs the fundamental-frequency range may be less than the piano, while in some large organs the fundamental-frequency range is C_0 to C_{10}.

Organs require large amounts of air for actuating the pipes because of the large size and number of pipes. The air pressure required is quite small, ranging from 3 inches of water in the early manually operated organs to 50 inches in some of the modern organs. In the early organs, large bellows operated by hand or foot were used. The air supply for modern organs is obtained from a centrifugal pump or fan driven by an electrical motor.

D. Lip-reed Instruments

Lip-reed instruments are actuated by a steady air stream supplied by the breath in which the lips throttle the steady air stream, thereby converting it into an alternating one. The frequency of the pulses correspond to the resonant frequency of the lips and an associated acoustical system. Modern examples of lip-reed instruments are the bugle, trumpet, cornet, trombone, French horn, and tuba.

1. Bugle. The bugle consists of a cupped mouthpiece coupled to a coiled tube with a slow rate of flare terminating in a bell-shaped mouth (Fig. 5.50). The length of the air column is fixed; therefore, the number of notes that can be sounded are limited. These notes are determined by the different resonant frequencies exhibited by the air column or horn (Sec. 4.11).

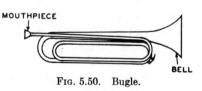

FIG. 5.50.   Bugle.

The notes of the conventional bugle are C_4, G_4, C_5, E_5, G_5, $B\flat_5$, and C_6.

The over-all length of the bugle is $22\frac{1}{2}$ inches.

In all the lip-reed brass wind instruments, the lips interrupt the steady air stream furnished by the lungs through the mouth and thereby furnish a complex sound wave to the flared horn. The action of the lips in throttling the air stream is as follows: The first position of the lips is shown in Fig. 5.51, $\theta = 0$ degrees. It will be noted the lips are closed. Under

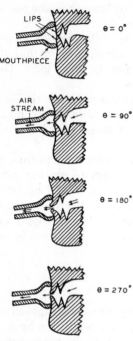

FIG. 5.51. The position of the lips and the particle velocities in a lip-reed instrument for a complete cycle. The arrows indicate the direction and magnitude of the air flow.

these conditions, the full pressure of the air acts upon the lips. This pressure opens the lips, as shown in Fig. 5.51, $\theta = 90$ degrees. The lips continue to open owing to the inertial energy of the lips and are shown at the maximum opening in Fig. 5.51, $\theta = 180$ degrees. At this point the velocity of the air through the lips is high. This high velocity of the air causes a reduction in air pressure upon the lips in accordance with Bernoulli's theorem. Furthermore, the force tending to close the lips is high, since the lips are at the maximum opening and therefore at maximum stress. These two effects conspire to reverse the motion of the lips, and the closing process begins. The lips are partially closed in Fig. 5.51, $\theta = 270$ degrees. The closing motion of the lips continues until the lips are closed, as shown in Fig. 5.51, $\theta = 0$ degrees, and the cycle of events is repeated at the resonant frequency of the system. The throttling action of the lips converts a steady air stream into a pulsating one of the saw-tooth type, which contains the fundamental and both even and odd harmonics (Sec. 6.3A). In this way a complex sound wave having a fundamental frequency of the fundamental vibration frequency of the lips is fed to the horn. In general, the fundamental resonant frequency of the lips must correspond to the resonant frequency of the horn, because the coupling between the lips and the horn is quite loose. However, it is possible to force the resonant frequency of the combination of the lips and the horn slightly up or down by a change of the tension in the lips and thereby change the resonant frequency of the system.

2. Trumpet. The trumpet consists of a cupped mouthpiece coupled to a coiled tube about 6 feet in length with a slow rate of flare terminating in a bell-shaped mouth (Fig. 5.52A). The first one-third of the tube is practically cylindrical, while the remainder is conical, save for the last 12 inches which flares rapidly into the bell-shaped mouth.

The action of the lips in generating the sound for actuating the horn is the same as that of the bugle, described in Sec. 5.4D1.

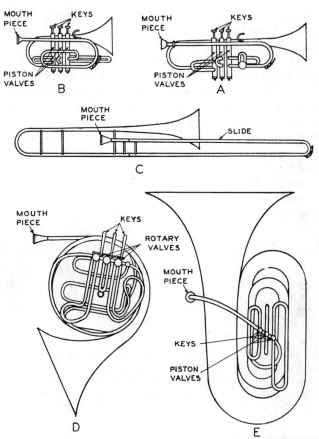

Fig. 5.52. *A.* Trumpet. *B.* Cornet. *C.* Trombone. *D.* French horn. *E.* Tuba.

Piston valves are provided for changing the length of the air column, as described in Sec. 4.11. The action of the valves is depicted in Fig. 5.53. Depressing the valve adds a length of tube. The trumpet is equipped with three valves, each of which adds a different length of tubing.

In this way, it is possible to obtain eight different lengths for the resonating tube (Sec. 4.11 and Fig. 4.28*A*). In this manner, it is possible to obtain a series of resonant frequencies corresponding to the musical scale.

A tuning slide or bit is provided so that the resonant frequencies of the trumpet can be made to coincide with other instruments.

A mute in the form of a pear-shaped piece of metal or plastic which fits into the bell is used to change the quality and attenuate the sound output.

The trumpet covers a fundamental-frequency range of about three octaves, from E_3 to $B\flat_5$.

The over-all length of the trumpet is $22\frac{1}{2}$ inches.

3. Cornet. The cornet consists of a cupped mouthpiece coupled to a coiled tube about 6 feet in length with a slow rate of conical flare terminating in a bell-shaped mouth (Fig. 5.52B). The appearance of the cornet is similar to that of the trumpet. The valve system is also similar, the essential difference being that the cornet is somewhat more compact and the bore is conical rather than cylindrical. The action of the lips in generating the signal for actuating the horn is the same as that of the bugle, described in Sec. 5.4D1.

The cornet covers a fundamental frequency range of about three octaves, from G_3 to $B\flat_5$.

The over-all length of the cornet is 14 inches.

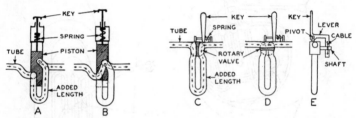

FIG. 5.53. Piston and rotary valves. *A.* Piston valve with key depressed which adds a length of air column. *B.* Piston valve in free position which shunts out added length of air column. *C.* Rotary valve with key depressed which adds a length of air column. *D.* Rotary valve in free position which shunts out added length of air column. *E.* Side view of rotary valve showing lever and rotating-shaft mechanism.

4. French Horn. The French horn consists of a coiled tube about 12 feet in length with a slow rate of conical flare terminating in a large bell-shaped mouth (Fig. 5.52D). The French horn is equipped with three additional lengths of tubing which may be added to the length of the horn by means of valves (Sec. 4.11).

The modern type of French horn employs rotary valves. The action of the rotary valves is depicted in Fig. 5.53. With this valve system and the three lengths of tubing, it is possible to obtain eight different lengths for the resonating tube (Sec. 4.11 and Fig. 4.28A). In this manner, it is possible to obtain a series of resonant frequencies corresponding to the musical scale.

The mouth of the French horn is large, which makes it possible for the player to insert his hand and thereby raise or lower the pitch or produce muted effects.

The fundamental frequency range of the French horn is over three octaves, from B_1 to F_5.

The over-all length of the French horn is 22¾ inches.

5. Trombone. The trombone consists of a cupped mouthpiece coupled to a cylindrical U-shaped tube, about 9 feet in length, doubled upon itself and connected to a conically tapered section terminating in a bell-shaped mouth (Fig. 5.52C). The telescopic section makes it possible to vary the length of the tube and thereby alter the resonant frequency (Sec. 4.11 and Fig. 4.28B). The trombone is the only instrument except those of the violin family in which it is possible to obtain a continuous frequency glide through the fundamental-frequency range. The pitch is determined by the position of the slide, the player's lips, and the air pressure; therefore, the player must have an accurate sense of pitch and the ability to produce the correct note almost instantly.

The action of the lips in generating the sound for actuating the horn is the same as that of the bugle, described in Sec. 5.4D1.

The fundamental-frequency range of the trombone is about two and one-half octaves, from E_2 to B_4.

The over-all length of the trombone is 45 inches.

6. Bass Trombone. The bass trombone resembles the conventional trombone in shape, the essential difference being that it is larger and covers a lower frequency range. The fundamental-frequency range of the bass trombone is about three octaves, from A_1 to $G\flat_4$.

7. Tuba. The tuba consists of a cupped mouthpiece coupled to a coiled tube about 18 feet in length with a slow rate of conical flare terminating in a large bell-shaped mouth (Fig. 5.52E).

Piston valves are provided for changing the length of the air column, as described in Sec. 5.4D2. The action of the valve is depicted in Fig. 5.53. The tuba is usually equipped with three valves, each of which adds a different length of tubing. In this way it is possible to obtain eight different lengths for the resonating tube (Sec. 4.11 and Fig. 4.28A). In some of the newer instruments a fourth valve and another length of tubing have been added, which makes it possible to obtain a larger number of different lengths for the resonating air column and a correspondingly greater number of different resonant frequencies.

The action of the lips in generating the sound for actuating the horn is the same as that of the bugle, described in Sec. 5.4D1.

The fundamental-frequency range of the tuba is about three octaves, from F_1 to F_4.

The over-all dimension of the tuba is 39 inches.

Besides the conventional tuba described above, the tuba is made in many sizes and forms. The largest is termed the sousaphone.

8. Electrical Analogy of Lip-reed Instruments. A schematic view, an electrical analogy, and the acoustical network of the vibrating system of a lip-reed musical instrument are shown in Fig. 5.54. The resonant frequency of the lips is determined by the inertance M_1 and the acoustical capacitance C_{A1} of the lips. The lips are pressed against the mouthpiece. The tension of the lips determines the acoustical capacitance C_{A1} and, hence, the resonant frequency of the lips, because the inertance M_1 of the lips is for all practical purposes fixed. The lips are brought into tune with the resonant frequency of the horn z_{AH}. The mutual coupling between the lips and the resonant tube represented by M aid in bringing the two resonances together. The throttling action of the lips converts a steady air stream into a pulsating one of the saw-tooth type, which contains the fundamental and all the harmonics (Sec. 6.3A). It will be seen that the electrical analogy is a conventional electronic oscillator, which converts a steady current into an alternating current. A vacuum-tube oscillator can be built to deliver the waveform produced by the lips. In the acoustical network the elements are the same as those in the electrical

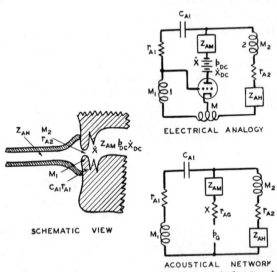

FIG. 5.54. Schematic view, electrical analogy, and acoustical network of a lip-reed type instrument. In the electrical analogy and acoustical network, p_{DC} = pressure of the air supply system; $\dot{X}_{DC}$ = direct volume current flow; M_1, C_{A1}, and r_{A1} = inertance, acoustical capacitance, and acoustical resistance of the lips; M_2 and r_{A2} = inertance and acoustical resistance of the aperture formed by the lips; z_{AH} = acoustical impedance of the horn; r_{AG} = acoustical resistance of the equivalent acoustical generator; p_G = pressure of the equivalent acoustical generator; z_{AM} = acoustical impedance of the mouth; M = mutual coupling between branches 1 and 2.

analogy, except that the electrical generator has been replaced by an acoustical generator p_G having an internal acoustical resistance r_{AG}. The waveform of the acoustical generator is the same as that of the electrical generator. The output contains the fundamental and all the harmonics. The amplitudes of these frequencies are modified by the resonant characteristics of the horn z_{AH}. Again it will be seen that the elements of the resonating system play an important part in determining the nature of the sound output.

E. Vocal-cord Reed Instrument or Human Voice

The voice mechanism, shown in Fig. 5.55, consists of three parts: the lungs and associated muscles for maintaining a flow of air; the larynx for converting the steady air flow into a periodic modulation; and the vocal cavities of the pharynx, mouth, and nose which vary the relative harmonic content of the output of the larynx. The vocal cords do not receive excitation at the frequency of vibration. The source of power is the steady air stream.

The voice mechanism is analogous to the electronic oscillator in that it

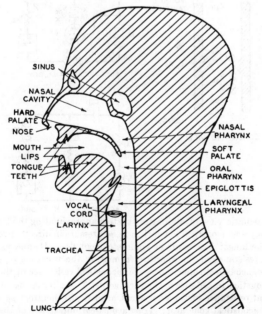

FIG. 5.55. Sectional view of the head showing the voice mechanism. (*After Olson, Acoustical Engineering, D. Van Nostrand Company, Inc., Princeton, 1957.*)

converts a direct-current flow into a pulsating flow. The elements of a simplified larynx are shown in Fig. 5.56. The electronic analogy is also shown in Fig. 5.56. The electronic system may be replaced by a generator having an internal electromotive force p_G and an internal resistance r_{AG}. M represents the mutual coupling between branch 1 and branch 2 of Fig. 5.56. The acoustical circuit under these conditions is depicted in Fig. 5.56. The frequency of the vibration is governed by all the elements of the vibrating system, that is, the acoustical capacitance C_{A1}, of the vocal cords incurred by tension, the inertance M_1, and the acoustical resistance r_{A1}, of the vocal cords, the inertance M_2, and acoustical resistance r_{A2}, of the aperture and the load acoustical impedance z_{AV}, due to the vocal cavities. A schematic view and the acoustical network of the vocal cavities can be seen in Fig. 5.57. This shows that the nature of the input acoustical impedance z_{AV} to the acoustical cavities is extremely complex. The inertances M_1, M_2, and M_3 and the acoustical capacitances C_{A1} and

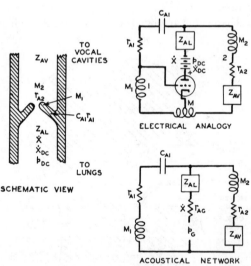

FIG. 5.56. Schematic view of a vibrating system approximating the larynx. In the electrical analogy and acoustical network, p_{DC} = pressure in direct-current air supply furnished by the lungs; X_{DC} = direct volume current flow; M_2 and r_{A2} = inertance and acoustical resistance of the aperture formed by the vocal cords; M_1, C_{A1}, and r_{A1} = inertance, acoustical capacitance, and acoustical resistance of the vocal cords; z_{AV} = input acoustical impedance to the vocal cords; $\dot{X}$ = alternating volume current; r_{AG} = acoustical resistance of the equivalent acoustical generator; p_G = pressure of the equivalent acoustical generator; z_{AL} = acoustical impedance of the trachea and lungs; M = mutual coupling between branches 1 and 2. (*After Olson, Acoustical Engineering, D. Van Nostrand Company, Inc., Princeton, 1957.*)

C_{A2} can be varied by changing the sizes of the apertures and the volumes of the cavities.

The oscillation of the vocal cords is of the relaxation type rather than the conventional sinusoidal variation. This is borne out by the rapid starting and stopping in the case of some sounds. The oscillator shown in Fig. 5.56 will produce waves of the relaxation type, provided that the circuit constants and the nonlinear elements are suitable. The wave shape of a relaxation oscillator corresponds to the general wave shape of the output of the vocal cords shown in Fig. 5.58. The output of the vocal cords was measured with a pressure microphone in the pharynx, with the mouth and nose cavities damped. A saw-tooth wave contains the fundamental and all the harmonics. Therefore, the generator p_G should produce the fundamental frequency and all the harmonics of the fundamental frequency.

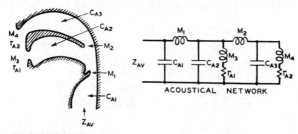

SCHEMATIC VIEW

FIG. 5.57. Schematic sectional view and the acoustical network of the vocal cavities. In the acoustical circuit, z_{AV} = input acoustical impedance to the vocal cavities; C_{A1} = acoustical capacitance of the laryngeal pharynx; M_1 = inertance of the narrow passage determined by the epiglottis; C_{A2} = acoustical capacitance of the mouth cavity; M_2 = inertance of the passage connecting the mouth and nasal cavities; M_3, M_4, r_{A1}, and r_{A2} = inertances and acoustical resistances of the mouth and nose openings and the air load upon these openings. (*After Olson, Acoustical Engineering, D. Van Nostrand Company, Inc., Princeton, 1957.*)

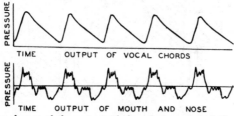

FIG. 5.58. Wave shapes of the output of the vocal cords and the mouth and nose for the vowel sound ĕ. (*After Olson, Acoustical Engineering, D. Van Nostrand Company, Inc., Princeton, 1957.*)

When the vocal cords are set into vibration as outlined above, the output of the larynx consists of a steady stream with superimposed impulses (Fig. 5.58). This pulsating air stream passes through the air cavities of the head. The harmonic content of the output is modified because of the discrimination introduced by the acoustical network of Fig. 5.57. The effect of the vocal cavities is illustrated in Fig. 5.58, which shows the wave shape of the sound output of the mouth and nose corresponding to the wave shape of the output of the vocal cords. When the shape of the vocal cavities is altered, the acoustical elements of the acoustical network of Fig. 5.57 are altered, which in turn alters the output harmonic content. These changes together with a change in the fundamental frequency of the vocal cords make it possible to produce an infinite number of different sounds. The tongue plays the major role in altering the shape of the vocal cavities. The shapes of the vocal cavities for four vowel sounds are shown in Fig. 5.59. It will be seen that the mouth

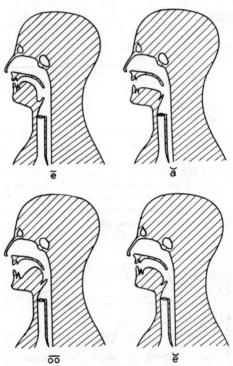

FIG. 5.59. Sectional views of the head showing the configuration of the mouth cavity for different vowel sounds. (*After Olson, Acoustical Engineering, D. Van Nostrand Company, Inc., Princeton, 1957.*)

opening, tongue, and epiglottis are the principle elements which are altered in these examples. Of course, the fundamental frequency of the vocal cords is also different in the four examples.

The true vowels and diphthongs are produced by the above outlined resonance method. The so-called unvoiced consonants, as, for example, S, are produced by air from the lungs passing over the sharp edges and through the narrow passages in the various parts of the mouth and nose. The vocal cords are not used in the production of these sounds. The voice consonants are produced by a combination of the two systems.

From the foregoing it will be seen that the voice mechanism consists of a number of acoustical elements which can be varied by the person at will to produce a wide variation of tones differing in frequency, quality, loudness, duration, growth, and decay.

The fundamental-frequency range of the human voice in singing is about two octaves. The fundamental-frequency ranges of the different voices are as follows: bass, E_2 to D_4; baritone, A_2 to G_4; tenor, D_3 to C_5; alto, G_3 to F_5; and soprano, C_4 to C_6. These are average values, and there is considerable variation among different individuals.

5.5 PERCUSSION INSTRUMENTS

As the name implies, in percussion instruments the vibrations are excited by a blow upon the vibrating system. The vibrating system may be a bar, rod, plate, membrane, or bell. There are two general classifications of percussion instruments, namely, definite and indefinite pitch.

A. Definite-pitch Instruments

Definite-pitch percussion instruments emit one or more musical tones which can be identified as possessing a definite pitch or frequency. These include the tuning fork, xylophone, glockenspiel, celesta, chimes, bells, and kettledrums.

1. Tuning Fork. The tuning fork, as the name implies, is used as a standard-frequency sound source. It consists of two cantilever bars attached to a common base (Fig. 5.60A). In the case of a single cantilever bar, a very massive base must be used in order to reduce loss of the vibration energy to the base, because the force exerted by the bar at the base is very large. However, in the

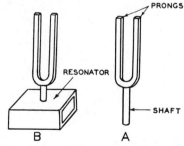

FIG. 5.60. *A.* Perspective view of a tuning fork. *B.* Perspective view of a tuning fork mounted on a resonating box.

tuning fork, since the two bars vibrate in opposition, the force exerted upon the external supporting means is small and a massive base is not required. This is the reason for the use of the double bar. The resonant frequency of a tuning fork may be determined from the dimensions of the prongs and Eq. (4.35).

The output of the tuning fork may be increased by mounting the base of the fork upon a hollow box open at both ends (Fig. 5.60B). The resonant frequency of the air column in the box is tuned to the resonant frequency of the fork. The vibration at the base of the fork is transmitted to the walls of the box, which vibration in turn is transmitted to the air column. The vibration of the air column produces radiation of sound. In this manner, the coupling between the fork and the air is improved and, thereby, a larger sound output is produced.

2. Xylophone. The xylophone consists of a number of resonant metal or wood bars with frequencies corresponding to the notes of a musical scale and arranged like the keyboard of a piano (Fig. 5.61). The bars are mounted in a horizontal position and supported upon soft material at the two nodal points. The vibration is the same as that of a free bar (Sec. 4.6B). The resonant frequency of a bar may be determined from the dimensions of the bar and Eq. (4.39). The fundamental frequencies of the different bars correspond to the tones of the musical scale. A

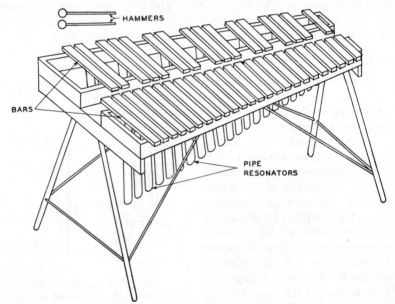

FIG. 5.61. Perspective view of a xylophone.

resonator in the form of a pipe is coupled to each bar to provide improved coupling between the bar and the air and thereby increase the sound output.   The bars are set into vibration by being struck by felt-covered hammers.

Xylophones are made in fundamental-frequency ranges of two to four octaves.   The usual frequency range is C_3 to E_7.

3. Marimba.   The marimba is essentially an enlarged version of the xylophone (see Sec. 5.5A2).   The fundamental-frequency range is usually about five octaves from F_2 to F_7.

4. Glockenspiel (Orchestra Bells). The glockenspiel consists of resonant bars with frequencies corresponding to the musical scale and arranged like the keyboard of a piano (Fig. 5.62). The bars are mounted in a horizontal position and supported upon soft material at the two nodal points. The vibration is the same as that of a free bar (Sec. 4.6B).   The resonant frequency of a bar may be determined from the dimensions of the bar and Eq. (4.39).   The fundamental frequencies of the different bars correspond to the tones of the musical scale. The bars are set into vibration by being struck by wooden hammers (Fig. 5.62).

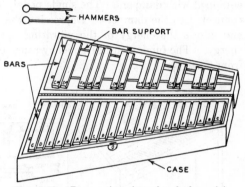

Fig. 5.62.   Perspective view of a glockenspiel.

The bell lyre is another form of the glockenspiel. A set of steel bars is mounted on a lyre-shaped frame (Fig. 5.63). It is played by striking the bars with a ball-shaped hammer. The bell lyre is used in marching bands. The frame is held in one hand and the hammer in the other.

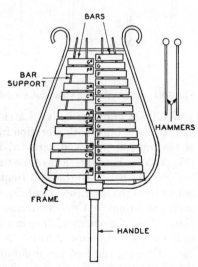

Fig. 5.63.   Front view of a bell lyre or glockenspiel.

The glockenspiel is made in various fundamental-frequency ranges up to three octaves, in the frequency range somewhere between C_3 to C_6.

5. Celesta. The celesta consists of a series of resonant steel bars with frequencies corresponding to the notes of a musical scale and actuated by being struck by hammers connected to keys forming a keyboard (Fig. 5.64). The vibration is the same as that of a free bar (Sec. 4.6B). The steel bars are suspended over wooden resonating boxes, which improve the coupling to the air and thereby increase the sound output. The action, shown in Fig. 5.64, is similar to that of a piano. The bars are equipped with dampers. The single pedal, termed a sustaining pedal, is connected to the dampers. Depressing the pedal removes the dampers and allows the energy of the vibrating bar to be dissipated as sound energy. The fundamental-frequency range of the celesta is four octaves, from C_4 to C_8.

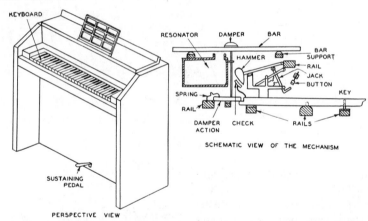

FIG. 5.64. A perspective view and a schematic view of the mechanism of a celesta.

6. Chimes (Tubular Bells). A chime of tubular bells consists of resonant tubes of brass with vibration frequencies corresponding to the notes of the musical scale and actuated by being struck by wooden mallets (Fig. 5.65). The vibrations of the tubes are the transverse modes of a free bar (Sec. 4.6B). The resonant frequency of a hollow tube may be determined from the dimensions and Eq. (4.35). The tubes are suspended by means of strings in a wooden frame and arranged like the keyboard of a piano. The fundamental frequencies of the different tubes correspond to the notes of the musical scale. A foot pedal is connected to a damping bar. Pressing the foot pedal damps the vibrations of the tubes. The tubular bells are designed as a substitute for real bells.

The chime of tubular bells is made in various fundamental-frequency ranges up to two octaves, in the frequency range somewhere between G_1 to G_3.

7. Bell. A bell is a metallic instrument shaped like an inverted cup with a flared mouth and containing a clapper which actuates the bell. The elements of the bell are shown in Fig. 5.66. The clapper is hung loosely inside the bell, and when the bell is rotated, the clapper strikes the bell. The fundamental frequency of the bell is a complex function

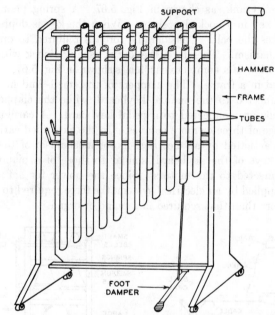

FIG. 5.65. Perspective view of orchestra chimes, or tubular bells.

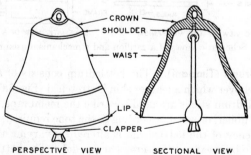

FIG. 5.66. Perspective and sectional views of a bell.

of the internal diameter, the wall thickness, and the density and moduli of the metal.

8. Carillon. A carillon consists of a set of fixed bells tuned to a musical scale and arranged to be actuated by being struck by hammers. The bells may be struck by hammers held in the hand of the carillonneur. The bells may be struck by the clapper in conjunction with a mechanism coupled to a keyboard, which is played by the carillonneur. In the classical carillon, the clapper of the bell is coupled to a key by means of a cable and bell crank, as shown in Fig. 5.67. A spring maintains the clapper and key in a cocked position. When the key is depressed, the clapper strikes the bell. Considerable force is required to operate the massive mechanism. Therefore, the key is usually struck with the fist. A schematic view of a complete carillon is shown in Fig. 5.67. The bells are supported in a frame. The clappers of the small- and medium-size bells are connected to the manual keyboard, while the clappers of the large-size bells are connected to a pedal keyboard. A carillon usually covers a range of three or more octaves. In the electrified carillons, the clappers are actuated by an electrical-motor mechanism of the solenoid type. The keys of the keyboard, similar to those of a piano, operate switches connected to the motors. Since the power for actuating the clapper is supplied by an electrical system, the force required to depress a key is no more than that required in a piano or organ.

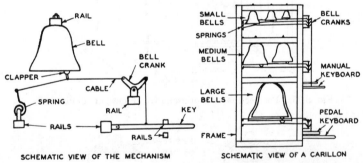

FIG. 5.67. Schematic views of a carillon and a mechanism used in carillons.

9. Kettledrums (Timpani). The kettledrum consists of a large hemispherical bowl over which a leather skin is stretched (Fig. 5.68). Various types of kettledrum sticks are used to strike the membrane, for example, sponge, felt, rubber, or wood, according to the tone required. The fundamental frequency of the kettledrum is altered by varying the tension of the membrane by means of the screws and pedal. The kettledrum emits a low-frequency sound of definite pitch. By means of the pedal, the pitch can be varied so quickly and accurately that a melody may be played. Timpani are usually made in two sizes. The fundamental-frequency ranges of the two timpani are as follows: the smallest $B\flat_2$ to F_3, and the largest F_2 to C_3. The diameters of the two timpani are 23 and 30 inches.

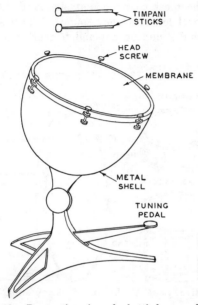

FIG. 5.68. Perspective view of a kettledrum and sticks.

B. Indefinite-pitch Instruments

Indefinite-pitch percussion instruments emit a musical noise which cannot be identified as possessing a definite pitch or frequency. These include triangles, drums, tambourines, cymbals, gongs, and castanets.

1. Triangle. The triangle consists of a steel rod bent into a triangular shape (Fig. 5.69). It is struck by means of a metal beater. The vibrations are a complex mixture of transverse, longitudinal, and torsional vibrations. Owing to the complex nature of the vibrations, the fundamentals and overtones form an in-definite noise mixture. The tri-angle is used to mark rhythms and to produce extraordinary effects.

2. Bass Drum. The bass drum consists of a hollow cylinder of metal or wood covered at each end

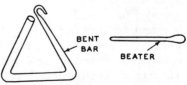

FIG. 5.69. Triangle and beater.

by a stretched membrane of parchment or skin (Fig. 5.70C). Means in the form of thumbscrews are provided for stretching the membrane. The vibration is complex, being a combination of the air column and the stretched membranes. The drum is played by being struck with a soft-headed stick. The bass drum is built in sizes ranging in diameters from 2 to 10 feet. The most common diameter is about 31 inches. The output is a powerful low-frequency sound. The bass drum is used to augment the general sound output and to mark rhythm.

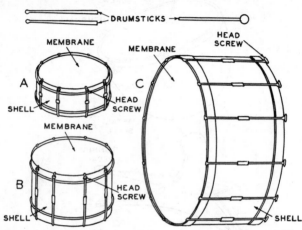

Fig. 5.70. Perspective views of drums. *A*. Snare drum. *B*. Military drum. *C*. Bass drum.

3. Military Drum. The military drum consists of a deep hollow cylinder of metal or wood covered at each end by a stretched membrane of parchment or skin (Fig. 5.70B). Means in the form of thumbscrews are provided for stretching the membrane. The drum is played by being struck with wood sticks. The military drum is about 16 inches in diameter and 12 inches in depth.

4. Snare Drum. The snare drum consists of a shallow hollow cylinder of metal or wood covered at each end by a stretched membrane of parchment or skin (Fig. 5.70A). Means in the form of thumbscrews are provided for stretching the membrane. Across the lower head, cords of catgut are stretched so that these cords are struck by the membrane when it vibrates. The result is a buzzing and rattling sound. The drum is played or excited by striking the top head by means of wood sticks. The tone of the snare drum is of indefinite pitch but is brilliant and crisp.

The diameter of the snare drum is about 14 inches.

5. Tambourine. A tambourine consists of a hoop of wood or metal with a single membrane of parchment or skin stretched over one end (Fig. 5.71). Small circular disks of metal are inserted in pairs in the hoop and loosely strung on wires. The tambourine is held in one hand and the membrane struck with the fingers or the palm of the other hand.

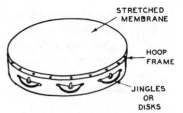

FIG. 5.71. Perspective view of a tambourine.

6. Cymbals. A cymbal is a circular disk of brass with a concave section at the center. The vibration is similar to that of a circular plate supported at the center (Sec. 4.8C).

The cymbals shown in Fig. 5.72B are equipped with handles. The two cymbals are held in each hand and struck together.

A single cymbal is sometimes supported in a fixed horizontal position on a drum or other support and struck with a drumstick.

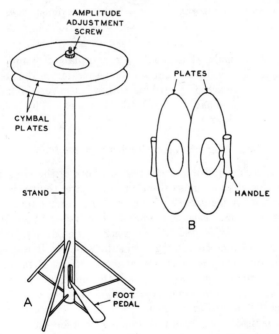

FIG. 5.72. Perspective views of cymbals. *A.* Sock- or foot-actuated cymbals. *B.* Hand-held cymbals.

A sock cymbal consists of a fixed and movable cymbal arranged so that the movable cymbal, connected to a pedal, is brought into contact with the fixed cymbal by depressing the pedal (Fig. 5.72A).

Cymbals are made in various sizes ranging in diameter from 2 to 20 inches.

The major portion of the sound energy of the cymbal lies above 8,000 cycles.

7. Castanets. Castanets are hollow shells of hard wood (Fig. 5.73). The castanets are held in the hand and clapped together. Castanets are used to add atmosphere in Spanish-type music.

Castanets are also mounted upon a handle (Fig. 5.73). Shaking the handle clacks the castanet.

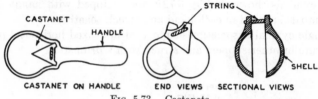

FIG. 5.73. Castanets.

8. Gong (Tam-tam). The gong is a round plate of hammered bronze with the edges turned up so that the shape resembles that of a shallow pan. The gong is excited by striking with a felt-covered hammer. The diameter of gongs range from 18 to 36 inches. The sound output resembles a heavy roar.

C. Electrical Analogies of Percussion Instruments

A schematic view and the mechanical network of the vibrating system of bar-type percussion musical instruments are shown in Fig. 5.74. These instruments include the following: the tuning fork, xylophone, marimba, glockenspiel, celesta, chimes, and triangle. A similar network would apply to the bell, cymbals, and gong. The bar is represented as a multiresonant system containing elements which will resonate at the fundamental frequency and all the overtones. These elements are represented as the masses m_1, m_2, . . . m_N, the mechanical resistance r_{M1}, r_{M2}, . . . r_{MN}, and the compliances C_{M1}, C_{M2}, . . . C_{MN}. In some instruments the bar radiates directly into the air. In some instruments a resonator is used to increase the coupling between the bar and the air which is represented as a quadripole z_M. The air load upon the quadripole is represented as the mass m_A, and the mechanical resistance r_{MA}.

The hammer is represented as the mass m_H. The force which imparts a velocity $\dot{x}$ to the hammer is represented as a battery of variable voltage. The force f_M causes the hammer m_H to move with a velocity $\dot{x}$. This corresponds to closing the switch on contact 1. The kinetic energy of the hammer is transferred to the bar, which causes it to vibrate. This corresponds to moving the switch to contact 3, so that contacts 1 and 3 are joined. The hammer then rebounds. This corresponds to moving the switch so that contacts 2 and 3 are joined. The resonant system transfers its vibrating energy to the surrounding air through the radiation mechanical resistances r_{M1}, r_{M2}, ... r_{MN}. If a resonator is used, the added radiation load is the quadripole z_M and its radiation resistance r_{MA}. From the mechanical network it will be seen that the ratio of the ampli-

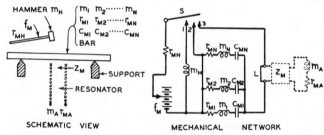

FIG. 5.74. Schematic view and mechanical network of a bar- or rod-type percussion instrument. In the mechanical network, f_M = driving force; r_{MH} = mechanical resistance of the generator; m_H = mass of the hammer; m_1, C_{M1}, r_{M1}, m_2, C_{M2}, r_{M2} ... m_N, C_{MN}, r_{MN} = lumped masses, compliances, and mechanical resistances representing the bar. In some instruments an additional resonator is used. This is shown dotted in the schematic view and mechanical network. If the resonator is used, the link L will be open in the mechanical network. z_M = quadripole representing the resonator. m_A and r_{MA} = mass and mechanical resistance of the air load terminating the resonator.

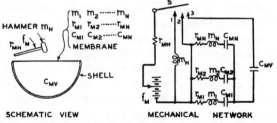

FIG. 5.75. Schematic view and mechanical network of a drum instrument. In the mechanical network, f_M = driving force; r_{MH} = mechanical resistance of the generator; m_H = mass of the hammer; m_1, C_{M1}, r_{M1}, m_2, C_{M2}, r_{M2} ... m_N, C_{MN}, r_{MN} = lumped masses, compliances, and mechanical resistance representing the membrane and air load; C_{MV} = compliance of the cavity.

tudes of the fundamental and the various overtones would be quite complex, owing to the complex nature of the striking force, the bar, and the coupling to the air.

A schematic view and the mechanical network of the vibrating system of drum-type percussion instruments are shown in Fig. 5.75. The system shown is the kettledrum, but the analogy would be similar for all drum-type instruments. The membrane is represented as a multiresonant system which will resonate with the remainder of the system at the fundamental frequency and all the overtones. These elements are represented as the masses m_1, m_2, . . . m_N, the mechanical resistances r_{M1}, r_{M2}, . . . r_{MN}, and the compliances C_{M1}, C_{M2}, . . . C_{MN}. The volume behind the membrane is represented as the compliance C_{MV}. The hammer is represented as the mass m_H. The force f_M which imparts a velocity $\dot{x}$ to the hammer m_H is represented as a battery of variable voltage. The force f_M causes the hammer m_H to move with a velocity $\dot{x}$. This corresponds to closing the switch on contact 1. The kinetic energy of the hammer is transferred to the membrane and cavity, which cause these elements to vibrate. This corresponds to moving the switch to contact 3 so that contacts 1 and 3 are joined. The hammer then rebounds. This corresponds to moving the switch so that contracts 2 and 3 are joined. The resonant system then transfers its vibrating energy to the surrounding air through the radiation mechanical resistances r_{M1}, r_{M2}, . . . r_{MN}. The mass in the vibrating system is relatively low, and the radiation mechanical resistance is relatively high. Therefore, the vibrations of the drum are very heavily damped, and the vibrations decay very rapidly. In general, the fundamental tone of the drum is most pronounced. This is due to the breakup of the membrane in the higher modes into sections of opposite phase which reduce the radiated energy (Sec. 4.7).

5.6 ELECTRICAL MUSICAL INSTRUMENTS

A tone of any pitch, timbre, and loudness may be generated by electrical means. The basic generator may be any of the following: an interrupted air stream, an electrically driven diaphragm, an electrical alternator and loudspeaker combination, and an electron-tube oscillator, amplifier, and loudspeaker combination. An example of the interrupted air stream is the siren. The electrically driven resonant diaphragm coupled to a horn is almost universally used for automobile horns. Electrical organs have been built, using miniature rotating electrical alternators. Others have used electrically driven resonant reeds. Electronic tubes have been used in electrical musical instruments in a number of ways. The strings in a piano have been coupled to electromagnetic or electrostatic transducers and the output has been amplified and fed to a loudspeaker. In other instruments, electronic oscillators are used to generate the tone without the use of any mechanical vibrating system as the fundamental resonant system. It is the purpose of this section to consider electrical musical instruments.

A. Siren

The simple siren consists of a rotating perforated wheel which interrupts an air supply. This is an interrupted air-stream form of generator and does not truly belong in the electrical class. Sectional and front views and the electrical analogy of the siren are shown in Fig. 5.76. The section is essentially the same as that of the interrupted air stream described in Sec. 1.4.

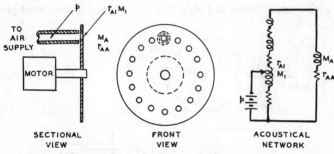

SECTIONAL VIEW   FRONT VIEW   ACOUSTICAL NETWORK

FIG. 5.76. Front and sectional views and the acoustical network of a siren. In the acoustical network, p = pressure in the direct-current air supply; r_{A1} and M_1 = acoustical resistance and inertance of the throttling aperture; r_{AA} and M_A = acoustical resistance and inertance of the air load.

B. Automobile Horn

The conventional automobile horn shown in Fig. 5.77 consists of an electrically driven diaphragm coupled to a horn. The driving system consists of an electromagnet, armature, and interrupter. In the repose

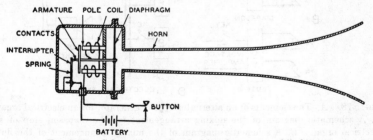

FIG. 5.77. Schematic sectional view of an automobile horn.

position the contacts of the interrupter are pushed together.   When the
button is pressed closing the circuit, a current flows in the electromagnet
which in turn exerts a pull on the armature.   As the armature is pulled
toward the poles of the electromagnet, the contacts of the interrupter
separate and the circuit is broken and there is no force exerted upon the
armature by the electromagnet.   Under these conditions, the restoring
force of the diaphragm drives the armature back so that the contacts are
again pressed together and the process is repeated.   The cycle of events
is repeated at the resonant frequency of the diaphragm, armature, and
horn combination.   The driving force exerted upon the armature is a
complex waveform.   As a result, the sound produced by the diaphragm is
rich in overtones.   Under these conditions, the resonant characteristics
of the horn play an important part in determining the quality.   Long
horns with a small throat and slow rate of flare produce the most pleasing

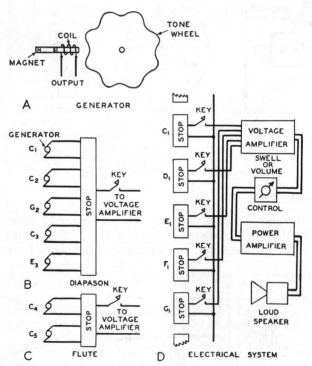

Fig. 5.78.   A. The elements of an alternator type of generator for an electrical organ.
B. A schematic diagram of the mixing arrangement for the diapason stop of an
electrical organ.   C. A schematic diagram of the mixing arrangement of the flute
stop of an electrical organ.   D. A schematic diagram of an electrical organ.

sounds because the overtone structure of the sound output resembles that of a musical instrument.

C. Electrical and Electronic Organs

One form of electrical organ which has been extensively commercialized consists of a system of electrical alternators, one alternator for each note. The alternator used in this electrical organ is shown in Fig. 5.78A. The rotation of the tone wheel produces a change in the flux, supplied by the permanent magnet, through the coil, which in turn produces the induction of an alternating voltage corresponding to the undulations in the disc. The outputs of the generators are mixed and amplified by means of vacuum-tube amplifiers (Sec. 9.5). The electrical variations are converted into the corresponding sound waves by the loudspeaker (Sec. 9.3). By means of the mixing or stop system, shown schematically in Fig. 5.78B and 5.78C, it is possible to obtain any combination of the harmonic series. A schematic diagram of an electrical organ is shown in Fig. 5.78D.

Vacuum-tube and transistor oscillators may be used to generate the tone in an electronic organ. A generic vacuum-tube oscillator network is shown in Fig. 5.79A. A generic transistor oscillator network is shown in Fig. 5.79B. Power from the output network is fed back to the input network. Oscillations occur when more power is developed in the output circuit than is required for the loss in the input circuit combined with suitable phase relations between the current and voltages in the input, feedback and output networks. Under these conditions, the action consists of regular surges of power at a frequency dependent upon the constants of the resonant elements in the output or input networks. The power output is the difference between the input power and the power dissipated in the system. The resonant elements may be inductance-capacitance, quartz, crystal, tuning fork, etc. Any fundamental frequency in the audio range may be obtained. Electronic systems can generate all the wave shapes shown in Figs. 6.7 to 6.12 as well as variations of these wave shapes. This means that the wave shape of any musical instrument

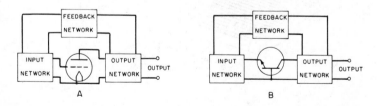

FIG. 5.79. *A*. Vacuum-tube oscillator network. *B*. Transistor oscillator network.

can be simulated.   Therefore, electronic oscillators offer great possibilities for use as tone generators in musical instruments.

A number of electrical organs employing electronic oscillators for generating the tone have been commercialized.   The principle is essentially the same as shown in Fig.5.78, save that an electronic oscillator is substituted for the generator.

An electrical organ with air-driven reeds and electrostatic pickups has also been commercialized.

Another form of electrical organ employs rotating electrostatic generators instead of the electromagnetic generators shown in Fig. 5.78A.

Most of the electrical and electronic organs consist of two manual keyboards and a pedal keyboard with a system of couplers and stops, as shown in Fig. 5.80.   The loudspeakers are usually housed in a separate cabinet. The entire electrical system, alternators, or oscillators, stop and coupler system, and vacuum-tube amplifiers are housed in a small cabinet integral with the console.

The fundamental objective in all electrical and electronic organs is to perform the functions of the pipe organ described in Sec. 5.4C.

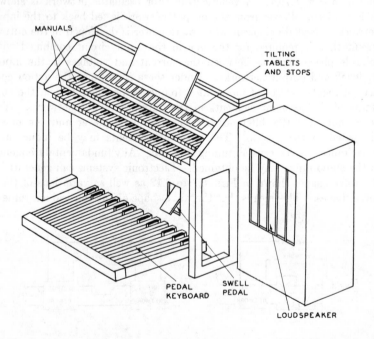

Fig. 5.80.   Perspective view of an electronic organ console and loudspeaker.

D. Electrical Piano

In the electrical piano, the system is essentially the same as the conventional piano, save that the soundboard has been replaced by electrical means. The vibrations of the strings are converted into electrical variations by means of the electromagnetic transducer shown in Fig. 5.81A or the electrostatic transducer shown in Fig. 5.81B. In these transducers the variation in distance between the string and pickup end of the transducer corresponds to the excursions of the vibrating string. In the electromagnetic transducer, the variation in distance produces a corresponding change in the magnetic flux, supplied by the permanent magnet, through the coil, which in turn leads to the induction of an alternating voltage corresponding to the vibrations of the string. In the electrostatic transducer, the electrical charge upon the capacitance between the fixed electrode and string remains constant. Therefore, a change in the capacitance, due to a variation in distance between the electrodes, produces an alternating voltage upon the fixed electrode corresponding to the vibrations of the string. The output of the transducer is amplified by means of a vacuum-tube amplifier (Sec. 9.5). The amplified electrical variations are converted into the corresponding sound waves by the loudspeaker (Sec. 9.3).

In the electrical piano the tonal qualities of a concert grand can be obtained in a very small instrument by suitable compensation of the electrical output. Another advantage of the electronic piano is the wide

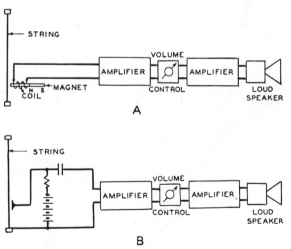

Fig. 5.81. Schematic views of the elements of electrical pianos. A. Electromagnetic pickup means. B. Capacitance pickup means.

dynamic range—the sound output can be adjusted so that it is suitable for the smallest apartment or the largest auditorium.

E. Electrical Guitar

The electrical guitar consists of six strings stretched between a tail-piece and the end of a fretted finger board and passing over a bridge coupled to a mechanoelectric transducer, an amplifier, and a loudspeaker (Fig. 5.82). The body is a small massive box which houses the transducer element. Vibrations of the strings are transmitted to the bridge. The motion of the bridge is transmitted to the mechanoelectric transducer. The transducer converts the mechanical vibrations into corresponding electrical variations. These variations are amplified by means of a vacuum-tube amplifier (Sec. 9.5). The output of the amplifier is coupled to the loudspeaker (Sec. 9.3). The loudspeaker converts the amplified electrical variations into the corresponding sound vibrations. Since the massive body of the electrical guitar is not an efficient soundboard, the output of the instrument without the electrical system is very small. However, with the electrical system, the output can be made comparable to that of an orchestra. The volume control makes it possible to adjust the level to any desired value. In some designs, a volume control is also placed on the body of the instrument, which makes it possible for the musician to make rapid changes in volume. The sound of the electrical guitar resembles that of a conventional instrument.

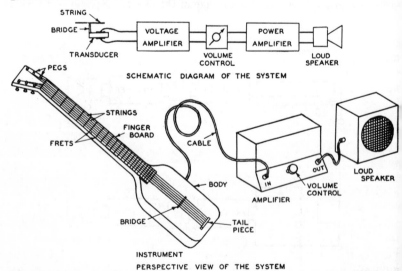

FIG. 5.82. Perspective view and schematic electrical diagram of electrical guitar.

F. Music Box

A music box is a mechanical musical instrument consisting of a set of resonant reeds tuned to the notes of a musical scale and a spring-driven rotating drum equipped with pins which set the reeds into vibration by mechanical contact as the drum rotates (Fig. 5.83). The sound output of the reeds is accentuated by mounting the reeds on a resonant sound box. The pins on the cylinder are arranged in a suitable sequence so that a tune is played as the pins pluck the reeds when the cylinder turns.

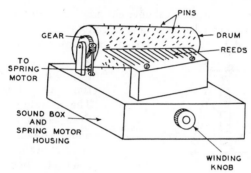

Fig. 5.83.   Perspective view of a music box.

G. Electrical Carillon

The electrical carillon consists of small vibrating reeds, springs, tubes, rods, bars, or vibrators of complex shapes which are struck by hammers actuated by keys forming a keyboard, an electrical means for converting the vibrations into the corresponding electrical variations, an amplifier, and a loudspeaker for amplifying and converting the electrical variations into powerful sound waves. The electrical pickup system is usually either electrostatic or electromagnetic, of the types described in Sec. 5.6D.

In one form, the actuating mechanism is similar to that of the celesta (Fig. 5.64). The hammers are usually made of metal so that the strike tones will be similar to that of conventional bells. In this type, the keyboard, vibrators, and electrical pickup system are located in a console cabinet.

In another form, an electromagnet mechanism is used to strike the vibrators. In this case, the vibrators and actuating mechanism are located in some convenient room with an electrical cable connected to a keyboard located in some other room. In this form, the housing for the keyboard is very small and may be placed in almost any location as, for example, alongside the organ console.

Some of the advantages of the electrical carillon as contrasted to the classical carillon, described in Sec. 5.5A8, are as follows: The space required for the vibrating system is relatively small. The keyboard may be located at any desired location. The loudspeakers may be located and arranged to direct the sound in the desired directions. Almost any tone structure may be obtained by a suitable mechanical vibrating system. The cost is relatively small.

H. Metronome

A metronome is a mechanical device consisting of a pendulum actuated by a clock mechanism driven by a spring motor (Fig. 5.84). An audible tick is produced by the pendulum at the extremities of the excursion.

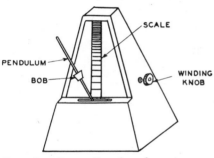

The movable bob on the pendulum is used to alter the time interval between ticks. The pendulum is graduated in ticks per minute. The metronome is useful for determining or establishing the time value of a beat. In using the metronome the movable bob on the pendulum is moved to the position giving the desired number of ticks per minute.

Fig. 5.84.   Perspective view of a metronome.

For example, if the metronome indication on a selection is ♩ = 84, the bob on the pendulum is set to 84. Under these conditions, a quarter note is played to each tick. If the setting is ♪ = 84, than a half note is played for each tick (see Sec. 2.3).

5.7 ORCHESTRAS

An orchestra is a company of musicians, under the direction of a conductor, performing upon string, wind, and percussion instruments. An orchestra differs from a band in that the main body of the tone is produced by the string instruments. There are several types of orchestras for rendering different types of music as, for example, symphonies and overtures, the accompaniment for operas and oratorios, music for theatrical performances, and popular or dance music.

The symphony orchestra is made up of four groups, namely, the strings, wood winds, brass, and percussion. The strings include the violin, viola, violoncello, and contrabass. The woodwinds include the flute, piccolo, oboe, bassoon, English horn, contra bassoon, clarinet, and bass clarinet.

The brass instruments include the trumpet, trombone, French horn, and tuba. The percussion instruments include the timpani, bass, military, and snare drums, tambourine, gong, celesta, glockenspiel, tubular chimes, castanets, xylophone, and triangle. In order to obtain sufficient sound output, suitable dramatic and artistic effects in large halls, the modern symphony orchestra must use 80 to 120 men. In some cases, with limited personnel or in an unusually large hall, it is possible to use good sound-reinforcing systems to produce the desired artistic effects.

The number of instruments of each kind in a typical symphony orchestra is shown in Table 5.1.

The arrangement of the instruments in a symphony orchestra is shown in Fig. 5.85.

Orchestras for the accompaniments for operas and oratorios and the music for theatrical performances may range in number from 5 to 50 instruments. The particular combination of instruments depends upon the type of music. The arrangement and the instruments in a general orchestra of this type is shown in Fig. 5.86B.

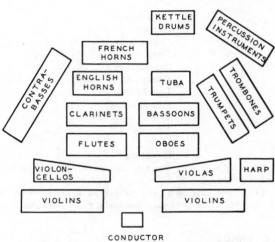

FIG. 5.85. A plan view of the arrangement of the conductor and the instruments in a symphony orchestra.

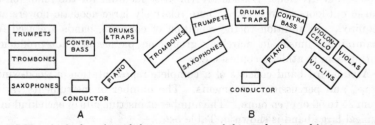

FIG. 5.86. A. A plan view of the arrangement of the conductor and instruments in a small dance band. B. A plan view of the arrangement of the conductor and instruments in a small orchestra.

Orchestras for playing popular and dance music range in number from 5 to 25. The arrangement and the instruments in an orchestra of this type are shown in Fig. 5.86*A*.

TABLE 5.1. THE ELEMENTS OF A SYMPHONY ORCHESTRA*

Instrument	Persons
First violins	16 to 20
Second violins	14 to 18
Violas	10 to 14
Violoncellos	8 to 12
Contrabasses	8 to 10
Flutes	2 to 3
Piccolos	1 to 2
Oboes	3
English horn	1
Bassoon	3
Contra bassoon	1
Clarinets	3
Bass clarinet	1
Trumpets	4
Trombones	4
French horns	4 to 12
Tuba	1
Timpani†	1
Harp	1
Percussion†	1 to 5

* A symphony orchestra may include other instruments as, for example, the piano and the organ.

† Percussion includes bass, military, and snare drums, tambourine, gong, celesta, glockenspiel, tubular chimes, castanets, xylophone, and triangle.

5.8 BANDS

A band is a company of musicians performing upon wind and percussion instruments, the main body of the tone being produced by the brass and wood-wind divisions. Bands are the best medium for the rendition of music outdoors in free space where relatively large acoustic powers are required. In addition to the rendition of concerts, bands are used for maintaining unison in marching groups on parade and for producing enthusiasm in athletic contests, etc.

A standard band consists of a complete representation of wood-wind, brass, and percussion instruments. The number of instruments varies from 25 to 50 or even more. The number of instruments of each kind in a typical large band is shown in Table 5.2.

The military band is practically the same as the standard band except that it may be augmented by a fife, drum, or bugle corps.

The concert band may include contrabasses, timpani, harps, and other instruments not suitable for marching.

A dance band is a term used synonymously with dance orchestra.

TABLE 5.2.   THE ELEMENTS OF A STANDARD BAND

Instrument	Persons
Flutes..................	4
Piccolo..................	1
Clarinets................	14
Oboes...................	2
Bassoons................	2
Sarrusophones...........	2
Saxophones..............	4
Cornets.................	4
Trumpets................	2
French horns............	4
Trombones..............	4
Tubas..................	6
Snare drum.............	1
Bass drum.............	1
Percussion*.............	1 to 5

* Percussion includes cymbals, triangle, bells, castanets, and xylophone.

5.9   ELEMENTS OF MUSICAL INSTRUMENTS

Action.   The action is a term used to designate the mechanism which translates the manual acts of the musician into motions of valves, hammers, strings, bars, air columns, and other devices of a musical instrument to produce the desired tonal effects.   The term is usually applied to keyboard instruments.

Air Column.   An air column is the air, within a pipe, tube, or horn of a wind instrument, which is set into motion when the instrument is actuated.

Bag.   A bag is the air-supply chamber in a bagpipe which serves as a buffer and reservoir between the breath and the pipes.

Bar.   A bar in musical instruments is a term used to designate a resonant piece of wood, metal, or other material in which the length is large in proportion to its breadth and thickness.   A musical tone is produced when the bars are excited by striking with a hammer.   Bars are used as the sound-producing element in xylophones, marimbas, glockenspiels, and celestas.

Bass Bar.   A bass bar is a strip of wood glued to the belly of a violin under the lowest string.

Baton.   A baton is a wand which the conductor waves in directing an orchestra to indicate by appropriate motions the rhythm, time, tempo, loudness, and other effects.

Beater. A beater is a steel bar used to strike the triangle.

Bellows. A bellows is a machine consisting of fluted collapsible sides which can be expanded to pull in air or contracted to expel air. It is used to supply air for musical instruments as, for example, organs and accordions. A bellows may also be used as a motor, in which case air under pressure is used to expand the machine.

Belly. The belly is the face of the body upon which the strings are stretched in viol and lute-type instruments.

Bit. A bit is a small length of tube used to shorten or lengthen the air column of a brass instrument and thereby alter the fundamental frequency of the instrument.

Blowhole. The blowhole is the hole in the side of a transverse flute across which the player blows.

Body. Body is a term used to designate the main portion of a musical instrument as, for example, the resonant box of the viol family or the central portion, excluding the mouthpiece and the mouth, of wood winds and horns.

Bow. A bow is a bowlike stick strung with a flat group of horsehair. The stretched horsehair is drawn across the strings of the viol family of instruments to actuate the strings. The iterated force is increased by applying rosin to the horsehair of the bow.

Bridge. A bridge is a transverse member, in contact with the body or soundboard, over which the strings are stretched. The bridge transmits the vibrations of the string to the body or soundboard, thereby increasing the output of the string. The bridge also determines the length of the string since it establishes one end of the vibrating string.

Button. Buttons are the keys that actuate the valves in an accordion. A button is a small post to which the tailpiece is attached in instruments of the viol class.

Cabinet. A cabinet is the case for a musical instrument, such as a piano, organ, phonograph, or radio receiver.

Catgut. Catgut is a term used for the gut strings of a musical instrument.

Celeste Pedal. See Soft Pedal.

Chanter. A chanter is a term used to designate the melody pipe in the bagpipe.

Choir Organ. The choir organ is the part of an organ which is controlled by the lowest manual.

Clapper. A clapper is the pendant, hammer, or other actuating means of a bell.

Closed Pipe. A closed pipe is an organ pipe or other pipe containing a resonant air column with a stopper.

Conductor. The conductor is the director of a chorus or orchestra.

Console. A console is the desklike case containing the keyboards and stop mechanisms of an organ.

Coupler. A coupler is a mechanism for joining the keys of two manuals of an organ so that the depression of the keys of one manual causes the actuation of the keys of the other manual. The pedal coupler joins the keys of the pedal keyboard to a manual.

Damper. The damper is the felt pad which is held against a piano string to increase the rate of decay of the vibration of the string. The damper is not in contact with the string when the key is depressed or when the sustaining pedal is depressed.

Damper Pedal. See Sustaining Pedal.

Drone. A drone is a term used to designate the pipe which sounds continuously in a bagpipe.

Drumstick. A drumstick is a stick or hammer used for beating a drum.

Echo Organ. The echo organ is a part of an organ enclosed in a box and located in a remote point to produce an echo effect. This division of the organ is controlled by the fourth manual.

Embouchure. Embouchure is a term used to designate the mouth-piece of some wind instruments.

F Holes. The F holes of a violin are two holes in the belly of a violin which couple the air chamber of the body to the outside air. The term is derived from the shape.

Finger Board. A finger board is the flat front of the neck in viol- and lute-type instruments over which the strings are stretched. The resonant frequency of the strings is altered by changing the length by pressing the string against the finger board by means of the fingers.

Finger Hole. A finger hole is a hole in the side of the tube comprising the resonating air column of wind instruments. Opening or closing the hole by means of the fingers changes the resonant frequency of the air column. Finger hole is sometimes used to designate openings controlled by a key, lever, and valve arrangement.

Fipple. A fipple is the block in whistle-type musical instruments which contains the mouthpiece, flue, hole, and lip.

Fipple Hole. The fipple hole is the aperture between the lip and the flue of whistle or flue organ pipe instruments. This aperture is the opening connecting the volume of the pipe, chamber, or resonator and the open air.

Flue. A flue is the passageway in a whistle or flue organ pipe for directing the air stream.

Fret. A fret is a crossbar or ridge of ivory, plastic, or metal inserted crossways in the finger board of string instruments for changing the length of the string by a discrete amount by pressing the finger against the string to bring it in contact with the ridge. The change in length of the string changes the resonant frequency.

Great Organ. The great organ is the part of an organ which is controlled by the second manual. This division of the organ controls the most powerful stops.

Hammer. A hammer consists of a heavy head fastened to a light shank. It is used to strike strings, bars, rods, bells, etc., to produce the musical tone. The hammer may be held in the hand or operated by means of a key and lever arrangement.

Head. In instruments of the viol class, the head is the upper part of the neck containing the peg box and tuning pegs. In a bow, the head is the outer end of the bow. In drums, the head is the stretched membrane. In a hammer, the head is the heavy portion which is fastened to the handle.

Heel. A heel is the part of the neck of a violin which is fastened to the body. A heel is the part of a bow which is held in the hand.

Horn. A horn is a tube of varying cross-sectional area with respect to the axis. It is also a general term applied to wind instruments.

Jack. The jack is the upright member which actuates the hammer in a piano.

Key. A key is the end of a lever system, actuated by the fingers, for performing the following operations in various musical instruments: for controlling the opening or closing of a hole in the side of a tube in a wind instrument; in conjunction with a valve for controlling the effective length of the air column in a wind instrument; for actuating a hammer which strikes a string or bar; in conjunction with a valve for controlling the air flow which actuates a reed or pipe; and for controlling the opening or closing of a switch in an electrical circuit.

Keyboard. A keyboard is a series of keys arranged in a row or some other geometrical pattern.

Knee Lever. A knee lever is used in an organ to control the volume or the stops.

Knob. A knob is a handle for actuating the stop mechanism.

Lip. Lip is the sharp edge against which the air stream in a flue pipe impinges and divides.

Loud Pedal. See Sustaining Pedal.

Manual. A manual is used to designate a keyboard for the hands as contrasted with one for the feet.

Membrane. A membrane is a stretched sheet of skin or parchment used as the principal sound emitter in drums, tambourines, and banjos.

Mouth. The mouth is a term used to designate the open end of a wind instrument.

Mouthpiece. A mouthpiece is the part of an instrument held against or by the lips. In air-reed instruments, the mouthpiece is a flat surface containing a hole against which the breath is directed. In mechanical-reed instruments, both single and double reed, the mouthpiece is the outer end of the reed which is held between the lips. In lip-reed instruments, the mouthpiece is a hemispherical cup against which the lips are pressed.

Mute. A mute is a wood, plastic, or metal clip attached to the bridge of a violin to attenuate the sound output and reduce the overtones. A mute is a cone or cylinder of metal, plastic, cardboard, or felt which is slipped into the mouth or bell of a horn to reduce the sound output and alter the quality.

Nut. A nut is the ridge between the bend and the finger board of string instruments over which the strings pass to the tuning pegs. The nut and the bridge determine the open length of the string.

Open Pipe. An open pipe is an organ pipe or other pipe containing a resonant air column without a stopper.

Pallet. A pallet is a valve that controls the admission of air to a pipe or reed of a pipe organ, reed organ, or accordion.

Pedal. A pedal is a foot-operated key. A pedal is also a series of keys operated by the feet. A pedal is also used to effect certain actions in harps, drums, cymbals, etc.

Pedal Harp. The pedals of a harp control the tension of the strings so that each string can produce three different tones.

Pedal Organ. The pedal organ is the part of an organ which is controlled by the pedal keyboard. This division controls the largest pipes of the lowest pitch.

Peg. The pegs are the thumb pins around which one end of the strings of the violin, guitar, ukulele, mandolin, banjo, etc., are wound. The tension in the string and the resultant pitch is changed by turning the peg.

Pipe. A pipe is a tube of constant cross section. A pipe is also used to designate a tube element in whistles, organs, and other instruments having an air column of constant cross section.

Piston. A piston is the valve in lip-reed instruments. A piston is a movable knob on an organ for changing the stop registration.

Plectrum. A plectrum is a small flat piece of metal, ivory, wood, or plastic held between the thumb and fingers and used to pluck the strings of the lyre, guitar, banjo, mandolin, or other plucked-string instruments. It is also made in the form of a ring or rod.

Port. A port is any opening in an air column or air chamber.

Rank. A rank is a row of organ pipes belonging to one stop.

Reed. A reed is a thin strip of metal, wood, cane, or plastic fixed at one end and free at the other. A reed is set into vibration by the actuating air stream which it modulates and thereby produces a sound. Lip reed is a term applied to designate an instrument in which the lip acts as the reed. Air reed is a term applied to an instrument in which air acts as the reed. Vocal-cord reed is a term applied to the voice mechanism in which the vocal cords act as reeds.

Reservoir. A reservoir is the air-supply chamber which serves as a buffer between the compressor and the reeds and pipes in an organ.

Resonator. A resonator is any vibratile element, as, for example, a string, a bar, a rod, an air column, or an air chamber, which exhibits the greatest response to the actuating force at certain discrete frequencies termed the resonant frequencies.

Rib. A rib is one of the vertical strips which join the belly and back of the instruments of the viol class.

Rosin. Rosin is a turpentine distillate in the solid state. It is applied to bows to increase the friction between the bow and the string of bowed-string instruments of the viol family.

Scroll. The scroll is a term used to designate the head of instruments of the viol family.

Slide. A slide is a U-shaped section of telescopic tubing used in lip-reed brass instruments for changing the length of the resonating air column. In the trombone, the instrument is played by varying the position of the U slide. In the valve-tuned instruments, it is used to make small adjustments in the pitch. A slide is also used to designate the valve which controls the air to a rank of pipes in an organ. The action of the valve is controlled by the stop knob.

Soft Pedal (Celeste Pedal). The soft pedal of a piano reduces the sound emitted by the piano, either by decreasing the length of the stroke of the hammer or by decreasing the number of strings which are struck when the pedal is depressed.

Soundboard. A soundboard is a thin sheet of wood coupled to the strings in string instruments to increase the coupling between the string and the air and thereby increase the sound output.

Sound Box. A sound box is an air-type resonator in the form of a hollow chamber with one or more openings. A sound box is used to augment the output of vibrating strings, bars, and rods in all manner of instruments.

Sound Hole. A sound hole is an aperture in a sound box.

Sound Post. A sound post is the wooden prop between the belly and back of instruments of the viol family. It is located under the bridge.

Stop. A stop is some mechanism which varies the resonant frequency of a string or tube. In an organ it is a mechanism for coupling various pipes together for simultaneous sounding.

String. A string is a tightly stretched thread of silk, gut, plastic, silver, bronze, brass, or steel which is thrown into vibration by some actuating means to produce a musical tone.

Sustaining Pedal. Loud Pedal. Damper Pedal. The sustaining pedal of a piano controls the string dampers. When the pedal is depressed, the dampers are removed and the vibration of the strings are damped only by the soundboard and end supports.

Swell Organ. The swell organ is the part of an organ enclosed within a box provided with shutters which are opened and closed by the swell pedal.

Swell Pedal. See Swell Organ.

Tailpiece. A tailpiece is a triangular-shaped member of ebony to which the lower ends of the strings are attached in the viol family of instruments.

Throat. The throat is a term used to designate the small end of a horn.

Tilting Tablet. A tilting tablet is a movable tablet which actuates the coupling mechanism in an organ. Tilting tablets are sometimes used for actuating the stop mechanism instead of the more conventional knob.

Tongue. The term tongue designates the vibrating end of a reed.

Touch. Touch is a term used to designate the resistance which the keys of an instrument, particularly the piano and organ, offer to motion.

Tremolant. A tremolant is a mechanism for modulating the amplitude or the frequency or both of a musical instrument.

Valve. A valve is a mechanism used for controlling the air stream or vibrating air column either by closing or opening a passage or by diverting it to another channel. In the pipe organ, reed organ, and accordion, a valve, operated by a key, is used to admit air to a pipe or reed and thereby actuate it. In lip-reed brass instruments, a valve, operated by a key, is used to change the length of the air column by introducing an additional length of tubing. In instruments with side holes, a valve, operated by a key, is used to open and close the holes and thereby alter the resonant frequency of the air column.

Ventil. A ventil is a valve. In the organ it refers to a valve which controls the air supply to a rank or a group of pipes in an organ.

Vocal Cavities. The vocal cavities are the resonators of the human-voice mechanism.

Vocal Cords. The vocal cords are the vibrating reeds of the larynx. The vocal cords are set into vibration by the air stream from the lungs. The vibration of the vocal cords throttle the air stream at the vibration frequency of the vocal cords. In this process a steady air stream is converted into a pulsating air stream. The pulsating air stream is a sound wave. The sound wave is sent into the vocal cavities where it is altered by the resonant properties of these cavities.

Wind Chest.   A wind chest is a reservoir for supplying air under pressure to the pipes or reeds of an organ.

Wrest Pin.   Wrest pins are the turnable pins around which the ends of the strings are wound in the harp and piano.   The strings can be tightened or loosened by turning the wrest pin.

CHAPTER SIX

Characteristics of Musical Instruments

6.1 INTRODUCTION

The two characteristics of music which are principally a function of the musical instrument are the tonal and dynamic. The tonal aspect depends upon the pitch and the timbre of the instrument. The pitch is in general determined by the fundamental frequency and fundamental-frequency range of the instrument. The timbre is the instantaneous-acoustical spectrum of the instrument. The timbre involves the frequencies and amplitudes of both the fundamental and the overtones. The dynamic aspect depends upon the absolute intensity level produced by the instrument and the dynamic range or intensity range.

Musical instruments and the voice produce fundamental frequencies and overtones of fundamental frequencies. The overtone structure is one of the characteristics which distinguishes various instruments and voices. If musical instruments produced the fundamental without overtones, each instrument would produce a pure sine wave, and it would, therefore, be the same as the output of all other instruments except for the possibility of a difference in frequency and intensity. The fundamental frequency is the lowest frequency component in a complex wave. When a musician speaks of the range of a voice or musical instrument, he usually means the frequency range of the fundamental frequency. In other words, the fundamental-frequency range of a musical instrument commands a certain section of the musical scale. In addition, each instrument or voice produces harmonics or overtones of the fundamental frequency. In general, the overtones cover a tremendous frequency range or section of the musical scale.

The dynamic aspect of music depends upon the intensity. The intensity ranges of musical instruments involve the absolute value of the upper and lower intensity ranges and the resultant dynamic range.

The intensity and timbre of the sound produced by a musical instrument are also governed by the directivity pattern. The directivity pattern depicts the sound output as a function of the angle with respect to some reference axis of the instrument. In general, the directivity pattern is complex, in that it is a function of both the angle and the frequency. Under these conditions, both the intensity and the timbre will vary as the frequency or angle of orientation of the instrument is altered.

The growth, decay, and steady-state characteristics and the duration of a tone produced by a musical instrument influence the tonal and dynamic characteristics of the tone.

It is the purpose of this chapter to consider the fundamental and overtone frequency ranges, the acoustic spectrums, the intensity ranges, the directional patterns, and the growth, decay, steady-state, and duration characteristics of various musical instruments.

6.2 FUNDAMENTAL AND OVERTONE FREQUENCY RANGES OF MUSICAL INSTRUMENTS

Musical instruments and the voice produce fundamental frequencies and overtones of fundamental frequencies. The overtone structure is one of the characteristics which distinguishes various instruments and voices. If musical instruments produced the fundamental without overtones, each instrument would produce a pure sine wave and would, therefore, be the same as the output of all other instruments except for the possibility of a difference in frequency and intensity. The fundamental frequency is the lowest frequency component in a complex sound wave. When a musician speaks of range of a voice or musical instrument, he usually means the frequency range of the fundamental frequencies which the voice or instrument is capable of producing. The fundamental-frequency ranges[1] of voices and various musical instruments are shown in Fig. 6.1. It will be seen that the fundamental-frequency range of each musical instrument commands a certain section of the musical scale. There may be some variations, from the frequency ranges shown in Fig. 6.1, among various instruments and voices, but, in general, these frequency ranges are typical. Comparing the frequency ranges of the fundamental of Fig. 6.1 with the entire frequency spectrum[2] of Fig. 6.2, it will be seen that the frequency ranges of the overtones of the instruments extend the frequency ranges by a factor of two or more octaves. Referring to Fig. 6.2, it will be seen that the low-frequency ranges of some instrument extend to a lower frequency than the fundamental-frequency ranges of Fig. 6.1.

[1] Olson, H. F., *Acoustical Engineering*, D. Van Nostrand Company, Inc., Princeton, 1957.
[2] Snow, W. B., *Jour. Acoust. Soc. Amer.*, Vol. 3, No. 1, Part 1, p. 155, 1931.

This is due to the fact that some instruments generate noises and sub-harmonics having frequency components below the true fundamental-frequency range. It will also be seen that the limits of the low-frequency range of some instruments in Fig. 6.2 are actually higher than those of Fig. 6.1. This is due to the fact that in the case of some of the lowest tones produced by some instruments the fundamental is so weak that it can be eliminated without discerning any change in the character of the tone. These facts relating to the low-frequency range are relatively unimportant. The important factor in the comparison between Figs. 6.1 and 6.2 is the great extension of the frequency range when the harmonics are included.

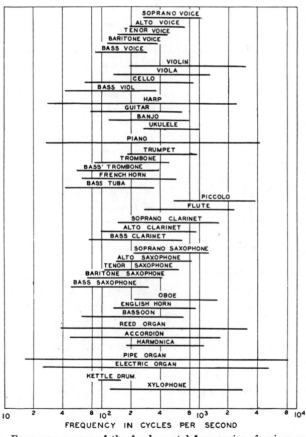

Fig. 6.1. Frequency ranges of the fundamental frequencies of voices and various musical instruments. (*After Olson, Acoustical Engineering, D. Van Nostrand Company, Inc., Princeton, 1957.*)

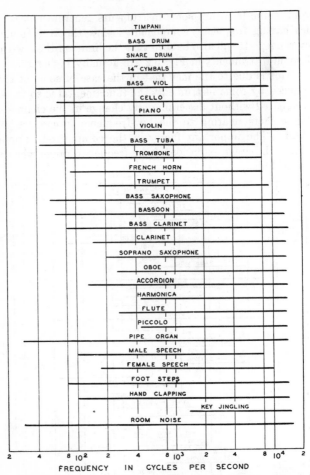

Fig. 6.2. Frequency ranges required for speech, musical instruments, and noises so that no frequency discrimination will be apparent. (*After Snow.*)

The ratios of the average sound pressure[3] per cycle to the average total pressure of the spectrum for speech, musical instruments, and orchestras are shown in Fig. 6.3.

The ratios of the peak pressures[4] to the average pressure of the entire spectrum for speech, musical instruments, and orchestras are shown in Fig. 6.4.

[3,4] Sivian, Dunn, and White, *Jour. Acoust. Soc. Amer.*, Vol. 2, No. 3, p. 330, 1931.

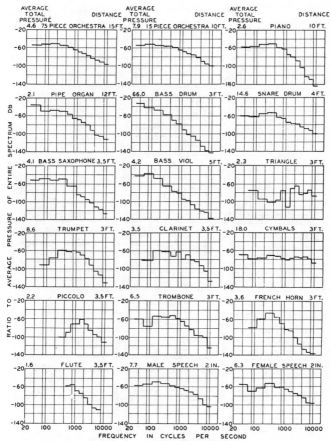

Fig. 6.3. Ratio of the average pressure per cycle to the average total pressure of the entire spectrum for speech, various musical instruments, and orchestras. The distance and average total pressure, in dynes per square centimeter, are shown above each graph. (*After Sivian, Dunn, and White.*)

The average sound-pressure characteristics give the average sound pressure per cycle over a long period of time. The peak-pressure characteristics give the ratio of the maximum peak of sound pressures that occur in a frequency band to the total average sound pressure. These bands are indicated by the steps 30 to 60, 60 to 120, 120 to 240, 240 to 500, 500 to 1,000, 1,000 to 1,400, etc. The peak levels may occur for only a small fraction of the time. From Fig. 6.3 it will be seen that the peak levels are many times as great as the average level. It is the peak pressure that must be considered in the design of sound-reproducing equipment.

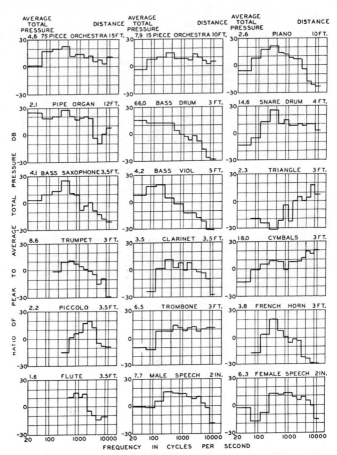

FIG. 6.4. Ratio of the peak pressure to the average total pressure of the entire spectrum for speech, musical instruments, and orchestras. The distance and average total pressure, in dynes per square centimeter, are shown above each graph. (*After Sivian, Dunn, and White.*)

The equipment must be capable of handling the peak power without overloading or distorting.

The average power gives the general impression of the balance of the music. The peak levels produce the dramatics and dynamics in musical renditions.

6.3 TIMBRE OF MUSICAL INSTRUMENTS

A sound wave may be represented by means of a graph in which the ordinates represent pressure or particle velocity in the sound wave and the abscissas represent time (Sec. 1.3). This type of representation of a sound wave may be obtained by the combination of a microphone amplifier and cathode-ray oscillograph (Fig. 6.5). The microphone converts the variations in pressure or particle velocity into the corresponding electrical variations (Sec. 9.2). The electrical variations are amplified by the vacuum-tube amplifier and applied to the vertical deflection system of the cathode-ray oscillograph (Sec. 9.5). The vertical deflections of the electron beam in the oscillograph correspond to the amplitudes of the pressure or particle velocity in the original sound wave. The horizontal deflection of the electron beam is produced by an oscillator which drives the electron beam at a constant rate in a horizontal direction. The electron beam impinges upon the fluorescent screen and produces a visible trace which depicts the sound wave in graphical form. This wave gives the cross section of the tone over a certain interval of time. If the wave on the oscillograph is produced by a musical instrument, the tonal structure of the musical instrument may be obtained from an analysis of this wave, because a complex wave may be resolved into components consisting of the fundamental and harmonics or overtones. The structure of complex waves will be described in the section which follows.

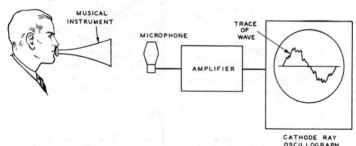

Fig. 6.5. Apparatus for depicting the wave shape of a sound wave.

A. Representation of Sound Waves

A complex wave may be considered to be composed of the fundamental and harmonics or overtones in the proper amplitude and phase relations. The composition of a complex wave is illustrated in Fig. 6.6. The waveform of the fundamental is shown in Fig. 6.6A. The waveform of the combination of the fundamental and the second harmonic, or first overtone, is shown in Fig. 6.6B. The waveform of the combination of the fundamental and the second and third harmonics, or the first and second overtones, is shown in Fig. 6.6C. The waveform of the combination of the fundamental and the second, third, and fourth harmonics is shown in Fig. 6.6D. The waveform of the combination of the fundamental and

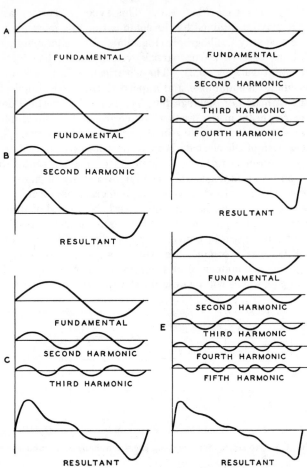

FIG. 6.6. *A.* Fundamental sine wave. *B.* Fundamental, second harmonic, and the resultant. *C.* Fundamental, second and third harmonics, and the resultant. *D.* Fundamental, second, third, and fourth harmonics, and the resultant. *F.* Fundamental, second, third, fourth, and fifth harmonics, and the resultant.

the second, third, fourth, and fifth harmonic is shown in Fig. 6.6*E.* If an infinite number of components with the appropriate amplitude and phase are used, the resultant will be a saw-tooth wave. This will be described later.

A complex wave may be expressed mathematically, as shown in Fig. 6.7. The first five components of the resultant wave of Fig. 6.6*E* are as follows:

$$P_1 = \sin \omega t \tag{6.1}$$
$$P_2 = \tfrac{1}{2} \sin 2\omega t \tag{6.2}$$
$$P_3 = \tfrac{1}{3} \sin 3\omega t \tag{6.3}$$
$$P_4 = \tfrac{1}{4} \sin 4\omega t \tag{6.4}$$
$$P_5 = \tfrac{1}{5} \sin 5\omega t \tag{6.5}$$

where P_1 = fundamental

P_2 = second harmonic

P_3 = third harmonic

P_4 = fourth harmonic

P_5 = fifth harmonic

$\omega = 2\pi f$

f = frequency

t = time

The resultant wave is given by

$$P_R = P_1 + P_2 + P_3 + P_4 + P_5 \tag{6.6}$$

This is the sine series

$$P_R = \sin \omega t + \tfrac{1}{2} \sin 2\omega t + \tfrac{1}{3} \sin 3\omega t + \tfrac{1}{4} \sin 4\omega t + \tfrac{1}{5} \sin 5\omega t \tag{6.7}$$

The structure of a sound wave produced by a musical instrument may be depicted by a spectrum graph. The spectrum of the resultant wave is shown in Fig. 6.7A. The relative amplitudes of the components of the resultant wave are depicted as a function of the frequency. The heights of the vertical lines are proportional to the amplitudes of the fundamental and harmonics. The position along the abscissa determines the frequency. For example, the spectrum shows that the frequency of the second harmonic is two times that of the fundamental and the amplitude is one-half that of the fundamental, the frequency of the third harmonic is three times that of the fundamental and the amplitude is one-third that of the fundamental, etc.

The first three components and the resultant of the combination of the first three components of the odd sine series are shown in Fig. 6.7B. The first three components of this wave are as follows:

$$P_1 = \sin \omega t \tag{6.8}$$
$$P_3 = \tfrac{1}{3} \sin 3\omega t \tag{6.9}$$
$$P_5 = \tfrac{1}{5} \sin 5\omega t \tag{6.10}$$

The resultant wave is given by

$$P_R = P_1 + P_3 + P_5 \tag{6.11}$$

This is the odd sine series

$$P_R = \sin \omega t + \tfrac{1}{3} \sin 3\omega t + \tfrac{1}{5} \sin 5\omega t \tag{6.12}$$

It will be seen that the wave of Fig. 6.7A is of the saw-tooth type, while the wave of Fig. 6.7B is of the rectangular type.

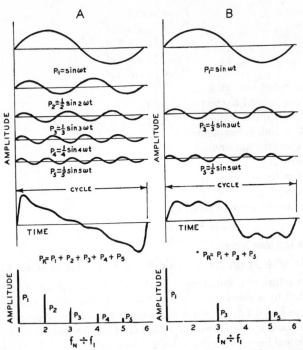

FIG. 6.7.   *A.* The fundamental, second, third, fourth, and fifth components of a wave, and the resultant wave consisting of the combination of these components of the sine series.   The spectrum depicts the relative amplitudes of the components of the resultant wave as a function of the frequency. f_1 = frequency of the fundamental, f_N = frequency of the Nth harmonic.   *B.* The fundamental, third, and fifth harmonic components of a wave, and the resultant wave consisting of these components of the odd sine series.   The spectrum depicts the relative amplitudes of the components of the resultant wave as a function of the frequency. f_1 = frequency of the fundamental, f_N = frequency of the Nth harmonic.

The output of musical instruments produces exceedingly complex waveforms.   A few illustrations of the simpler complex waves will now be given.

The waveform of the series

$$P_R = \cos \omega t + 0.7 \cos 2\omega t + 0.5 \cos 3\omega t \tag{6.13}$$

is shown in Fig. 6.8*A*.   The spectrum of this wave is also shown in Fig. 6.8*A*.

The waveform of the series

$$P_R = \sin \omega t - 0.5 \sin 3\omega t + 0.33 \sin 5\omega t - 0.25 \sin 7\omega t \tag{6.14}$$

is shown in Fig. 6.8*B*.   The spectrum of this wave is also shown in Fig. 6.8*B*.

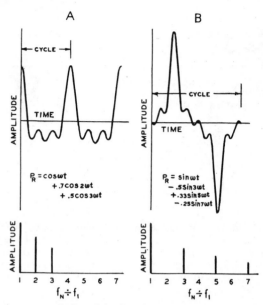

FIG. 6.8.   *A.* A wave consisting of the first three components of a cosine series.   The spectrum depicts the relative amplitudes of the components of the wave as a function of the frequency. f_1 = frequency of the fundamental, f_N = frequency of the Nth harmonic.   *B.* A wave consisting of the first four components of an odd sine series with alternate signs.   The spectrum depicts the relative amplitudes of the components of the resultant wave as a function of the frequency. f_1 = frequency of the fundamental, f_N = frequency of the Nth harmonic.

The resultant waveforms of Figs. 6.7 and 6.8 illustrate how different wave shapes are developed from various complexions of the overtone structure.

The phase of the components plays an important part in determining the shape of the resultant wave.   In Fig. 6.9A the equation of the wave is given by the series

$$P_R = \sin \omega t + \tfrac{1}{3} \sin 3\omega t + \tfrac{1}{5} \sin 5\omega t \qquad (6.15)$$

In Fig. 6.9B the equation of the wave is given by the series

$$P_R = \sin \omega t - \tfrac{1}{3} \sin \omega t + \tfrac{1}{5} \sin 5 \omega t \qquad (6.16)$$

The phase of the third harmonic of Eq. (6.15) differs in phase by 180 degrees from the third harmonic of Eq. (6.16).   It will be seen that the shapes of the two waves are entirely different.   However, the spectrums are the same.   In general, except for very large phase shifts or very intense sounds, phase plays a relatively small role.

The wave shape and the equation of a saw-tooth wave are shown in

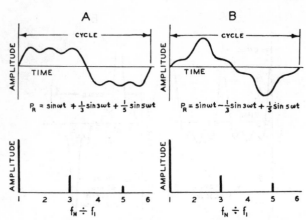

FIG. 6.9. *A.* A wave consisting of the first three components of the odd sine series, the equation of the wave, and the spectrum. f_1 = frequency of the fundamental, f_N = frequency of the Nth harmonic. *B.* A wave consisting of the first three components of the odd sine series with alternate signs, the equation of the wave, and the frequency spectrum. f_1 = frequency of the fundamental, f_N = frequency of the Nth harmonic.

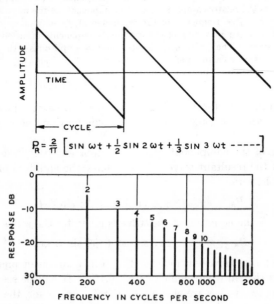

FIG. 6.10. A saw-tooth wave, the equation of the wave, and the spectrum for a fundamental frequency of 100 cycles, that is, $t = \frac{1}{100}$ second for a complete cycle.

Fig. 6.10. An infinite number of components in the sine series is required to produce a saw-tooth wave shape. The spectrum containing the components of the wave up to 2,000 cycles, for a saw-tooth wave having a fundamental frequency of 100 cycles, is shown in Fig. 6.10. The spectrum depicts the amplitudes and frequencies of the fundamental and the harmonics. It will be seen that this wave contains the fundamental and all the harmonics. Spectrums will be used in this chapter to depict the timbre or overtone structure of the sounds produced by musical instruments.

The wave shape and the equation of a triangular wave are shown in Fig. 6.11. An infinite number of components in the odd cosine series is required to produce the triangular wave shape. The spectrum con-

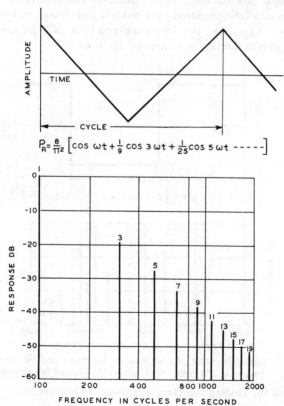

$$P_R = \frac{8}{\pi^2} \left[\cos \omega t + \frac{1}{9} \cos 3\omega t + \frac{1}{25} \cos 5\omega t \; ----- \right]$$

FIG. 6.11. A triangular wave, the equation of the wave, and the spectrum for a fundamental frequency of 100 cycles, that is, $t = \frac{1}{100}$ second for a complete cycle.

taining the components of the wave up to 2,000 cycles, for a triangular wave having a fundamental frequency of 100 cycles, is shown in Fig. 6.11. It will be seen that this wave contains the fundamental and the odd harmonics. The even harmonics are missing.

The wave shape and the equation of a rectangular wave are shown in Fig. 6.12. An infinite number of components is required to produce the rectangular wave shape. The spectrum containing the components of the wave up to 2,000 cycles, for a rectangular wave having a fundamental frequency of 100 cycles, is shown in Fig. 6.12. It will be seen that this wave contains the fundamental and the odd harmonics. The even harmonics are missing.

From the preceding examples of complex waves, it will be seen that the spectrum depicts the relative amplitudes and the frequency of the components of the fundamental and harmonics or overtones. It is this complexion of a tone produced by a musical instrument that determines the timbre. Therefore, the frequency spectrum will be used in this chapter to depict the timbre of musical instruments.

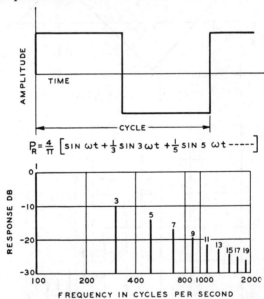

$$P_R = \frac{4}{\pi} \left[\sin \omega t + \frac{1}{3} \sin 3\omega t + \frac{1}{5} \sin 5\omega t \ ----- \right]$$

Fig. 6.12. A rectangular wave, the equation of the wave, and the spectrum for a fundamental frequency of 100 cycles, that is, $t = \frac{1}{100}$ second for a complete cycle.

B. Methods for Testing the Timbre of Musical Instruments

The cathode-ray oscillograph provides a means for depicting the wave shape of a sound wave. A schematic diagram of the apparatus for depicting the shape of sound waves is shown in Fig. 6.5. The system includes a microphone for converting sound waves into the corresponding electrical waves and an amplifier for increasing the output to an amplitude suitable for operating the cathode-ray oscillograph. The horizontal deflection of the electron beam which produces the trace on the cathode-ray oscillograph is deflected at a constant velocity by an internal oscilator. The output of the amplifier is fed to the vertical deflection system of the cathode-ray oscillograph. Under these conditions, the vertical deflection of the electron beam, which produces the trace on the cathode-ray oscillograph, is proportional to the sound pressure or particle velocity in the sound wave. The net result is that the output of the musical instrument produces a sound wave in air which is translated by the equipment into the corresponding graphical wave upon the screen of the cathode-ray oscillograph. This wave can be analyzed into components consisting of the fundamental and the harmonics, as outlined in the preceding section. However, this type of analysis is laborious and complex. It is more convenient to employ an analyzer which depicts the amplitude of the components of a wave with respect to frequency in a graph similar to that of the spectrums of Figs. 6.7 to 6.12.

A schematic diagram of the apparatus for obtaining a spectrum of the components of a complex wave is shown in Fig. 6.13. The sound wave in air is converted into the corresponding electrical wave by the microphone. The electrical output of the microphone is amplified by means of a vacuum-tube amplifier and fed to the analyzer. The analyzer is a scanning heterodyne type of instrument which automatically separates the frequency components of a complex wave and simultaneously measures their frequency and magnitude. The results of the analysis are depicted on the screen of a cathode-ray tube with a screen of the long-persistence variety, thereby providing a continuous over-all view of the spectrum. The analyzer consists of a narrow band-pass heterodyne filter which sweeps through a frequency range of 20 to 20,000 cycles once

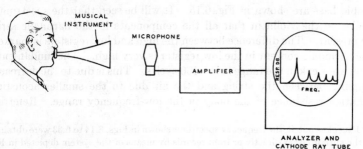

FIG. 6.13.   Apparatus for depicting the acoustic spectrum of a sound wave.

every second. The output of the heterodyne filter is fed to an amplifier which drives the vertical system of the cathode-ray tube. When a component is passed by the heterodyne filter, a vertical deflection of the cathode-ray tube is produced. Thus it will be seen that, as the wave components are progressively tuned through, they are spectrographically displayed on the cathode-ray tube as sharp peaks. The height of each deflection indicates the relative magnitude of the component. The position along the horizontal logarithmic frequency scale indicates the frequency of the component. Since the scanning time for one complete sweep is 1 second, the tone to be analyzed must persist for 1 second without any appreciable change in order to obtain a spectrum. For sound in which the timbre varies over a period of a second, it is necessary to record the tone on a disk phonograph or magnetic tape record. The section of the tone to be analyzed is reproduced and fed to the analyzer. The record is reproduced repeatedly until a complete spectrum has been obtained.

C. Acoustic Spectrums of Musical Instruments[5]

1. String Instruments. *a. Violin.* The violin is the outstanding example of the bowed-string musical instrument (Sec. 5.3B1). The acoustic spectrums for the four open strings of a violin are shown in Fig. 6.14. The difference between the quality of the low register and the high register is quite marked. The low registers exhibit more harmonics. As in all string instruments, it will be seen that all the components in the harmonic series are present in the violin (Sec. 4.5). This is probably the reason that the violin produces a beautiful tone. It resides in the fact that the violin produces all the partials with an even amplitude distribution. The quality is uniform through the frequency range even though there is a wide variation in quality from the low registers to the high registers.

b. Double bass. The acoustic spectrums for the four open strings of the double bass are shown in Fig. 6.15. It will be seen that the spectrum is similar to the violin in that all the components in the harmonic series are present. The difference between the high and low registers is marked. The harmonic content in the low register is very high. The fundamental tones are weak in the range below 100 cycles. This is due to the decreased coupling between the string and the air due to the smaller acoustical radiation resistance of the body in the low-frequency range. Referring

[5] Except as noted, the frequency spectrums shown in Figs. 6.14 to 6.33 were obtained by the author either directly or from records by means of the system depicted in Fig. 6.13. The level of 0 decibels in these diagrams is arbitrary.

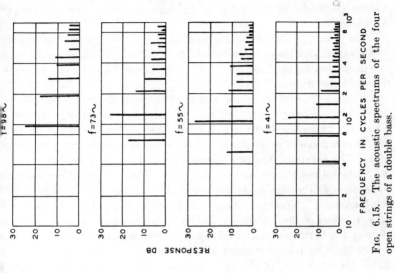

Fig. 6.15. The acoustic spectrums of the four open strings of a double bass.

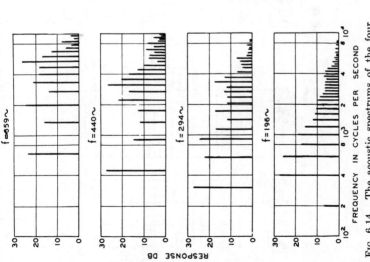

Fig. 6.14. The acoustic spectrums of the four open strings of a violin.

to Fig. 4.15 it will be seen that the radiation resistance of a vibrating surface decreases as the ratio of the dimensions to the wavelength decreases.

Considering the size of the double bass, the sound output of the instrument is relatively low. Nevertheless, the double bass plays an important role in reinforcing the bass portion of an orchestra which, under any condition, is usually weak.

c. Piano. The piano is the outstanding example of a struck-string musical instrument (Sec. 5.3C1). The acoustic spectrums of three notes of a piano are shown in Fig. 6.16. As in other string instruments, all the components in the harmonic series are present in the piano (Sec. 4.5). The overtone structure depends to some extent upon the velocity with which the hammer strikes the string. The velocity of the hammer depends upon the velocity of the key at the moment that it "throws off" the hammer. The pianist can produce different tone qualities but only as a function of the intensity. In the high-frequency range, the amplitude of the harmonics decreases rapidly with frequency. The piano has a very uniform quality in going from one tone to the next. On the other hand, there is a marked difference between the tones in the low registers and the high registers. The low tones are richest in harmonics, because

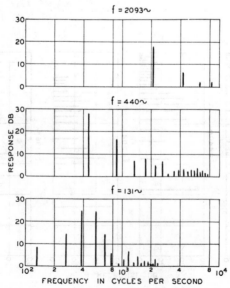

Fig. 6.16.   The acoustic spectrums of three tones of a piano.

the low tones are not produced with the same efficiency as the higher tones. This is caused by the decreased coupling between the sound-board and the air in the low-frequency range due to the relatively small acoustical-radiation resistance. Referring to Fig. 4.15, it will be seen that the acoustical-radiation resistance of a vibrating surface decreases as the ratio of the dimensions to the wavelength decreases. It also follows that, in the larger grand pianos with correspondingly larger soundboards, the low tones will be reproduced with greater efficiency, because as the soundboard is made larger the coupling between the strings and the air is improved, owing to the larger acoustical radiation resistance.

The sound-output characteristic of a piano tone consists of growth and decay, but no steady-state condition. There is considerable variation in the frequency spectrum during the growth and decay intervals. The acoustic spectrum shown in Fig. 6.16 represents the frequency components in the output during a short interval of time in the decay period.

d. Guitar. The guitar is an example of a plucked-string instrument. The acoustic spectrums of two open strings of the guitar are shown in Fig. 6.17. As in other string instruments, all the components in the harmonic series are present in the guitar (Sec. 4.5). As in the piano, the complexion of the overtone structure depends upon the intensity with which the instrument is plucked, that is, the deflection of the string from the normal repose position when it is released. The sound-output characteristic of a guitar tone consists of a growth and decay, but no steady-state condition. There is considerable variation in the acoustic spectrum

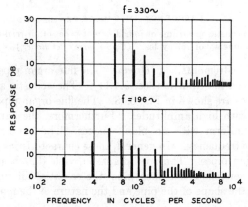

FIG. 6.17. The acoustic spectrums of two tones of a guitar.

during the growth and decay intervals. The acoustic spectrum shown in Fig. 6.17 represents the frequency components in the output during a short interval of time in the decay period.

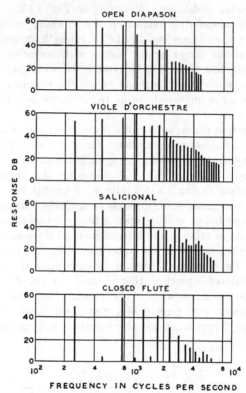

FIG. 6.18. The acoustic spectrums of four different types of air-reed organ pipes for a fundamental frequency of 262 cycles. (*Upper three after Boner.*)

2. Air-reed Instruments. *a. Organ.* The flue organ pipe is an example of an air-reed instrument (Sec. 5.4A1). The acoustic spectrums[6] of four organ pipes are shown in Fig. 6.18. The flue organ pipe is very rich in harmonics with high amplitudes. Furthermore, the harmonic structure is very uniform with no blank spaces. The overtones decrease uniformly with frequency. In general, these characteristics are exhibited by all flue organ pipes. The open diapason, viole d'orchestre, and salicional are all versions of an open pipe, the essential difference in these three being in the shape of the pipe and the nature of the mouth termina-

[6] Boner, C. P., *Jour. Acoust. Soc. Amer.*, Vol. 10, No. 1, p. 32, 1938.

tion. Since these three are open pipes, the resonant frequencies correspond to the fundamental and all the harmonics (Sec. 4.10). As a result, the acoustic spectrum of these pipes contains the fundamental and both even and odd harmonics. The shape of the pipe and the nature of the mouth termination modify the radiation characteristics as a function of the frequency. The net result is a difference in the complexion of the overtone structure in the three pipes, as shown in Fig. 6.18. Referring to the spectrum of the closed flute pipe shown in Fig. 6.18, it will be seen that the sound output contains the fundamental, the odd harmonics, and some weak even harmonics. The reason for the presence of even harmonics is due to the fact that the flipping air reed generates the fundamental and all the harmonics. Therefore, the fundamental and all the harmonics are radiated, but only the fundamental and the odd harmonics are reinforced by the pipe, because the resonant frequencies of a closed pipe correspond to the fundamental and odd harmonics (Sec. 4.10). As a result, the closed flute pipe contains the fundamental, strong odd harmonics, and weak even harmonics, as shown in Fig. 6.18.

b. *Flute.* The flute is another example of an air-reed instrument (Sec. 5.4*A*5). As contrasted to most other musical instruments, the

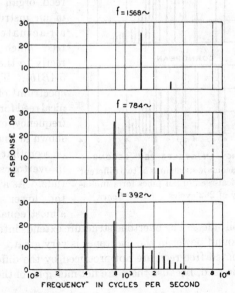

FIG. 6.19. The acoustic spectrums of three tones of a flute.

fundamental carries a preponderant amount of the acoustical-energy output. The acoustic spectrums of three tones emitted by a flute are shown in Fig. 6.19. The lower registers are richer in harmonics. In the case of the highest tone, the overtones are practically nonexistent and the notes emitted are very clear and clean. The large percentage of sound power in the fundamental together with the small number of overtones is the reason that the flute has the thinnest and purest tone of all the musical instruments.

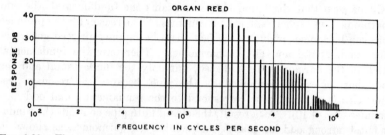

FIG. 6.20.   The acoustic spectrum of a mechanical reed alone, of the type used in reed organ pipes, for a fundamental frequency of 262 cycles.   (*After Boner.*)

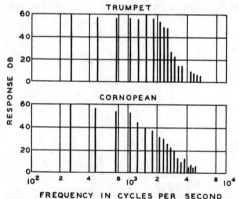

FIG. 6.21.   The acoustic spectrums of two different types of mechanical-reed organ pipes for a fundamental frequency of 262 cycles.   (*After Boner.*)

3. **Mechanical-reed Instrument.** *a. Organ.* The reed organ is an example of an instrument with an air-actuated mechanical reed which is coupled directly to the open air (Sec. 5.4*B*1*b*). The acoustic spectrum[7] of a free mechanical reed for a fundamental frequency of 262 cycles is shown in Fig. 6.20. It will be seen that the reed is rich in overtones of high amplitude. As a matter of fact, the lower overtones are almost equal to the fundamental in amplitude. The overtone structure extends into the ultrasonic range. The sound power in the overtones is very great. The overtones of the air-actuated free reed are not produced by the different modes of vibration of the reed, because these are not multiples of the fundamental,

[7] Boner, C. P., *Jour. Acoust. Soc. Amer.*, Vol. 10, No. 1, p. 32, 1938.

as shown in Sec. 4.6*A*, but by the nonlinear characteristics of the throttling action of the reed upon the air stream. The throttling action of the reed converts the steady air stream into a pulsating one of the saw-tooth type which contains the fundamental and all the harmonics (Sec. 6.3*A*).

The acoustic spectrums[8] for trumpet and cornopean types of reed organ pipes are shown in Fig. 6.21. Comparing Figs. 6.20 and 6.21, it will be seen that the addition of a pipe to a reed modifies the overtone structure. Nevertheless, the overtones, particularly of a low order, exhibit relatively large amplitudes. This accounts for the reedy sound of these pipes.

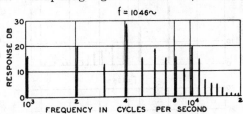

FIG. 6.22. The acoustic spectrum of a tone of a harmonica.

b. Harmonica. The harmonica is an example of a mechanical-reed instrument in which the reed feeds directly into the air (Sec. 5.4*B*1*d*).

The acoustic spectrum of a tone produced by a harmonica is shown in Fig. 6.22. As in the case of the free-reed organ, the harmonica is rich in overtones of high amplitude which extend into the ultrasonic region. Furthermore, the acoustic spectrum contains the fundamental and both even and odd harmonics.

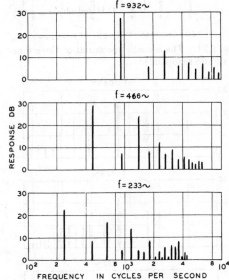

FIG. 6.23. The acoustic spectrums of three tones of a clarinet.

c. Clarinet. The clarinet is an example of an instrument with a single mechanical reed coupled to a cylindrical tube terminating in a flared mouth (Sec. 5.4*B*1*f*). The acoustic spectrums of three tones of the clarinet are shown in Fig. 6.23. The acoustic spectrum contains the

[8] Boner, C. P., *Jour. Acoust. Soc. Amer.*, Vol. 10, No. 1, p. 32, 1938.

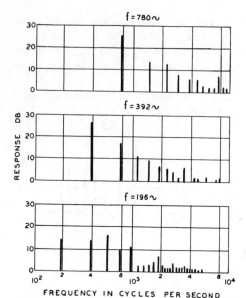

FIG. 6.24.   The acoustic spectrums of three tones of an alto saxophone.

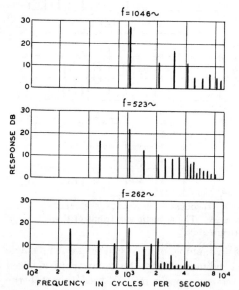

FIG. 6.25.   The acoustic spectrums of three tones of an oboe.

fundamental and both even and odd harmonics. It will be seen that most of the energy resides in the fundamental. Therefore, the sound output is clear and bright. It will be seen that the even harmonics are suppressed. This is a characteristic of a cylindrical pipe closed at one end (Secs. 4.10 and 5.4$B1f$). The tone of the clarinet is sometimes incisive, but it can also be mournful. It produces tones of considerable sound power in the lower registers.

d. Saxophone. The saxophone is an example of an instrument with a single mechanical reed coupled to a conical tube terminating in a flared mouth. (Sec. 5.4$B1h$). The acoustic spectrums of three tones of a saxophone are shown in Fig. 6.24. There is considerable energy in the overtones. Apparently there is no fixed pattern in the overtone structure. The tone of the saxophone resembles both the brasses and wood winds. Its tone is full and rich combined with a large sound power output.

e. Oboe. The oboe is an example of an instrument with a double mechanical reed coupled to a conical tube terminating in a flared mouth (Sec. 5.4*B2a*). The acoustic spectrums of three tones of the oboe are shown in Fig. 6.25. The major portion of the energy resides in the fundamental and overtones in the frequency range from 500 to 1,500 cycles. The tone of the oboe is bright and reedy and sometimes like a flute in brilliance. The low tones are rich in powerful harmonics and therefore are reedy in character. The tone can also be very incisive.

f. Bassoon. The bassoon is an example of an instrument with a double mechanical reed coupled to a conical tube (Sec. 5.4*B2d*). The acoustic spectrums of three tones of the bassoon are shown in Fig. 6.26. The fundamental and lower harmonics are low in intensity in the low registers. The tone is usually called weird and grotesque. The dominant region is around 500 cycles.

4. Lip-reed Instruments.
a. Trumpet. The trumpet is an example of an instrument with a lip reed, coupled to a tube with a combination of cylindrical and conical tube terminating in a flared mouth (Sec. 5.4*D2*.) The acoustic spectrums of three tones of the

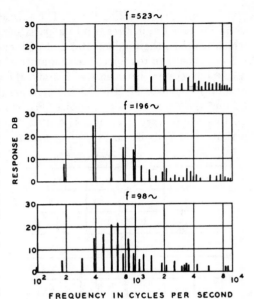

FIG. 6.26. The acoustic spectrums of three tones of a bassoon.

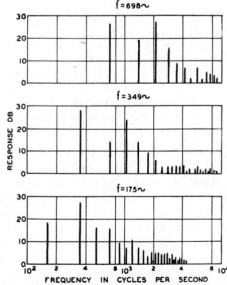

FIG. 6.27. The acoustic spectrums of three tones of a trumpet.

trumpet are shown in Fig. 6.27.   The throttling action of the lips converts the steady air stream into a pulsating one of the saw-tooth type
which contains the fundamental and both even and odd harmonics (Sec.
6.3A).   The horn reinforces the fundamental and overtones by providing
a coupling system between the lips and the air (Sec. 4.11).   The sound
output of the instrument is rich in overtones throughout the range.   The
tone of the trumpet is characterized by its golden clarity and brilliance.
Its tone can be very strident.   It can also be very soft and rich.   Under
certain conditions it can be very expressive.

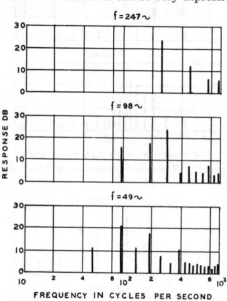

FIG. 6.28.   The acoustic spectrums of three tones of a tuba.

b. Tuba.   The tuba is
the largest of the lip-reed-
type instruments (Sec.
5.4D7).   The acoustic spectrums of three tones of the
tuba are shown in Fig. 6.28.
The principal sound output
occurs in the range from
100 to 300 cycles.   The
tuba produces a large output in the low-frequency
region.

5. **Human Voice.**   The
human voice possesses the
greatest flexibility of all
musical instruments (Sec.
5.4E).   The several reasons
for this characteristic are as
follows: The throttling action of the vocal cords converts a steady air stream
supplied by the lungs into
a pulsating one of the saw-tooth type which contains the fundamental and
both even and odd harmonics (Sec. 5.4 and Fig. 5.58 and Sec. 6.3A).   The
fundamental frequency of this wave can be varied by the person at will over
a range of about two octaves.   The harmonic content of the output of the
vocal cords can also be varied to some extent.   The output of the vocal
cords passes into the vocal cavities.   The vocal cavities possess several
different discrete resonant frequencies which can be varied by the person
in many ways (Sec. 5.4E).   Some of the tones emitted by the vocal cords
are accentuated, and some of the tones are attenuated.   Thus it will be
seen that the quality of the human voice can be varied in a complex
fashion.   The output can also be varied over a wide range.   Almost all the

emotions can be expressed by the voice. The reason for this is due to the great flexibility of the control of the frequency, the timbre and the output in the voice.

The acoustic spectrums of bass, tenor, alto, and soprano voices are shown in Fig. 6.29. It will be seen that the patterns of the acoustic spectrums are similar but shifted with respect to frequency.

The sound-output characteristic of the human voice producing a vowel sound consists of the following periods: growth, steady state, and decay. There is considerable change in the acoustic spectrum during all these periods. However, the steady-state period exhibits the least change in the frequency components during the sounding period. The acoustic spectrum shown in Fig. 6.29 represents the output for a short interval of time during the steady-state period. The growth, steady-state, and

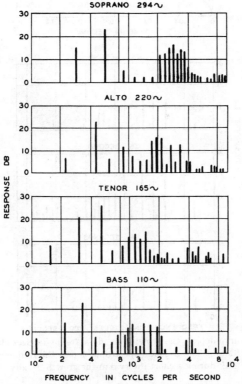

Fig. 6.29. The acoustic spectrums of four voices producing the vowel sound "ah."

decay sound output time characteristics are shown in Fig. 6.30. The acoustic spectrums of a tenor voice for a short interval of time during the growth, steady-state, and decay periods are shown in Fig. 6.30. It will be seen that the acoustic spectrum changes during the production of the vowel sound. This is true of all vowel sounds.

Since the acoustic spectrums of speech and music change with time, systems which will depict the structure in a continuous manner are useful for certain types of investigations. One of these systems is termed visible speech. Visible speech is an electronic method of changing spoken words into visible patterns that someone may learn to read. A schematic arrangement of the apparatus for depicting speech in visible patterns is

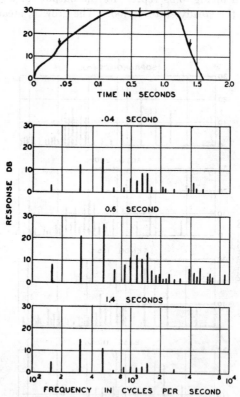

FIG. 6.30. The curve depicts the growth, steady-state, and decay characteristics of the vowel sound "ah." The acoustic spectrums depict the frequency structure of the vowel sound "ah" during a short interval in the growth, steady-state, and decay periods.

shown in Fig. 6.31.   Speech is picked up by the microphone and converted into the corresponding electrical variations.   These variations are amplified and limited in amplitude so that the amplitude range is confined within relatively narrow limits.   The output of the amplifier is coupled to 12 bandpass filters.   Each filter covers a frequency band of 300 cycles.   The entire frequency range covers the band from 150 to 3,750 cycles.   The output of each filter is coupled to a lamp.   When a lamp is illuminated, it produces a trace on the moving belt of phosphor.   With this apparatus, a complex sound wave is divided into 12 discrete frequency bands.   The portions of the frequency range, with intensity sufficient to produce illumination on the phosphor screen, will leave a trace.   A complex wave or a series of complex waves will leave patterns on the moving belt of phosphor.   Each vowel and consonant sound produces a unique and distinguishable pattern.   Under these conditions, speech picked up by the microphone may be read from the moving belt of phosphor.   Music or any other sound may also be picked up by the microphone and portrayed on the screen.   Some of the uses of the visible-speech apparatus are as follows: visual hearing for the deaf, teaching of the deaf to speak, speech correction, and aid in the study of vocal music.

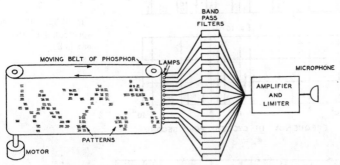

Fig. 6.31.   The elements of a system for producing visible speech.   (*After Potter.*)

6. Percussion Instruments.   *a. Bell.*   The bell is an example of a percussion instrument of the definite-pitch type (Sec. 5.5A7).   The acoustic spectrums of two bells are shown in Fig. 6.32.   When a bell is struck by a clapper, the acoustic spectrum is a function of the time.   Some of the partials[9] disappear after a very short interval, and other partials which do not occur at first begin to appear.   The explanation is that energy in one mode of vibration is converted into another mode.   An examination

---

[9] Curtiss and Giannini, *Jour. Acoust. Soc. Amer.*, Vol. 5, No. 2, p. 159, 1933.

of the acoustic spectrum of Fig. 6.32 shows that the overtones are not harmonics of the fundamental. It is for this reason that when bells are used in a musical instrument, as, for example, the carillon, the music appears to be discordant, particularly if more than one bell is sounded at the same time.

b. Cymbal. The cymbal is an example of a percussion instrument of the indefinite-pitch type (Sec. 5.5*B*6). The acoustic spectrum of a cymbal is shown in Fig. 6.33. The output of the cymbal extends into the ultrasonic region. Furthermore, the overtones are not harmonics. In addition, there is no definite pattern in the acoustic spectrum. It is for this reason that the cymbal does not possess any definite pitch. The cymbal is used to produce brilliant crashes and for reinforcing the tempo in bands and orchestras. The outstanding characteristic is the brilliant sound which is due to the large output in the high-frequency range.

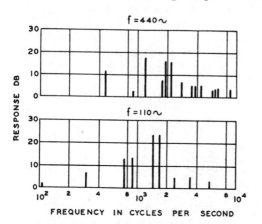

Fig. 6.32. The acoustic spectrums of two different bells.

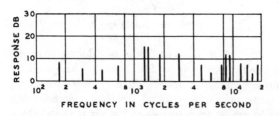

Fig. 6.33. The acoustic spectrum of a 12-inch-diameter cymbal.

6.4 INTENSITY RANGES OF MUSICAL INSTRUMENTS

The tonal aspects of music depend upon pitch and timbre. The dynamic aspect of music depends upon intensity. The intensity and change in intensity play an important role in musical renditions, particularly in obtaining emotional effects. To obtain these effects, the musician must be able to control the intensity of intonation and to produce artistic deviations in intensity.

The intensity ranges of typical musical instruments, all reduced to a distance of 10 feet, are shown in Fig. 6.34. The upper and lower ranges are subject to considerable variation. For example, the maximum output depends upon the individual musician, because some are able to produce much greater outputs than others. The lower limit also depends upon the particular individual playing the instrument. Actually, there is really no lower limit for many of the instruments. The lower limits shown in Fig. 6.34 are the lower limits used in actual musical renditions.

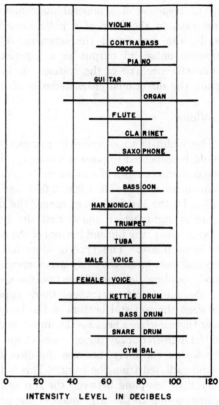

FIG. 6.34. The intensity ranges for various musical instruments at a distance of 10 feet. 0 decibels = 0.000204 dyne per square centimeter, or 10^{-16} watt per square centimeter.

6.5 DIRECTIONAL CHARACTERISTICS OF MUSICAL INSTRUMENTS[10]

As in the case of most acoustical radiating systems, the intensity of the sound radiated from musical instruments varies with the direction of observation with respect to a reference axis of the system. In general, this radiation pattern varies with the emitted frequency. Thus, it will be seen that the magnitude of the sound output, as well as the quality, of a musical instrument will vary with the orientation of the instruments with respect to the listener. It is customary to depict the sound out-

[10] Except as noted, the directional patterns shown in Figs. 6.35 to 6.44 were obtained by the author.

put or response of a musical instrument, with respect to some reference axis, in the form of a polar directional characteristic. In other words, the directional characteristic of a musical instrument is the response or sound output as a function of the angle with respect to some reference axis of the system. It is the purpose of this section to depict the directional characteristics of musical instruments.

A. Violin

The violin is an exceedingly complex vibrating system; therefore, it would be expected to possess a complex directivity pattern. The directional characteristics of a violin in the plane normal to the strings, for the frequencies of 200, 500, 1,000, 2,000, and 4,000 cycles, are shown in Fig. 6.35. In the low-frequency range, the directional pattern resembles a figure of eight, which shows that the system is an acoustical doublet. That is to say, the top and bottom of the body vibrate with approximately the same phase. This type of vibration is to be expected, because the top and bottom of the body are connected by the post, which provides good coupling between these two surfaces in the low-frequency range. In the mid-frequency range, there appears to be a phase difference between the top and bottom of the body, as well as a phase difference over these surfaces because the directional pattern exhibits several lobes. In the high-frequency region, most of the radiation appears to come from the top of the body, because the directional pattern shows very little sound radiated from the back. This is to be expected because there is very little coupling between the top and bottom in the high-frequency region, owing to the large mass of the post and bottom, with the result that the vibrations transmitted to the bottom are very small.

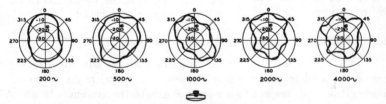

FIG. 6.35.   The directional characteristics of a violin for five different frequencies.

B. Viola, Violoncello, and Double Bass

The directivity patterns of the viola, violoncello, and double bass are similar to the violin except that the frequency for each pattern is lowered by the ratio of the linear dimensions of the instruments.

C. Piano

The directional characteristics of a grand piano in the plane of the strings, for the frequencies 100, 400, 1,000, 2,000, and 4,000 cycles, are shown in Fig. 6.36. The listening positions for the piano usually do not deviate very much from this plane. These characteristics were obtained with the large lid in the open position. In the low-frequency range, the sound output does not vary to any marked degree with the direction. In the mid- and high-frequency region, the lid acts as a reflector for the sound emitted by the strings and soundboard. As a consequence, the major portion of the sound is radiated in the direction opposite the reflector. In concerts, the piano is usually placed so that the lid will reflect the sound into the audience. In sound reproduction, the microphone is usually placed opposite the reflecting lid so that the lid will reflect the sound toward the microphone.

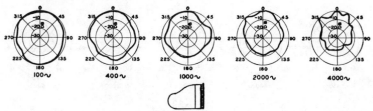

FIG. 6.36. The directional characteristics of a grand piano for five different frequencies.

D. Clarinet

The directional characteristics of a clarinet, for the frequencies 1,000, 2,000 and 4,000 cycles, are shown in Fig. 6.37. Since the mouth is very small, the clarinet does not show any marked directivity except at the very high frequencies (Sec. 4.12E). There is some radiation from the holes which introduces small sharp lobes in the directivity pattern. Since these vary rapidly in direction as well

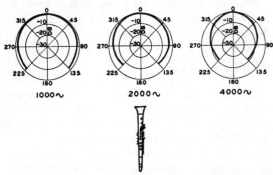

FIG. 6.37. The directional characteristics of a clarinet for three different frequencies.

as with the particular hole opening, the curves of Fig. 6.37 are smoothed out and do not show these variations.

E. Saxophone

The directivity patterns of the soprano saxophone are similar to the clarinet except that they are shifted downward in frequency due to the larger mouth.

The directional characteristics of an alto saxophone, for the frequencies 150, 400, 1,000, 2,000, and 4,000 cycles, are shown in Fig. 6.38. The saxophone consists of a tube with a conical flare ending in a relatively small bell. Therefore, the high-frequency radiation pattern is very narrow. In the mid-frequency range, the pattern is shifted owing to the turn at the mouth and the relatively large diameter at the bend.

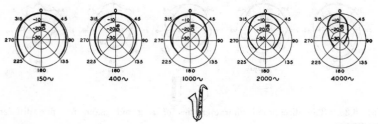

FIG. 6.38. The directional characteristics of an alto saxophone for five different frequencies.

F. Oboe, English Horn, and Bassoon

Since the mouths of the oboe, English horn, and bassoon do not differ materially from the clarinet, the average directional characteristics of these instruments are similar to those of the clarinet.

G. Flute and Piccolo

The open ends of the flute and the piccolo are small compared to the wavelength up to a very high audio-frequency range. Therefore, the sound radiation which issues from the mouth does not vary appreciably with the direction up to 10,000 cycles (Sec. 4.12B). There is considerable sound radiation from the embouchure and some of the open holes. Under these conditions, the system consists of multiple sound sources, and the directivity pattern exhibits a large number of lobes (Sec. 4.12A). However, the average pattern may be said to be nondirectional.

H. Trumpet, Cornet, and Trombone

The directional characteristics[11] of a trumpet, for the frequencies 220, 480, 920, 1,840, and 4,000 cycles, are shown in Fig. 6.39. The directional pattern of the trumpet becomes sharper as the frequency increases (Sec. 4.12E). All the radiation issues from the mouth of the horn. Therefore, the directional pattern is very smooth and free of secondary lobes. Over a total angle of 90 degrees there is very little frequency discrimination up to 4,000 cycles.

The directional characteristics of the cornet are practically the same as the trumpet because the dimensions and shape of the mouths of these two instruments are almost the same.

The directional characteristics of the trombone are similar to those of the trumpet save that the frequencies shown in Fig. 6.39 are shifted downward by the ratio of the linear dimensions of the bells of the two horns (Sec. 4.12E).

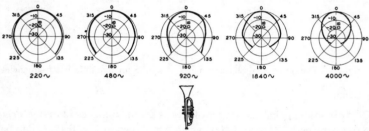

Fig. 6.39. The directional characteristics of a trumpet for five different frequencies. (*After Martin.*)

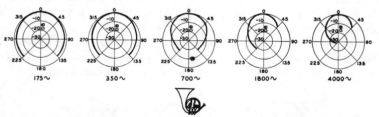

Fig. 6.40. The directional characteristics of a French horn for five different frequencies. (*After Martin.*)

I. French Horn

The directional characteristics[12] of the French horn, for the frequencies 175, 350, 700, 1,800, and 4,000 cycles, are shown in Fig. 6.40. The

[11,12] Martin, D. L., *Jour. Acoust. Soc. Amer.*, Vol. 13, No. 3, p. 309, 1942.

directivity pattern is similar to the trumpet except that it is shifted downward in frequency owing to the larger mouth.

J. Tuba

The directional characteristics of a tuba, for the frequencies 100, 400, 1,000, 2,000, and 4,000 cycles, are shown in Fig. 6.41. The tuba consists of a tube with conical bore ending in a large bell-shaped mouth. The diameter at which the conical taper ends and the bell-shaped mouth begins is relatively large. The directivity pattern in the high-frequency range is governed by this conical taper. Therefore, the directivity pattern becomes quite narrow in the high-frequency range.

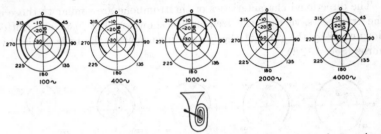

FIG. 6.41.   The directional characteristics of a tuba for five different frequencies.

K. Drum

The directional characteristics of a bass drum 26 inches in diameter, for the frequencies 60, 120, and 400 cycles, are shown in Fig. 6.42. The bass drum is a two-sided radiator; therefore, the directivity pattern is governed by the dimensions and separation, as well as the phase of the two vibrating diaphragms. For this reason, the directivity patterns are complex with secondary lobes (Sec. 4.12A).

The directional patterns of all two-sided drums are similar to those of the bass drum of Fig. 6.42. As in most radiating systems, the effects are shifted in frequency as the inverse of the linear dimensions.

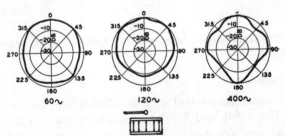

FIG. 6.42.   The directional characteristics of a bass drum 26 inches in diameter for three different frequencies.

L. Voice

The directional characteristics[13] of the human voice in a horizontal plane passing through the mouth, for the frequencies 100, 400, 1,000, 4,000, and 10,000 cycles, are shown in Fig. 6.43. Similar characteristics in a symmetrical bisecting vertical plane passing through the mouth, for the frequencies 100, 400, 1,000, 4,000, and 10,000 cycles, are shown in Fig. 6.44. It will be noted that there is very little frequency discrimination over a total angle of 90 degrees in the forward direction. However, beyond this angular range there is considerable loss of high-frequency radiation. This is in accordance with experience, in which one observes that there is considerable frequency discrimination as the head of a speaker or singer is turned more than 45 degrees from the normal, or 0-degree position.

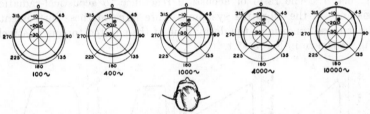

FIG. 6.43. The directional characteristics of the human voice in a horizontal plane passing through the mouth for five different frequencies. (*After Dunn and Farnsworth.*)

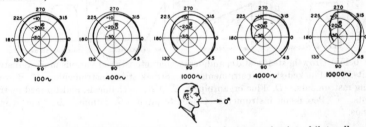

FIG. 6.44. The directional characteristics of the human voice in a bilaterally symmetrical vertical plane passing through the mouth for five different frequencies. (*After Dunn and Farnsworth.*)

6.6 GROWTH AND DECAY CHARACTERISTICS OF MUSICAL INSTRUMENTS

The growth of a tone emitted by a musical instrument involves the time required for the tone to build up to some fraction of its ultimate value. The decay of a tone emitted by a musical instrument involves the time required for the tone to fall in intensity from one arbitrary level to some other arbitrary level. The growth and decay characteristics affect the pitch, loudness, and timbre of a tone.

[13] Dunn and Farnsworth, *Jour. Acoust. Soc. Amer.*, Vol. 10, No. 3, p. 184, 1939.

The growth and decay characteristics of musical instruments in various classifications are shown in Fig. 6.45. The growth and decay characteristics of individual instruments in the different classifications vary from instrument to instrument. The characteristics also vary with frequency. The characteristics depicted in Fig. 6.45 may be said to be typical of each classification.

The growth and decay characteristics of plucked-string instruments, as, for example, the guitar, harp, mandolin, ukulele, zither, and harpsichord, are shown in Fig. 6.45A. The vibrating system of plucked-string instruments is actuated by pulling the string to one side and then releasing it. The potential energy stored in the displaced string and the remainder of the vibrating system is liberated very suddenly as the string is released. Therefore, the build-up time of the emitted tone is very short, as shown in Fig. 6.45A. The ratio of acoustical reactance to acoustical-radiation resistance in the vibrating system is quite large in these instruments. Under these conditions, the decay time of the emitted tone is relatively long, as shown in Fig. 6.45A.

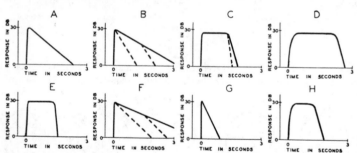

Fig. 6.45. Growth, steady-state, and decay characteristics of various musical instruments. *A.* Plucked-string instruments. *B.* Struck-string instruments. *C.* Bowed-string instruments. *D.* Flue organ pipe. *E.* Air-, mechanical-, and lip-reed instruments. *F.* Percussion instruments of definite pitch. *G.* Drums. *H.* Voice vowel sounds.

The growth and decay characteristics of a struck-string instrument, as for example, the piano,[14] are shown in Fig. 6.45B. The vibrating system consists of a relatively heavy string coupled to a large soundboard. The system is actuated by a hammer striking the string. The kinetic energy of the rapidly moving hammer is transferred to the vibrating system in a very short interval of time. Under these conditions, the build-up time of the emitted tone is very short. Although the soundboard is very large in area, the effective mass of the string and soundboard is also very large. Therefore, the ratio of the acoustical reactance to the acoustical-radiation

[14] Fletcher, H., *Amer. Jour. Phys.*, Vol. 14, No. 4, p. 215, 1946.

resistance is large. As a consequence, the decay time of the emitted tone is long. However, when the key is released or the damping pedal pressed down so that the damper engages the string, the decay time is shortened, as shown by the dotted lines of Fig. 6.45B.

The growth and decay characteristics of bowed-string instruments, as for example, the violin, viola, violoncello, and double bass, are shown in Fig. 6.42C. The build-up time of the emitted tone is quite short because it is a forced vibration produced by the nonlinear characteristics of the bow against the string. In the decay cycle, if the string is free, the decay time is relatively long. However, if the bow is in contact with the string, the decay time is quite short, as shown by the dotted line of Fig. 6.45C.

The growth and decay characteristics of a large flue organ pipe are shown in Fig. 6.45D. The ratio of acoustical reactance to acoustical resistance in a low-frequency organ pipe is very large. The acoustical impedance of the generator is relatively large. Under these conditions, the growth and decay times of the emitted tone will be long.

The growth and decay characteristics of air-, mechanical-, and lip-reed instruments are shown in Fig. 6.45E. Most of these instruments operate in the mid audio-frequency range. Furthermore, the resonance system is the air in a cylindrical, conical, or flared tube. As a consequence, the ratio of acoustical reactance to acoustical-radiation resistance is smaller than in most musical instruments. Under these conditions, the growth and decay times of the tones emitted are relatively short.

The growth and decay of the percussion instruments of the definite-pitch type, as for example, the xylophone, marimba, celesta, glockenspiel, chimes, and bell, are shown in Fig. 6.45F. The vibrating elements in these instruments are excited by being struck with hammers. The kinetic energy of the rapidly moving hammer is transferred to the vibrating system in a very short interval of time. Under these conditions, the build-up time of the emitted tone is very short. The vibrating system of these instruments is very massive and the acoustical-radiation resistance very small. These conditions conspire to make the decay time exceedingly long. In instruments equipped with dampers, the decay time is shortened when the damper engages the vibrator, as shown by the dotted lines of Fig. 6.45F.

The growth and decay characteristics of drums are shown in Fig. 6.45G. The build-up time is very short because the light vibrating system is struck by a heavy fast-moving hammer. The mass of the vibrating system is relatively small. The acoustical-radiation resistance is relatively large owing to the large vibrating surfaces. Therefore, the ratio of acoustical reactance to acoustical resistance is relatively small, which means that the decay time will be relatively short.

The growth and decay characteristics of the vowels[15] produced by the human voice vary over very wide ranges. However, on the average the growth and decay times are relatively long, as shown in Fig. 6.45H. These conditions are due to the highly resonant nature of the vocal cavities, that is, the large ratio of acoustical reactance to acoustical-radiation resistance. The duration of the vowel sound shown is relatively long. The duration of the average vowel sound ranges from 0.2 to 1 second.

There is a special case of growth and decay, namely, portamento. Portamento is the passage from a tone of one frequency to another tone of a different frequency in one continuous glide through all the intervening frequencies. The musical instruments capable of executing a portamento are of the continuously variable-pitch type as, for example, the violin family, the trombone, and the human voice. In going from one tone to a second tone, the second tone is approached by the method of successive approximations. For this reason it is difficult to define the duration of the portamento. In general, the average time of the portamento is 0.13 second.

6.7 STEADY-STATE CHARACTERISTICS OF MUSICAL INSTRUMENTS

The steady-state characteristic of a tone emitted by a musical instrument involves that portion of the complete sound-output time characteristic in which the sound output does not vary appreciably with respect to time. Referring to the sound-output-time characteristics of Fig. 6.45C, D, E, and H, it will be seen that the bowed-string, flue-organ, and air-, mechanical-, and lip-reed instruments, and the human voice exhibit steady-state output during a portion of the sounding time. On the other hand, examining the sound-output-time characteristics of Fig. 6.45A, B, F, and G, it will be seen that plucked-string, struck-string, and percussion instruments do not produce a steady-state output during any portion of the sounding time.

6.8 DURATION CHARACTERISTICS OF MUSICAL INSTRUMENTS

The duration of a tone is the length of time that a tone emitted by a musical instrument lasts or persists without an interruption or discontinuity in the sound output. The duration characteristics influence to some degree the pitch, loudness, and timbre of a tone.

From the standpoint of duration of a tone, musical instruments may be classified as follows: fixed duration, variable but fixed maximum duration, and unlimited duration.

The musical instruments in which the emitted tones are of fixed duration are as follows: harp, zither, guitar, ukulele, mandolin, banjo, harpsi-

[15] Drew and Kellogg, *Jour. Acoust. Soc. Amer.*, Vol. 12, No. 1, p. 95, 1940.

chord, piano, dulcimer, xylophone, marimba, glockenspiel, celesta, chimes, bells, kettledrum, triangle, drums, tambourine, gong, cymbal, and castanets. In all these instruments the output rises very rapidly after the system has been actuated and then decays relatively slowly after the maximum output has been attained (see Sec. 6.6 and Fig. 6.45). The sound output may be varied by changing the actuating force, but the growth and decay characteristics are fixed. In some instruments, damping means are provided to increase the decay rate and in this way change the duration. This expedient merely changes the duration to another fixed value.

Musical instruments in which the emitted tones are of a variable but fixed maximum duration are as follows: violin, viola, violoncello, double bass, flute, piccolo, fife, accordion, harmonica, clarinet, saxophone, oboe, English horn, oboe d'amore, bassoon, bugle, trumpet, cornet, French horn, trombone, and tuba. In the bowed-string instruments the maximum duration is determined by the elapsed time for a complete sweep of the bow in one direction, because a change in the direction of motion of the bow causes a break or discontinuity in the sound output. In the air-, mechanical-, and lip-reed instruments, and the human voice the maximum duration is determined by the length of time a person can blow without taking a breath of air, because inhaling a breath of air causes a cessation of the pressure of the air supply and a corresponding break in the sound output. In the accordion, the maximum duration is determined by the elapsed time for a maximum excursion of the bellows.

Musical instruments in which the emitted tones are of unlimited duration are as follows: free-reed organ, organ, bagpipe, and electrical organ. In these instruments a continuous tone, without a break of discontinuity in the sound output, can be maintained for an unlimited time.

Properties of Music

7.1 INTRODUCTION

Music is the art of producing pleasing, expressive, or intelligible combinations of tones. Noise is any undesired sound. Therefore, the demarcation between music and noise is not a definite line because both music and noise have certain features in common. The physical properties of a sound wave are frequency, intensity, waveform, and time. The psychological characteristics of sound which depend upon the physical properties are pitch, loudness, time, and timbre. Every sound in nature may be described by means of these four attributes of sound. The musical mind must be capable of discerning and comprehending these four. From a musical standpoint there is a further delineation which involves four fundamental sensory capacities, namely, the sense of tone quality, the sense of consonance, the sense of volume, and the sense of rhythm. The sensory capacities are based on the physical properties of a sound wave. The relation is, in general, very complex. For example, it is possible to enjoy and perform good music without any insight of its true nature. This is very strikingly demonstrated in the case of musical instruments which were developed by very crude empirical means. It is only within the last two decades that modern acoustics has been used to describe the action of musical instruments. The relation between the sensory capacities and the physical characteristics in the reproduction of sound is extremely involved and has led to certain anomalies which have not been explained. For example, it has been found that the introduction of certain types of distortion in reproduced sound improves the rendition. The nature of music is further complicated by the large number of instruments with widely different shapes, different modes of action, different methods of actuation, different types of sound quality, and different intensities of sound output.

From the above it is obvious that the subject of the properties of music and musical instruments is extremely complex. It is impossible to resolve all the physical characteristics and psychological attributes of music in a single chapter. However, the fundamental aspects relating to these fields may be adequately described in condensed form. It is the purpose of this chapter to describe the human hearing mechanism and the physical and psychological characteristics of speech and music.

7.2 HUMAN HEARING MECHANISM[1-3]

The ultimate useful destination of all desirable original and reproduced sound is the human ear. Therefore, it seems logical to describe the physical structure of the human hearing mechanism as well as the characteristics of hearing.

The human hearing mechanism shown in Fig. 7.1 may be divided into three parts: the outer ear, the middle ear, and the inner ear. The outer ear consists of the external ear, or pinna, and the ear canal which is terminated in the eardrum, or tympanic membrane. Behind the eardrum is the middle ear, a small cavity in which three small bones—the hammer, the anvil, and the stirrup—form the elements of a system for transmitting vibrations from the eardrum to an aperture termed the oval window of the inner ear. The cavity in the middle ear is filled with air by means of the equalizing tube, termed the Eustachian tube, leading to the nasal pharynx. The casing of the inner ear, or cochlea, is a bony structure of a spiral form (two and three-quarter turns).

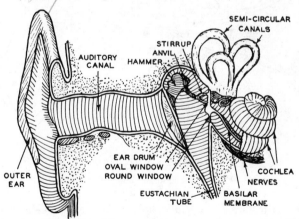

FIG. 7.1. Sectional and perspective views of the human hearing mechanism. *Olson, Acoustical Engineering, D. Van Nostrand Company, Inc., Princeton, 1957.*)

[1] Stevens and Davis, *Hearing*, John Wiley & Sons, Inc., New York, 1938.

[2] Wever, *Theory of Hearing*, John Wiley & Sons, Inc., New York, 1949.

[3] Fletcher, *Speech and Hearing in Communication*, D. Van Nostrand Company, Inc., Princeton, 1953.

A cross-sectional view of the cochlea is shown in Fig. 7.2.   The cochlea is divided into three parts by the basilar membrane and Reissner's membrane.   These three parallel canals are wound into a spiral.   On one side of the basilar membrane is the organ of Corti, which contains the nerve terminals in the form of small hairs extending into the canal of the cochlea.   These nerve endings are stimulated by vibrations in the cochlea.   When a sound wave traverses the longitudinal length of the cochlea, there is relative motion between the basilar membrane and the tectorial membrane which causes the hair cells to stimulate the nerve endings at the base.   The nerve fibers run from the hair cells through small holes into the central core of the cochlea.

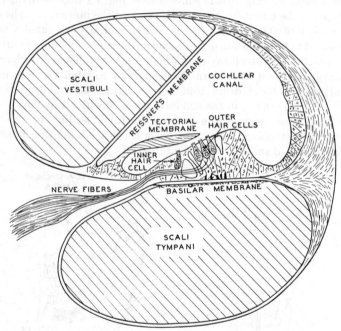

Fig. 7.2.   Sectional view of the cochlea.

A schematic view of the human hearing mechanism is shown in Fig. 7.3 When a sound wave enters the ear canal, it impinges upon the eardrum. The eardrum vibrates with a motion corresponding to the undulations in the sound wave.   The motion of the eardrum is transmitted to the oval window of the cochlea by the lever system of the middle ear.   The vibrations of the oval window are transmitted into the fluid of the cochlea back of the oval window.   The canal of the cochlea is about 35 millimeters, or

1.4 inches, in length. There are 4,000 nerve fibers running from the cochlea to the brain. There are about five hair cells for each nerve fiber. Each nerve fiber is enclosed in a sheath like that of an insulated wire. The 4,000 nerve fibers form a single cable a little over a millimeter in diameter. The cable of nerves passes through the temporal bone to the base of the brain. In the brain the nerve fiber endings are spread over an area.

The cochlea is a frequency-selective mechanism. For example, the portion of the cochlea nearest the oval window is excited by high frequencies, the mid-portion of the cochlea is excited by medium frequencies, and the portion farthest removed from the oval window is excited by low

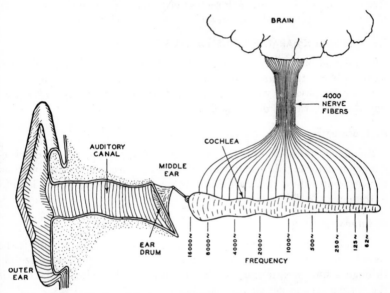

Fig. 7.3.   Schematic view of the human hearing mechanism showing the outer ear, the middle ear, the cochlea, and the nerve fibers leading to the brain.

frequencies. A frequency scale can be established along the cochlea, as shown in Fig. 7.3. When a certain part of the cochlea is excited by a sound wave, it appears that the hair cells are bent, which sets off a nerve impulse. The nerve does not transmit a continuous wave but an impulse. The number of impulses transmitted along a fiber per unit of time depend upon the intensity of the sound. The greater the intensity, the greater the excitation of the hair cells, and a correspondingly greater number of nerve impulses are sent to the brain. In the above mechanism, it will be seen that the cochlea is in effect a sound analyzer. The various frequencies in a complex sound wave are sorted out by the frequency-selective properties of the cochlea. The excited portions of the cochlea actuate corresponding nerve fibers, which in turn send impulses to the brain. The exact nature of the frequency-selecting system is not too well known or understood. However, it is known that the system is highly selective. Unlike the slow response of electrical or electronic systems with a high-order frequency selectivity, the response of the ear is so fast that only a few cycles are required to bring it to full sensitivity. The ear is capable of distinguishing 1,500 separate frequencies. No known electrical system with the rapid response of the ear is capable of resolving this number of frequencies in the audio-frequency range. Since the length of this selective system is 1.4 inches, each section of response is confined to a length of less than a thousandth of an inch.

7.3 PHYSICAL CHARACTERISTICS OF MUSIC

The medium of transmission from the musician to the listener is sound waves. From the standpoint of music, sound waves may be defined in terms of six physical variables, namely, frequency, intensity, waveform, duration, growth and decay, and vibrato.

A. Frequency

Frequency is the number of cycles occurring per unit of time. Frequency is determined by counting the waves per second. A complete set of recurrent values of a periodic quantity comprises a cycle. The unit of frequency is the cycle per second.

B. Intensity

The intensity of a sound field in a specified direction at a point is the sound energy transmitted per unit of time in the specified direction through a unit area normal to this direction at the point. The unit is the erg per second per square centimeter.

C. Waveform

A complex sound wave is made up of the fundamental frequency and overtones. The overtone or harmonic structure is expressed in the number, intensity, distribution, and phase relations of the components. The nature of a complex sound wave and the overtone structure of a number of different waveforms have been considered in Chap. 6.

D. Duration

Duration is the length of time that a tone persists or lasts. The unit is the second or some submultiple of a second. In musical notation the duration is indicated by the kind of note to be played (see Sec. 2.3). Physically, it is a relatively simple matter to measure the duration of a tone as well as its frequency and intensity during the time the tone exists.

6. Growth and Decay

Most of the growth and decay characteristics exhibited by musical instruments are exponential functions. For example, in the growth of a tone, the sound pressure p, in dynes per square centimeter, produced by the musical instrument is given by

$$p = p_0(1 - \epsilon^{-kt}) \tag{7.1}$$

where p_0 = ultimate or steady-state sound pressure, in dynes per square centimeter

k = constant of the instrument

t = time, in seconds

In the decay of a tone, the sound pressure p, in dynes per square centimeter, produced by the musical instrument is given by

$$p = p_0\epsilon^{-kt} \tag{7.2}$$

where p_0 = sound pressure, in dynes per square centimeter, for $t = 0$

The growth of a tone involves the time required for the tone to build up to some arbitrary fraction of its ultimate intensity. The arbitrary fraction for the growth of a tone is $(\epsilon - 1/\epsilon)$, where $\epsilon = 2.718$, the base of the natural system of logarithm. The decay of a tone involves the time required for a tone to fall in intensity to some arbitrary fraction of the reference intensity. The arbitrary fraction for the decay of a tone is $1/\epsilon$. The growth and decay of tones depend upon the type of generator. In some, as, for example, the piano, the build-up time of a tone is short, while the decay time is long. In the pipe organ, the growth and decay time of a tone are both relatively long.

7. Vibrato

Vibrato is a term used to designate primarily a frequency modulation of the musical tone. The vibrato is also accompanied by an amplitude modulation at the modulating frequency as well as a pulsating change in the timbre.

The vibrato is used as an artistic embellishment by singers. The average rate of the frequency modulation in the vibrato is 7 cycles per second.

Tremolo is an amplitude modulation. Therefore, it is a special case of the vibrato. For example, in the organ it is produced by a mechanical means which modulates the air supply to the pipes.

Pitch vibrato is practically a pure frequency modulation.   The violin is the most common example of an instrument in which the pitch vibrato is employed.

As in the case of the voice, the optimum frequency of amplitude modulation of the tremolo and the frequency modulation of the pitch vibrato are of the order of 7 cycles.

7.4 PSYCHOLOGICAL CHARACTERISTICS OF MUSIC

The psychological characteristics of music may be classified as follows: tonal, dynamic, temporal, and qualitative.   The tonal characteristic involves pitch, timbre, melody, harmony, and all forms of pitch variants. The dynamic characteristic depends principally upon the loudness.   The temporal characteristic involves time, duration, tempo, and rhythm. The qualitative characteristic involves timbre or the harmonic constitution of the tone.   In a musical performance, all four of the characteristics may be employed in a well-balanced rendition, or one or more may be emphasized.   These divisions are somewhat intangible, and a more concrete division is desirable for outlining the psychological characteristics of music.   These are as follows: pitch, loudness, timbre, duration growth and decay, consonance, volume, rhythm, presence, and vibrato

A. Pitch

Pitch is defined as a sensory characteristic arising out of frequency, which may assign to a tone a position in a musical scale.   Since it is subjective in character, the measurement of pitch can be made only by an average of judgment tests.

The lower limit of pitch is the lowest frequency which gives us a sensation of tone.   The lower limit depends upon the individual and a number

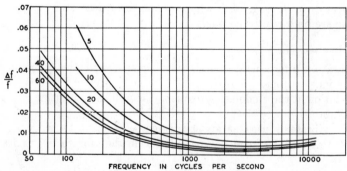

Fig. 7.4.   The variation of $\Delta f/f$ with frequency.   Numbers on the curve indicate level above the threshold of hearing. f is the frequency, Δf is the change in frequency (*After Shower and Biddulph.*)

f physical factors, such as the intensity and the waveform of the sound.
Jnder very favorable conditions most individuals can obtain tonal charac-
eristics as low as 12 cycles.

The upper limit of pitch is the highest frequency which can be heard.
The upper limit of pitch depends upon the individual.   The upper limit
decreases with increase in age.   The average for a person under forty
years of age with good hearing, unimpaired by disease or injury, is about
15,000 cycles.

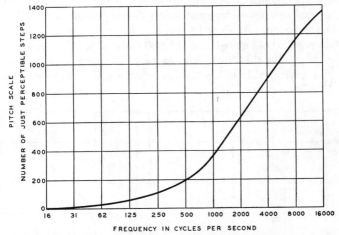

Fig. 7.5.   The cumulative number of just-perceptible steps in frequency as a function
of the frequency.   For example, there are 30 perceptible steps in frequency between 62
and 125 cycles, there are 280 perceptible steps in frequency between 1,000 and 2,000
cycles, etc.   (*After Lewis.*)

The difference in pitch which an individual can detect is termed pitch
discrimination.   Pitch discrimination is tested by sounding two frequencies in rapid succession and gradually reducing the difference until the
observer is unable to distinguish a difference.   The measured pitch
discrimination[4] as a function of the intensity and the frequency is shown
in Fig. 7.4.   It will be seen that the ear is most sensitive to frequency
changes at the higher frequencies.

The number of just-noticeable differences[5] in pitch which an individual
can hear as a function of the frequency under the most favorable conditions is depicted in Fig. 7.5.   It will be seen that the total number of just-
noticeable differences in pitch throughout the hearing range is about
1,400.   The number of tones in the equally tempered scale is only 120.

[4] Shower and Biddulph, *Jour. Acoust. Soc. Amer.*, Vol. 3, No. 2, Part 1, p. 275, 1931.
[5] Lewis, D., *Univ. Iowa Studies in Psychol. of Music*, Vol. IV, p. 346, 1937.

Thus it will be seen that the pitch capabilities of the ear are not fully realized in music.

It is obvious that a tone must persist for a certain length of time in order to establish pitch. If the duration of a tone is very short, say about a millisecond, it appears as a click. If it is somewhat longer, it appears as a noise with some attributes of pitch. As the length of the sounding time of the tone is increased, it finally becomes a tone of definite pitch. The number of cycles[6-8] required to ascribe a definite pitch to a tone depends upon the frequency, as shown in Fig. 7.6. It will be seen that the number of cycles required to establish the pitch of a tone increases with the frequency. At 50 cycles it is 3 cycles, at 400 cycles it is 7 cycles, at 1,000 cycles it is 12 cycles, and at 8,000 cycles it is 145 cycles. It may be noted in passing that the time required to ascribe a definite pitch does not vary a great deal from low to high frequencies. The average time is about 13 milliseconds.

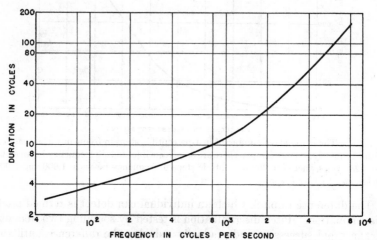

FIG. 7.6. The duration of a tone in cycles required to ascribe a definite pitch. (After Burck, Kotowski, and Lichte; von Bekesy and Turnbull.)

The pitch of a tone depends upon the loudness.[9,10] In these tests the loudness of a tone is varied and compared with a reference tone. The results of these tests are shown in Fig. 7.7. It will be seen that the pitch of a tone may be varied as much as a full tone by a change in loudness.

[6] Burck, Kotowski, and Lichte, *Ann. Physik.*, Vol. 16, p. 433, 1936.

[7] Bekesy, G. von, *Physik. Z.*, Vol. 30, p. 721, 1929.

[8] Turnbull, W. W., *Jour. Expl. Psychol.*, Vol. 34, p. 302, 1944.

[9] Snow, W. B., *Jour. Acoust. Soc. Amer.*, Vol. 8, No. 1, p. 14, 1936.

[10] Stevens, S. S., *Jour. Acoust. Soc. Amer.*, Vol. 6, No. 3, p. 150, 1935.

Another interesting aspect of pitch concerns the subject of discordant sounds. If two tones sounded separately appear to differ slightly in pitch, one cannot conclude that they will be discordant. Furthermore, if two tones when sounded separately appear to have the same pitch, it cannot be said that the combination will be pleasant. As to whether a combination will be harmonious or not depends upon the frequency and not the pitch of the components.

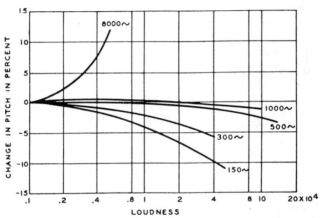

FIG. 7.7. The change in pitch in per cent with loudness for various frequencies as indicated on the curves. (*After Stevens.*)

B. Loudness

The dynamic aspects of music depend primarily upon the intensity. Loudness of a sound is the magnitude of the auditory sensation produced by the sound. The units on the scale of loudness should agree with common experience in the estimates made upon sensation magnitude.

The sone is the unit of loudness. By definition, a pure tone of 1,000 cycles per second 40 decibels above a listener's threshold produces a loudness of one sone. The loudness of another sound that is judged by the listener to be n times that of one sone is n sones. The loudness level of a sound, in phons, is numerically equal to the sound pressure level in decibels, relative to 0.0002 dynes per square centimeter of a free progressive sound wave of 1,000 cycles per second which is judged to be equally loud. A scale showing the relation between loudness, in sones, and the loudness level, in phons, is shown in Fig. 7.8.

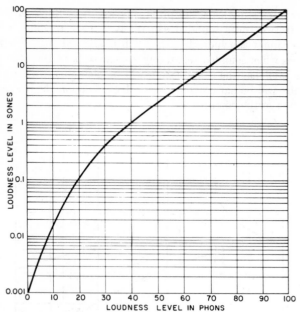

FIG. 7.8.   Loudness versus loudness level.   (*After Fletcher.*)

The loudness[11] of pure tones of various frequencies is shown in Fig. 7.9.
For tones between 800 and 2,000 cycles the loudness is the same for the
same pressure.   The difference is small up to 8,000 cycles.   For higher
frequencies than this the loudness decreases as the frequency increases.
Further, it will be seen that for a 50-cycle tone the intensity required to
reach the threshold of hearing is 250,000 times that required for the
reference 1,000-cycle tone.

Loudness may be depicted in another way, by a series of contours[12] of
equal loudness, as shown in Fig. 7.10.   The loudness level of other tones
was determined with a 1,000-cycle tone as the reference tone.   The loudness level of other tones is the intensity level of the equally loud 1,000-
cycle tone.   These characteristics show that the ear is most sensitive in
the region between 3,000 and 4,000 cycles.   The sensitivity of the ear
decreases above and below this frequency.   The lowest contour in Fig.
7.10 is the threshold of hearing.   The average normal ear cannot hear a
tone below this intensity.   The upper contour is the threshold of feeling.
A tone having an intensity greater than this level will be painful.

The minimum perceptible change in intensity level[13] which the ear can

[11] Fletcher, H., *Jour. Acoust. Soc. Amer.*, Vol. 9, No. 4, p. 275, 1938.

[12] Fletcher and Munson, *Jour. Acoust. Soc. Amer.*, Vol. 5, No. 2, p. 82, 1933.

[13] Fletcher, *Speech and Hearing in Communication*, D. Van Nostrand Company, Inc.,
Princeton, 1953.

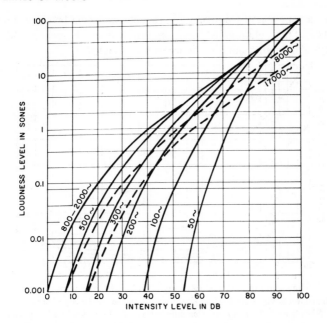

FIG. 7.9.   The relation between the intensity level and the loudness of pure tones of the frequencies indicated.   0 decibels = 0.000204 dyne per square centimeter.   (*After Fletcher.*)

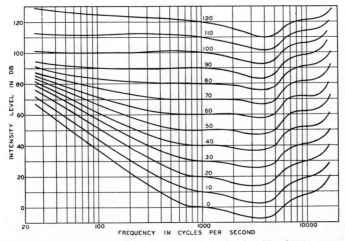

FIG. 7.10.   Contour lines of equal loudness for normal ears.   Numbers on curves indicate loudness level in phons.   0 decibels = 0.000204 dyne per square centimeter.   (*After Fletcher and Munson.*)   Also see Fig. 9.77.

detect as a function of frequency for various sensation levels is shown in Fig. 7.11. These characteristics show that the ear is most sensitive to intensity-level changes at the higher sensation levels.

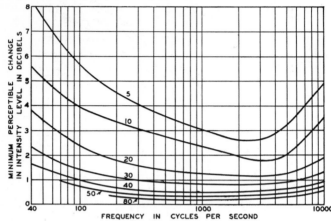

FIG. 7.11. The minimum perceptible change in intensity level of pure tones as a function of the frequency. Numbers on the curves indicate level above threshold. (*After Fletcher, Speech and Hearing in Communication, D. Van Nostrand Company, Inc. Princeton, 1953.*)

C. Timbre (Quality)

Timbre is the most important fundamental attribute of all music. Timbre is that characteristic of a tone which depends on its harmonic structure. The timbre of a tone is expressed in the number, intensity, distribution, and phase relations of its components. Timbre, then, may be said to be the instantaneous cross section of the tone quality. For practical considerations, timbre is the tonal spectrum. It ranges from a pure tone through an infinite number of changes in complexity up to a pitchless sound such as thermal noise. Timbre may be described as the characteristic which enables one to judge that two tones with the same pitch and loudness are dissimilar. For example, the frequency spectrums of the tones produced by the A string and the D string of a violin both sounding the fundamental tone A or 440 cycles are shown in Fig. 7.12. These two tones, emanating from the same instrument, have the same pitch and loudness and still appear to the listener to be two entirely different sounds. The reason for this fact is explained by the difference in the acoustic spectrums of the two tones. Referring to Fig. 7.12, it will be seen that the complexion of the acoustic spectrums of the two tones is entirely different. This accounts for the fact that these two tones appear to be entirely different to the listener.

Timbre may be said to be the characteristic which enables the listener to recognize the kind of musical instrument which produces the tone. Chapter 6 gave the tone structure of typical musical instruments. An examination of these spectrums shows that the essential difference between the musical instruments resides in the overtone structure. It is in terms of timbre that we differentiate the tonal character of musical instruments, of voices, of animal sounds, of insect sounds, and of warning signals.

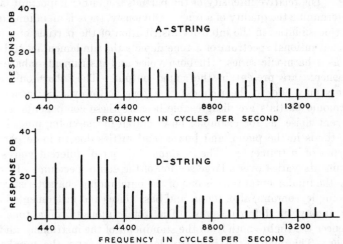

Fig. 7.12.   The acoustic spectrums of the A and D strings of a violin for the tone 440 cycles.

The timbre or quality also depends upon the intensity of the sound. This may be due in part to the fact that the ear is nonlinear and produces new overtones or alters the existing ones.   For example, when a pure tone of suitable intensity is impressed upon the ear, a series of harmonics or overtones of the original frequency is heard.   Furthermore, when two loud tones are sounded together, a group of tones is heard, consisting of the sums and differences of the two primary tones and their harmonics. These phenomena show that the ear is a nonlinear system.   The sensation levels[14] of the fundamental at which the various harmonics first become detectable are shown in Fig. 7.13.   The subjective effects of the harmonics generated in the ear are more pronounced at the lower frequencies.   Furthermore, the harmonics appear at a lower level at the lower frequencies.

In the consideration of timbre of a tone there are six physical character-

[14] Wegel and Lane, *Phys. Rev.*, Vol. 23, No. 2, p. 266, 1924.

istics which determine the quality, namely, the number of partials, the distribution of the partials, the relative intensity of the partials, the inharmonic partials, the fundamental tone, and the total intensity. The number of overtones present determines the richness of the tone. This may vary from a pure tone to a tone with a tremendous number of partials. The largest factor in determining whether a tone is beautiful or ugly is the distribution of the partials in the audio-frequency range. The beauty of a tone depends upon the location of the partials in the harmonic series. The relative intensity of the partials is another important factor in determining the quality of a tone. Obviously, there is an infinite number of possibilities in the intensity distribution of the partials of a tone. The conventional spectrum of a tone depicts the fundamentals and partials as a harmonic series. In both voice and instrument, inharmonic components are present in the sound output. The inharmonic components may add to the richness of a tone. However, in most cases inharmonic partials are disagreeable because these components usually represent noise, as for example, bow scratch in the violin, pings in the harp, thuds in the piano, and buzzes and rattles due to loose parts in all types of instruments. The fundamental tone of different voices and instruments varies over a large section of the audio spectrum. In some cases, the fundamental tone is one of the outstanding characteristics, as, for example, soprano, alto, tenor, and bass voices and instruments in the same family of different sizes. The total intensity of a tone also influences the timbre both from the standpoint of the instrument and the listener. The higher the intensity level, the greater the number of partials which are generated by the instrument. In general, the higher the intensity level, the greater the number of partials which the ear can discern.

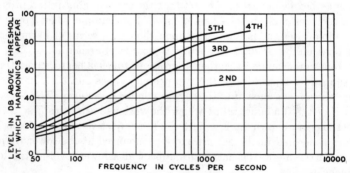

Fig. 7.13. The level above threshold at which harmonics are generated in the ear at various frequencies. The numbers on the curve indicate the order of the harmonic. (*After Wegel and Lane.*)

). Duration (Time)

The duration of a tone affects the pitch, loudness, and quality.   Dura-
ion is involved in the temporal aspects of music.   The aspects of dura-
ion or time are as follows: duration of the tone, tempo, time, and pauses.
The duration is the length of time that a tone persists or lasts.   In
nusical notation, the duration is indicated by the kind of note to be
olayed.   Tempo is a term used to designate the rate of movement of the
music.   Time is the division of music into portions marked by a regular
eturn of the accent.   A pause or rest is a cessation of all or a part of the
nusical activity.   It is obvious that duration directly involves the length
of time that a note persists.   Tempo, time, and pauses indirectly affect
he duration of a tone because of the variation in the overlap and spacing
of the tones as these factors are altered.

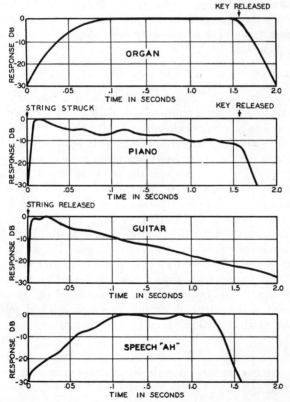

Fig. 7.14.   The growth and decay as a function of the time of an organ, piano, and
guitar tone and the speech sound "ah."

E. Growth and Decay[15] (Attack and Release)

The growth of a tone involves the time required for a tone to build up to some fraction of its ultimate intensity level (Sec. 7.3E). The decay of a tone involves the time required for a tone to fall in intensity from one arbitrary level to some other arbitrary level (Sec. 7.3E). The growth and decay affect the pitch, loudness, and timbre of a tone. The growth and decay characteristics of an organ, piano, guitar, and speech are shown in Fig. 7.14. It will be seen that the growth and decay times of the organ tone are relatively long. The growth time of the tone of a piano is short, but the decay time is long. The growth time of the tone of a guitar is very short, but the decay time is extremely long. In the case of most of the vowel sounds in speech, the growth and decay times are relatively long, owing to the highly resonant vocal cavities.

There is a special case of growth and decay, namely, portamento. Portamento is the passage from a tone of one frequency to another tone of a different frequency in one continuous glide through all the intervening frequencies.

F. Vibrato[16]

Vibrato is primarily a frequency modulation of a musical tone. It is also accompanied by an amplitude modulation at the modulating frequency as well as a pulsating change in the timbre. The vibrato is used as an artistic embellishment by musicians. It has been found that the average rate of the vibrato or frequency modulation of a good singer is 7 cycles per second. The vibrato is the most important of the musical ornaments, because it occurs so frequently and contributes so much to musical aesthetics. The vibrato is universally employed in all good singing. The vibrato is used in playing many musical instruments. It is somewhat different from the tremolo, which is primarily a pulsating change in intensity. Seashore states that, "a good vibrato is a pulsation of pitch, usually accompanied by a synchronous pulsation of loudness and timbre, of such extent as to give a pleasing flexibility, tenderness and richness of tone." Without doubt, a large part of the beauty of the vibrato is due to the artistic deviation from a precise and uniform tone and without which it would be somewhat mechanical.

G. Beats

When two tones impinge upon the ear, various interaction phenomena are produced. One of these is termed beats. There are two kinds of

---

[15] Fletcher, H., *Amer. Jour. Phys.*, Vol. 14, No. 4, p. 215, 1946.

[16] Seashore, *Psychology of Music*, McGraw-Hill Book Company, Inc., New York 1938.

beats, namely, beats of imperfect unisons and beats of mistuned consonances. In the first kind, beats arise owing to a small difference in frequency, for example, the simultaneous production of tones of 200 and 202 cycles. The beating note is equal to the difference frequency. In the case of mistuned consonances, beats appear when two tones depart slightly from the simple interval relationship; for example, the simultaneous production of tones of 302 and 200 cycles or 403 and 300 cycles, etc.

The character of the beat note changes as the frequency of the beat note or the rate of beating is varied. Slow beats or low-frequency beats appear as a slow rise and fall in loudness of the sound at the frequency of the beat note. When the beat frequency increases, the beats appear as pulses. High-frequency beats appear as a rough complex tone.

H. Consonance and Dissonance

When two or more tones are sounded simultaneously and the result is pleasing to the ear, the resultant sound is termed consonant. When such a combination of tones is not pleasing to the ear, the resultant sound is termed dissonant. Consonance depends upon the degree of coincidence of sound waves. The subject of pleasing combinations of tones was discussed in the section on scales. The factors which determine consonance are smoothness, purity, and blending.

Smoothness is determined by the prominence of the beats between the combinations which form the tone. Sounds are dissonant when there is a roughness or harshness. Dissonance appears to be a disagreeable sensation produced by beats. When two tones are sounded together, the combination is disturbed by beats of the upper partials so that the effect is a pulsing of sound which produces a roughness. It is the prominence of the beats that determines the roughness of a tone. This roughness persists until the interval becomes so large that the beats disappear. On this basis we would expect that the combination of tones in which the ratio of frequencies can be expressed by the ratio of two numbers neither of which is large would be the most pleasing, for example, 2:1, 3:2, 5:3, 5:4, 4:3, etc.

The relative purity is determined by the harmonic content. The greater the harmonic content for the same fundamental frequency, the less the subjective purity.

Blending is a question of whether two tones belong together. For example, notes of the interval of the seventh do not seem to belong together. Tones may not seem to belong together if they are too close together because this causes roughness. On the other hand, tones may not seem to blend if they are too far apart. It appears that blending rests to a large extent upon smoothness and purity.

From a psychological point of view there are three factors that determine consonance or dissonance,[17] namely, smoothness, purity, and blending. These three factors have been employed to determine the order of merit of interval in the consonance-dissonance series. The results are depicted in Fig. 7.15. It will be seen that order of merit decreases as the two numbers in the ratio become larger.

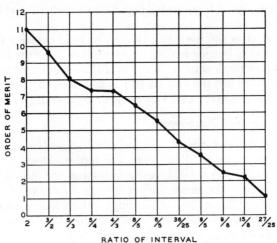

FIG. 7.15.   Consonance-dissonance characteristic for various interval ratios.   (*After Malmberg.*)

I. Volume

Volume is the psychological term applied to describe the musical characteristics of a tone involving the frequency, intensity, duration, harmonic content, spatial complexions of the sound sources, and boundary conditions of the environment.   Volume should not be confused with intensity. It is a much more comprehensive term.   Volume, however, is a function of intensity.   Increasing the intensity level of reproduced sound appears to contribute to spaciousness.   The addition of tones increases the volume.   An increase in reverberation time increases the volume.   The expedient of adding synthetic reverberation is employed in broadcasting and the recording of phonograph records to give the reproduced sound the effect of the expanse of a concert hall even though the sound is reproduced in a small room.   The greater the number of harmonics, the greater the volume.   The volume is increased by resonance or hang-over.   The warble of the frequency of a tone as in the vibrato tends to increase the largeness or volume as compared with a single

[17] Malmberg, G. F., *Univ. Iowa Studies in Psychol. of Music*, Vol. VII, p. 93, 1918.

tone. The volume is proportional to the duration of a tone. An organ always appears to have more volume than a piano. The spatial distribution of individual tones affects the volume. For example, if a single loudspeaker is compared with a large number of similar loudspeakers distributed around the room with the same power going into the single loudspeaker as in the case of the multiple arrangement, the volume will appear to be greater for the multiple sources. In general, in the case of reproduced music, an increase in volume enhances the aesthetic and artistic effects of the rendition. From the foregoing, it will be seen that volume is a somewhat intangible psychological characteristic of music. It plays a major role in determining the acoustical merit of a concert hall. If a concert hall does not possess volume as described above, the hall will not be considered to be outstanding.

J. Rhythm

A series of pulses with varied spacing in time and of different intensities possess a subjective quality termed rhythm. Uniform successions of sounds such as the tick of a clock or the exhaust of a motorcar or radio-telegraph dots and dashes are often irresistibly grouped into a rhythmic measure. In a six-cylinder motorcar, the exhaust pattern may be as follows:

$$\begin{array}{ccccccccccccccc} | & | & | & | & | & & | & | & | & | & | & & | & | & | & | \\ 1 & 2 & 3 & 4 & 5 & & 6 & 1 & 2 & 3 & 4 & 5 & & 6 & 1 & 2 & 3 \end{array}$$

where the numbers represent the series of cylinder exhausts, the lengths of the vertical lines represent the intensities of the exhaust sounds emitted by the different cylinders, and the spacings between the lines indicate the intervals between sounds emitted by the different cylinders. This sort of rhythm in a motorcar is not pleasant because it appears to be limping. However, it is possible subjectively to group pulses of equal intensity and equally spaced in time into a series of pulses. The frequency of a low-frequency sine wave may be determined by actual counting of the impulses by grouping into five or six impulses and then counting the group. So one should sound a word of caution: a motor may not be limping even though it appears so, because it may be entirely subjective. In this manner of grouping, it is possible to perceive or grasp many combinations of pulses as individual ones. Individual sounds are grouped in series of pulses, a series of combination of pulses, etc. This grouping or rhythmic periodicity is instinctive. Children perceive rhythm before they know the meaning.

Rhythm in music is a regular occurrence of stressed and relaxed pulses. The term rhythm means a repetition of groups of sounds at regular intervals. Therefore, there must be at least two similar groups in order to

establish rhythm.   Two examples of simple rhythmic patterns are shown
in Fig. 7.16.   In these illustrations it will be seen that the same patterns
are repeated again and again.   For further considerations of rhythm,
see Sec. 2.4.

FIG. 7.16.   Examples of simple rhythmic patterns.

K. Presence and Absence

Presence is a psychological term used to describe a type of intimacy in
live and reproduced sound.   Presence is a function of the amplitude-fre-
quency characteristic, the transient characteristics, the distortion characteristics, the spatial arrangement, and the directional patterns of the
sound sources and the reverberation and the resonance characteristics of
the enclosed space or room.   Presence in a symphony orchestra is
obtained by the relative amplitudes of the various instruments.   Reproduced speech is said to possess presence when the speaker appears to be in
the same room.   If the reproduced speech appears to come from a point in
front of the loudspeaker, the reproducing system is said to have presence.
If the speech comes from behind the loudspeaker and is muffled, the reproduced speech is said to lack presence.   When either live or reproduced
sound is incisive, crisp, and clear, it will in general exhibit considerable
presence.   The reverberation-frequency characteristic has a marked
effect upon presence.   Excessive reverberation in the low-frequency
range reduces presence.   A uniform directional pattern in the directivity
characteristic of a loudspeaker enhances the presence.   A uniform ampli-
tude-frequency characteristic in a reproducing system improves presence,
while frequency discrimination detracts from the presence.   A poor
transient-response characteristic detracts from the presence of the reproduced sound.   Nonlinear distortion in a sound-reproducing system
reduces the effects of presence.

Absence is a psychological term sometimes used to indicate a lack of
presence.

7.5 MEASURES OF MUSICAL APTITUDES[18]

It is somewhat paradoxical that talent in a subject fraught with as many
intangibles as music may be measured with considerable accuracy.   The
reason for this fact is that it is possible to measure the relative aptitudes

[18] Seashore, *Psychology of Music*, McGraw-Hill Book Company, Inc., New York,
1938.

for the perception of pitch, loudness, time, timbre, rhythm, and tonal memory. A person must possess considerable aptitude in each of these factors in order to possess sufficient talent to justify a serious program in training for a musical career.

A series of disk phonograph records[19] have been developed and recorded and are commercially available for use with a conventional phonograph to measure musical talent. These provide a means for scoring the following: pitch, loudness, time, timbre, rhythm, and tonal memory. The pitch record presents a series of two tones which differ in pitch. In each case, the listener indicates if the first tone is higher or lower in pitch than the second. The loudness record presents a series of two tones which differ in loudness. In each case, the listener indicates if the first tone is louder or weaker than the second. The time record presents a series of two tones which differ in length. In each case, the listener indicates if the first tone is longer or shorter than the second. The timbre record presents a series of two tones which are either the same or different in timbre. The listener is asked to indicate whether the first is the same as the second or different. The rhythm record presents a series of two rhythmic patterns. The listener indicates whether the first is the same as the second or different. In the tonal-memory record a series of tones are played twice. In each case in the second playing one note is changed. The listener is asked in each case to indicate which note is changed, that is, 1, 2, 3, 4 or 5, etc. Means have been devised for scoring each talent.

The tests do not measure training or achievement in music. They do not measure intelligence, feeling, or the will to work. As an instrumental means of testing and scoring musical talents in each musical category they are exceedingly accurate.

7.6 HEARING LOSS

The hearing loss of an ear at a given frequency is the ratio of the threshold of audibility for that ear to the normal threshold of audibility at the same frequency. Hearing loss is expressed in decibels. A loss in hearing may be produced by a severe disease either in the ear or outside the ear which in some way affects the hearing mechanism or by some accident which injures some part of the hearing mechanism. The most common cause of deafness or hearing loss is senility. The average hearing characteristics of various investigators[20,21] for different age groups are given in Fig. 7.17.

[19] Seashore, Measures of Musical Talents. Revised by Seashore, Lewis, and Saetveit. RCA Victor Album, E65. RCA Victor Division, Radio Corporation of America, Camden, N.J.

[20] Steinberg, Montgomery, and Gardner, *Jour. Acoust. Soc. Amer.*, Vol. 12, No. 2, p. 291, 1940.

[21] Bunch, C. C., *Arch. Otolaryngol.*, Vol. 9, p. 625, 1929.

It will be seen that the loss in hearing acuity occurs in the high-frequency ranges.

The type of hearing loss shown in Fig. 7.17 is not a flat loss in hearing because the loss increases with frequency. The question arises as to what effect the type of hearing loss shown in Fig. 7.17 has upon a musician. The acuity of hearing cannot be classed as a talent like a sense of pitch, loudness, time rhythm, etc. On the other hand, it does affect the individual's perception of the higher overtones. It also upsets the balance of high- and low-frequency sound energies. The musical talents which are least affected by a hearing loss are pitch, loudness, duration, time, vibrato, and rhythm. The talents which are most effected by a hearing loss are timbre, growth and decay, consonance, volume, and presence.

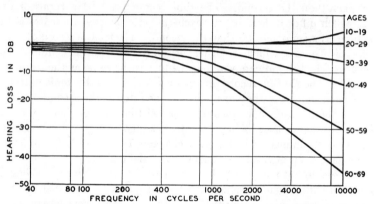

FIG. 7.17. Hearing-loss frequency characteristics for various age groups. These characteristics are averages of both men and women and different observers.

The acuity of hearing is measured by an audiometer. The audiometer consists of an audio oscillator for generating pure tones, an attenuator calibrated in decibels, and a telephone receiver (Fig. 7.18). The usual test tones are 128, 256, 512, 1,024, 2,048, 4,096, and 8,192 cycles per second. The reference level is the normal threshold of hearing, as outlined in Sec. 7.4B and Fig. 7.10. This level is the zero level of the audiometer. The person to be tested wears the earphone in the normal manner, and the level at which the sound is no longer audible is noted on the

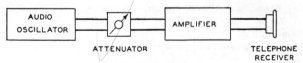

FIG. 7.18. Elements of an audiometer.

attenuator.   For a person with normal hearing, this will occur when the attenuator setting is zero.   This indicates that there is no hearing loss. For a person who is hard of hearing, a higher sound level will be required before the sound becomes audible.   The difference in level between the normal threshold level and the threshold level for the person being tested is the hearing loss.   These results are plotted on a graph, with the hearing loss in decibels as the ordinates and the frequency as abscissa.

CHAPTER EIGHT

Theater, Studio, and Room Acoustics

8.1 INTRODUCTION

The acoustical performance of enclosures is exceedingly complex because so many parameters are involved, as, for example, the cubical content, the shape of the enclosure, the configuration of the boundaries, the absorption of the boundaries, the furnishings, the characteristics of the source of sound, and the audience or other ultimate sound-pickup means. In order to achieve good acoustical performance, the following conditions should be realized: the sound should be loud enough; the ambient noise should be low; the room should be free of echoes, flutter, and resonance; the boundaries and source of sound should be arranged so that the reflected sound is uniform throughout the space; and the reverberation time should be appropriate for the particular source of sound, be it speech or music.

The advent of sound-reproducing means has changed the problems involving architectural acoustics. Before the introduction of sound-reproducing systems, the major concern was the optimum reverberation time and the proper geometrical configuration for the best artistic effects in music and the maximum intelligibility in speech, as outlined in the preceding paragraph. By means of sound-reproducing systems, speech can be rendered intelligibly, where before it was either too weak to be heard above the general noise level or too reverberant. Furthermore, these instruments have opened a field for all manner of artistic effects never before possible.

The theaters which suffer most from insufficient loudness are, of course, the large enclosed theater and the open-air theater. Sound-reproducing systems have opened new vistas in musical renditions both by reproduction and reinforcement. In certain instances the volume range of an orchestra is inadequate for full artistic appeal or to utilize the full capabilities of the hearing range. In these cases, means are required for augmenting the intensity of the original sound. The systems for accomplishing this objective are termed sound-reinforcing systems.

The acoustic problems involving the reproduction of sound motion pictures are quite unlike those of stage presentations. The acoustics of radio broadcasting differ from those of the stage and sound motion pictures in that the action cannot be seen. Therefore, sound carries the entire load of the transmission of intelligence. Television acoustics are the most complex of all because they involve a part of stage, sound motion pictures, and radio techniques, as well as entirely new problems. It is evident that reproduced sound offers greater possibilities for obtaining the proper artistic effects by the use of the following expedients: incidental sound, a wide volume range, the control of the reverberation or room characteristics, and various sound effects.

For large outdoor gatherings, such as state occasions and athletic events in large stadiums and parks, sound-reproducing systems are employed to amplify the speaker's voice.

In department stores, hotels, hospitals, schools, and factories, sound-reproducing systems are employed to transmit sound from a central point to several independent rooms or stations. The systems for accomplishing this objective have been termed general-announce or call systems.

It is the purpose of this chapter to outline the applied phases of theater, studio, and room acoustics and the applications of the collection and dispersion of sound.

8.2 DISPERSION OF SOUND
A. Sound Absorption and Reverberation[1-6]

When a source of sound is started in a room, the energy does not build up instantly, owing to the finite velocity of propagation of a sound wave. Each pencil of sound sent out by the source is reflected many times from the partially absorbing walls of the room before it is ultimately dissipated. A two-dimensional schematic version of the growth of sound in a room, when there is a steady source of sound S operating, is shown in Fig. 8.1. Figure 8.1A shows the direct sound D and the first pencils of reflected sound R_1, R_2, R_3, and R_4 which arrive at the observation point O. The energy in the direct sound is greater than the reflected sound because

[1] Sabine, *Collected Papers in Acoustics*, Harvard University Press, Cambridge, Mass., 1922.

[2] Watson, *Acoustics of Buildings*, John Wiley & Sons, Inc., New York, 1923.

[3] Knudsen and Harris, *Acoustical Designing in Architecture*, John Wiley & Sons, Inc., New York, 1950.

[4] Sabine, *Acoustics and Architecture*, McGraw-Hill Book Company, Inc., New York, 1932.

[5] Rettinger, *Applied Architectural Acoustics*, Chemical Publishing Company, Inc., Brooklyn, 1947.

[6] Olson, H. F., *Acoustical Engineering*, D. Van Nostrand Company, Inc., Princeton, 1957.

of the smaller distance and the absence of absorption by the walls which
occurs in the reflected sound. The increments of direct and reflected
sound energy as a function of the time are shown in the graph of Fig.
8.1A. Some of the second reflections such as R_5, R_6, R_7, and R_8 are
shown in Fig. 8.1B. The reflected sound suffers a reduction in energy at

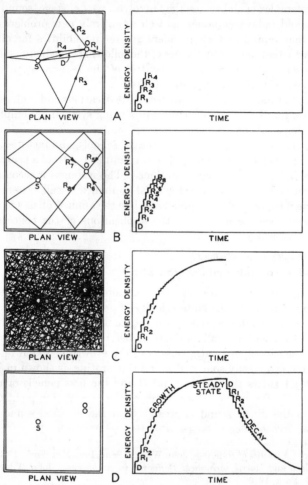

FIG. 8.1. A two-dimensional version of the growth and decay of the sound in a room·
S is the source of sound and O the observation point. A. The direct sound energy
D and the reflected sound energy R_1, R_2, R_3, and R_4 from the four walls. B. The
addition of the second reflections R_5, R_6, R_7, and R_8. C. A large number of reflections which approximate steady-state conditions. D. The decay of sound energy
after the source has stopped.

each reflection; therefore, the succeeding increments of reflected sound become smaller and smaller and are ultimately dissipated. A steady-state condition obtains when the energy absorbed by the walls equals the energy delivered by the sound source. The sound-growth characteristic approaches asymptotically the ultimate energy density, as shown in Fig. 8.1C.

As in the case of the growth of sound, some time is required for the sound energy to be completely absorbed after the source is stopped. The decay of sound is depicted in Fig. 8.1D. After the source has stopped, the first drop in sound energy is that due to the direct sound D. Then follows the reflected sound R_1, R_2, R_3, etc. The complete growth and decay characteristic of sound energy in a room is depicted in the graph of Fig. 8.1D.

The decay characteristic is exponential. The ideal exponential decay characteristic, with decibels as the ordinates and time as the abscissa, would be a straight line, as shown in Fig. 8.2A. Owing to the characteristic modes of vibration in a room, the measured growth and steady-state

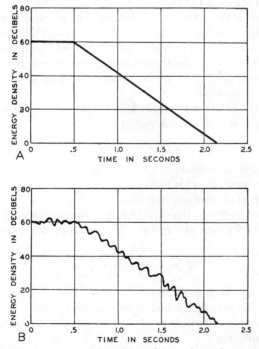

FIG. 8.2. The decay characteristics of sound in a damped room. A. The ideal characteristic. B. The measured characteristic.

and decay characteristics are not so smooth as the ideal characteristics. A typical decay characteristic is shown in Fig. 8.2B. It will be seen that average decay is an exponential characteristic.

The sound-energy density at any time after the source has stopped is given by

$$E = E_0 \epsilon^{-\frac{cAt}{4V}} \tag{8.1}$$

where E = sound-energy density, after a time t seconds, after stopping the source, in ergs per cubic foot

A = total number of absorption units, in sabins (see definition below)

$$E_0 = \frac{4P_0}{cA} \tag{8.2}$$

where P_0 = rate at which sound is generated by the source, in ergs per second

t = time, in seconds

c = velocity of sound, in feet per second

V = volume of the room, in cubic feet

The acoustic absorptivity (or absorption coefficient) of a surface is the ratio of the flow of sound energy into the surface on the side of incidence to the incident rate of flow. The sabin is a unit of equivalent absorption and is equal to the equivalent absorption of one square foot of a surface of unit absorptivity, that is, of one square foot of surface which absorbs all incident sound energy.

From Eq. (8.1) the time[7] required for the sound in a room to decay to one millionth of its original intensity is

$$T = 0.050 \frac{V}{A} \tag{8.3}$$

where T = time, in seconds

V = volume, in cubic feet

A = total absorption, in sabins

Later work has shown that Eq. (8.3) is unsatisfactory for large rooms or rooms with very large absorption. The equation developed by Eyring[8] is

$$T = \frac{0.05V}{-S \log_\epsilon (1 - a_{av})} \tag{8.4}$$

where V = volume, in cubic feet

S = total area of the boundaries in square feet

a_{av} = average absorption of the boundaries per square foot, in sabins

[7] Franklin, W. S., *Phys. Rev.*, Vol. 16, p. 372, 1903.

[8] Eyring, C. F., *Jour. Acoust. Soc. Amer.*, Vol. 1, No. 2, p. 217, 1930.

A tabulation of sound-absorption coefficients for various building materials and objects is shown in Table 8.1. The coefficients in this table were obtained on small samples in chambers having long reverberation

TABLE 8.1. ABSORPTION COEFFICIENTS OF VARIOUS ACOUSTICAL MATERIALS, BUILDING MATERIALS, AND OBJECTS

Material	Thickness, inches	Mounting	Frequency						Author
			128	256	512	1,024	2,048	4,096	
			Coefficient						
Corkoustic B-5	1½	2	0.18	0.41	0.70	0.51	0.58	0.65	A.M.A.
Cushiontone A-3	⅞	2	0.17	0.51	0.73	0.95	0.75	0.72	A.M.A.
Sanacoustic pad, with metal facing	1⅛	3	0.25	0.56	0.99	0.99	0.91	0.82	A.M.A.
Fibretex	1 3/16	2	0.16	0.49	0.56	0.78	0.84	0.78	A.M.A.
Acoustex 4OR	¾	2	0.09	0.17	0.59	0.90	0.75	0.73	A.M.A.
Fiberglass tile Type A	1	2	0.17	0.44	0.91	0.99	0.82	0.77	A.M.A.
Acoustone F	1 3/16	1	0.12	0.31	0.85	0.88	0.75	0.75	A.M.A.
Acousti-Celotex C-4	1¼	2	0.25	0.58	0.99	0.75	0.58	0.50	A.M.A.
Draperies hung straight, in contact with wall, cotton fabric, 10 ounces per square yard	...	..	0.04	0.05	0.11	0.18	0.30	0.44	P.S.
The same, velour, 18 oz. per square yard	...	..	0.05	0.12	0.35	0.45	0.40	0.44	P.S.
The same as above, hung 4 inches from wall	...	..	0.09	0.33	0.45	0.52	0.50	0.44	P.S.
Felt, all hair, contact with wall	...	..	0.13	0.41	0.56	0.69	0.65	0.49	P.S.
Rock wool	1	..	0.35	0.49	0.63	0.80	0.83	.....	V.K.
Carpet, on concrete	0.4	..	0.09	0.08	0.21	0.26	0.27	0.37	B.R.
Carpet, on ⅛-inch felt, on concrete	0.4	..	0.11	0.14	0.37	0.43	0.27	0.27	B.R.
Concrete, unpainted	...	..	0.010	0.012	0.016	0.019	0.023	0.035	V.K.
Wood sheeting, pine	0.8	..	0.10	0.11	0.10	0.08	0.08	0.11	W.S.
Brick wall, painted	...	..	0.012	0.013	0.017	0.020	0.023	0.025	W.S.
Plaster, lime on wood lath on wood studs, rough finish	½	..	0.039	0.056	0.061	0.089	0.054	0.070	P.S.

Individual object			Absorption units, square foot (sabins)						
Audience, per person, man with coat	...	..	2.3	3.2	4.8	6.2	7.6	7.0	B.S.
Auditorium chairs, solid seat and back	...	..	0.15	0.22	0.25	0.28	0.50	.....	P.S.
Auditorium chairs, upholstered	...	..	.....	3.1	3.0	3.2	3.4	.....	F.W.

Abbreviations in the above table are as follows: A.M.A., Acoustical Materials Association; W.S., Wallace Sabine; P.S., P. E. Sabine; F.W., F. R. Watson; V.K., V. O. Knudsen; B.R., Building Research Station, England; B.S., Bureau of Standards.

Mountings in the above table are as follows:
1. Cemented to plaster board.
2. Nailed to 1- by 2-inch furring 12 inches apart.
3. Attached to metal supports applied to 1- by 2-inch wood furring.
4. Laid on 24-gauge sheet iron, nailed to 1- by 2-inch wood furring 24 inches apart.

times.   In general, these measurements do not agree with those obtained
under actual conditions in practice.   That is, field measurements yield
smaller values than laboratory measurements.   However, the values of
Table 8.1 show the relative absorption coefficients of the various materials.   For a complete résumé of this subject, see the anniversary issue
of the *Journal of the Acoustical Society of America*, Vol. 11, No. 1, Part 1,
July, 1939.

The reverberation time of a room may be computed from Eqs. (8.3)
and (8.4).   Referring to these equations it will be seen that the reverberation time of a room may be reduced by the introduction of absorbing
means such as carpets, tapestries, and commercially developed materials
(see Table 8.1).

The effect of reverberation time upon the perception of sound in a room
is depicted in Fig. 8.3.   A sound source produces a series of sound pulses
of constant intensity but of different time lengths.   These sound pulses
may be considered to represent somewhat idealized sounds of speech and

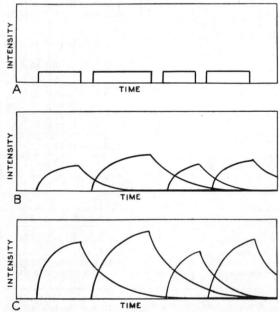

FIG. 8.3.   A series of four pulses of sound of different time lengths produced by a
source under different environments.   *A*. Source of sound and observation point
located in free space, that is, no reverberation.   *B*. Source of sound and observation
point located in a room of moderate reverberation time.   *C*. Source of sound and
observation point located in a room of long reverberation time.

music. The intensity as a function of the time of the sound pulses at an observation point with both the observation point and the source of sound located in free space, that is, no reflecting boundaries is shown in Fig. 8.3A. It will be seen that the sound at the observation point consists of a series of discrete pulses. To illustrate the effects of reverberation of sound in a room, the system consisting of the sound source emitting the same pulses and the observation point are moved into a room of moderate reverberation time. The results are shown in Fig. 8.3B. Owing to the finite decay time of sound in a room, the sound energies of the various sounds overlap. This overlapping becomes more pronounced as the reverberation time of the room is increased, as depicted in Fig. 8.3C. As a result of this overlapping, the intelligibility of speech is impaired. Under conditions of very long reverberation time, the sounds of speech may become practically unintelligible. The reason being that instead of a single speech sound at a time as in the case of free space, two or more speech sounds are heard at the same time. Under these conditions, the overlapping speech sounds mask and interfere with each other, which results in a reduction of the intelligibility. On the other hand, reverberation of an optimum value in music is desirable, because the prolongation and blending of musical tones due to reverberation produce a more pleasing musical performance. The optimum reverberation time of theaters, halls, studios, and other rooms will be considered in detail in subsequent sections.

When a source of sound operates in a room there are two sources of sound-energy density at the observation point, namely, the direct and the reflected sound (Fig. 8.1).

A nondirectional sound source, that is, a sound source which radiates equally in all directions, is depicted in Fig. 8.4A. The energy density, in ergs per cubic centimeter, due to direct sound radiation at the observa-

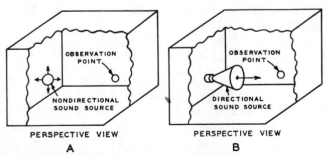

FIG. 8.4. Perspective views of a room containing a sound source and an observation point. A. A nondirectional sound source. B. A directional sound source.

tion point is

$$E_D = \frac{p_0^2 x_0^2}{r^2 \rho c^2} \qquad (8.5)$$

where p_0 = sound pressure, in dynes per square centimeter, obtained at a distance x_0 in free space

   r = distance between the sound source and the observation point, in centimeters

   ρ = density of air, in grams per cubic centimeter

   c = velocity of sound, in centimeters per second

The sound-energy density due to generally reflected sound is a function of the absorption characteristics of the room, the volume of the room, the power output of the sound source, and the time. The sound-energy density, in ergs per cubic centimeters, due to the generally reflected sound is given by

$$E_R = \frac{16\pi p_0^2 x_0^2}{\rho c^2 a S} \left\{ 1 - \epsilon^{\frac{cS[\log_\epsilon (1-a)]t}{4V}} \right\} (1-a) \qquad (8.6)$$

where a = average absorption per unit area, absorption coefficient

   S = area of the absorbing material, in square centimeters

   V = volume of the room, in cubic centimeters

   t = time, in seconds

   c = velocity of sound, in centimeters per second

   ρ = density of air, in grams per cubic centimeter

   p_0 = sound pressure in dynes per square centimeter, obtained at a distance x_0 from the sound source operating in free space

The total sound-energy density at the observation point will be the sum of the direct and generally reflected sound and may be expressed by

$$E_T = E_D + E_R \qquad (8.7)$$

In general, the sound source is directional, as shown in Fig. 8.4B. Under these conditions the energy density due to generally reflected sound is given by

$$E_R = \frac{4\Omega p_0^2 x_0^2}{\rho c^2 a S} \left\{ 1 - \epsilon^{\frac{cS[\log_\epsilon (1-a)]t}{4V}} \right\} (1-a) \qquad (8.8)$$

where Ω = effective solid angle of radiation, in steradians, and all the other constants are the same as those of Eq. (8.6)

The ratio of the generally reflected to the direct sound is a measure of the effective reverberation at the observation point and is given by

$$\frac{E_R}{E_D} = \frac{4r^2\Omega\left\{ 1 - \epsilon^{\frac{cS[\log_\epsilon (1-a)]t}{4V}} \right\}(1-a)}{aS} \qquad (8.9)$$

If the sound continues until the conditions are steady, Eq. (8.9) becomes

$$\frac{E_R}{E_D} = \frac{4r^2\Omega}{aS}(1 - a) \tag{8.10}$$

From Eqs. (8.9) and (8.10) it will be seen that the received reverberation or the effective reverberation at the observation point can be reduced by decreasing the distance between the source and observation point, by decreasing the solid angle Ω, by increasing the sound absorption of the walls, by increasing the volume, or by decreasing the time the source is excited.

B. Reproduction of Sound in Rooms[9]

Sound-reproducing systems are used in all manner of theaters and rooms to reinforce the original sound or reproduce recorded sound. As a matter of fact, the number of cases in which some sort of sound-reproducing system is not used is so small as to be almost negligible. Therefore, the main portion of this chapter will be concerned with theater, studio, and room acoustics as related to the reproduction of sound.

C. Reverberation Time of a Theater, Hall, or Room for the Reproduction of Sound

The optimum reverberation time of a theater, hall, room, or any enclosure for the reproduction of sound as a function of the volume of the room for different frequencies is shown in Fig. 8.5. It will be noted that the reverberation time increases at the lower frequencies so that the aural rate of decay of pure tones will be approximately the same for all frequencies (see Sec. 7.4B and Fig. 7.10). It has also been found to be desirable to increase the reverberation time in the frequency range above 4,000 cycles by a few per cent.

D. Sound-reinforcing Systems[10]

Sound-reinforcing systems are used to augment the sound output of a speaker, singer, or musical instrument. By means of sound-reinforcing systems, speech can be rendered intelligibly, where before it was either too weak to be heard above the general noise or too reverberant. In certain instances, the volume range of an orchestra is inadequate for full artistic appeal or to utilize the full capabilities of the hearing range. In these cases, means are required for augmenting the intensity of the original sound.

[9, 10] Olson, H. F., *Acoustical Engineering*, D. Van Nostrand Company, Inc., Princeton, 1957.

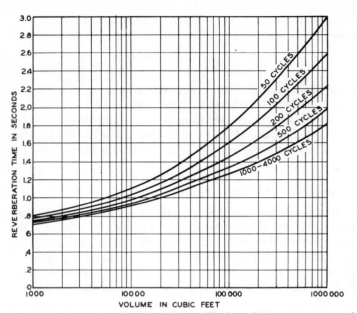

Fɪɢ. 8.5.   Optimum reverberation time of an auditorium, theater, or room as a function of the volume for different frequency ranges.

Two typical sound-reinforcing systems are shown in Fig. 8.6.   The sound-reinforcing system consists of a microphone, an amplifier, and a loudspeaker (see Sec. 9.12).   In Fig. 8.6*A* the loudspeakers are located in a cluster above the stage.   In Fig. 8.6*B* the loudspeakers are distributed around the walls of the room.   Either system may be used with equally good results.   The single loudspeaker station located above the original sound source produces somewhat more realistic results since the original and the augmented sound originate in the same general direction. In the system with the distributed loudspeakers it is somewhat easier to obtain good results because the feedback difficulties are reduced.   The latter arrangement is used for schoolrooms and churches.

In theaters the microphones for collecting the sounds are usually concealed in the footlight trough.   By employing directional loudspeakers, the sound level at the microphones due to the loudspeakers is lowered, which reduces the possibility of oscillations due to acoustic feedback or regeneration in the reproducing system.   In large theaters, having an expansive stage, the pickup distance will be very large.   Consequently, the sound which reaches the microphones from the original source will be small and will require considerable amplification, which increases the tendency for feedback.   In cases where difficulties are

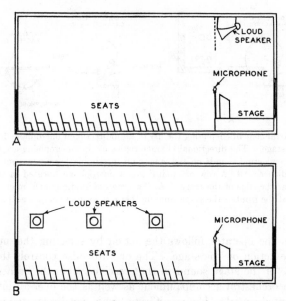

FIG. 8.6. Arrangements of the loudspeakers for a sound-reinforcing system in a large room as, for example, a schoolroom, a church, an auditorium, etc. *A.* A horn loud-speaker cluster located above the stage. *B.* A large number of direct-radiator permanent-magnetic dynamic loudspeakers distributed around the walls of the room.

experienced, owing to acoustic feedback, a reduction in acoustic coupling between the loudspeaker and the microphone can be obtained by employing directional microphones. Furthermore, the stage collecting system should not be responsive to sound originating in the orchestra or audience. For these applications, the unidirectional microphone has been found to be very useful, as will be seen from a consideration of the directional characteristics of this microphone shown in Fig. 8.7*A*. Where the stage is very deep, so that for action near the rear of the stage the distance from the action to the microphones becomes so large that it is impossible to obtain good sound pickup, another set of microphones may be located below a grill in the stage, as shown in Fig. 8.7*B*. These microphones are only used to pick up action at the rear of the stage, because the noise of footsteps would prevent the pickup of action directly over the microphones.

In order to "cover" the action from any part of the stage, several microphones are employed, usually spaced at intervals of 10 feet. The output of each stage microphone and orchestra microphone is connected to a separate volume control on the mixer panel. This mixer and volume-control system is located in the monitoring box. By means of

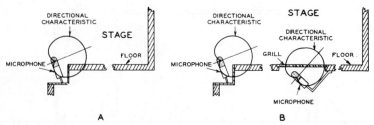

Fig. 8.7. Arrangements depicting the use of a unidirectional microphone for pickup of sound on a stage. The directional characteristics of this microphone are particularly adapted for collecting sounds on the stage and discriminating against sounds coming from the audience.   A. A row of unidirectional microphones located in the footlight trough or near the edge of the stage.   B. Two rows of unidirectional microphones, one row located at the front of the stage and another row located below a grill at the center of the stage.

this system the operator follows the action by selecting the microphone nearest the action on the stage. The operator also controls the ratio of the volume of the stage sound to that received from the orchestra when there is an orchestral accompaniment as well as the over-all intensity of the augmented sound. The monitoring box is usually located in the rear balcony, the position which is the most susceptible to the augmented sound and one which also furnishes a good view of the action.

In the sound-reinforcing system shown in Fig. 8.6B the sound from the loudspeakers will arrive at the ears of the listeners ahead of the direct sound from the person speaking on the stage. Thus the illusion of the sound emanating from the speaker on the stage is destroyed. The reason for this state of affairs is due to the precedence or delay effect.

Subjective experiments[11] have established that if there are several separated sources of sound, identical in content and amplitude but displaced with respect to time, the sound will appear to come from the source which leads the other in time. The experiment which illustrates this phenomenon is shown in Fig. 8.8. The same signal is reproduced from loudspeakers 1 and 2. The signal from loudspeaker 1 can be delayed by means of the delay system. For each value of delay, the ratio of the voltage input to the two loudspeakers is varied until it is impossible to distinguish which loudspeaker appears to be the source. The results of this test are shown by the graph of Fig. 8.8. This experiment shows that there can be considerable unbalance before the sound ceases to appear to come from undelayed source. With the same intensity of sound emanating from both loudspeakers, the source always appears to be the undelayed loudspeaker. The delay phenomena plays an important role in stereophonic systems.

[11] Haas, H., *Akrustika*, Vol. 1, No. 2, p. 49, 1951.

A sound-reinforcing system[12] employing a delay system is shown in Fig. 8.9. Audio delayers for use in sound reinforcing systems are described in Sec. 8.3J. The total time delay between loudspeaker 1 and loudspeakers 2 or 3 is made greater than the time for sound to travel in air from loudspeaker 1 to loudspeakers 2 or 3. The total time delay between loudspeaker 1 and loudspeakers 4 or 5 is made greater than the time required for sound to travel in air from loudspeaker 1 to loudspeakers 4 or 5, etc. If a system of the type shown in Fig. 8.9 is used, the sound will always appear to emanate from the person on the stage. Furthermore, the intelligibility will be superior to the conventional system of Fig. 8.6B without delay.

An architectural enclosure equipped with a highly sophisticated acousto-electronic system[13] which will provide the appropriate acoustics for all manner of programs ranging from speech to musical programs of all kinds

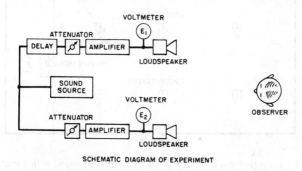

SCHEMATIC DIAGRAM OF EXPERIMENT

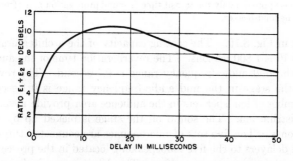

FIG. 8.8. Arrangement of the apparatus for demonstrating the precedence effect. The graph shows the ratio of the voltages E_1 and E_2 as a function of the delay for the conditions when it is impossible to distinguish which loudspeaker is the source.

[12] Olson, H. F., *Jour. Acoust. Soc. Amer.*, Vol. 31, No. 7, p. 872, 1939.
[13] Olson, H. F., *Jour. Audio Eng. Soc.*, Vol. 13, No. 4, p. 307, 1965.

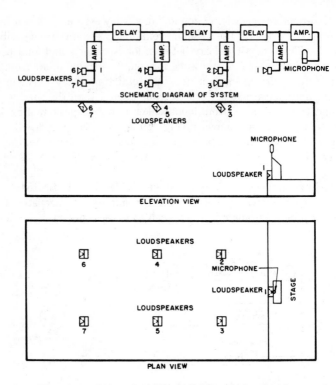

FIG. 8.9. Elevation and plan view of an auditorium equipped with a sound-reinforcing system with progressive delay between the microphone and loudspeakers. The
delay time is approximately the transit time of sound propagation in air from the microphone to the loudspeaker.

is shown in Fig. 8.10. The seating capacity of the architectural enclosure
of Fig. 8.10 is 1,800 persons. The reverberation time of the audience volume in the mid audio-frequency range is 0.7 second. The reverberation
time of the stage in the mid audio-frequency range is 0.4 second. The
large number of loudspeakers in the audience area provide a very uniform
sound distribution. The sound on the stage is picked up by directional
microphones. The five sets of microphones are connected through amplifiers and delayers to the five loudspeakers, located in the proscenium arch
to supply the auditory perspective. The delay system is designed so that
the sound appears to originate from the source on the stage (see Sec. 8.3J).
Electronic synthetic reverberators supply the reverberation envelope for
the sound reproduced from the stage (Sec. 8.3K). The electronic synthetic
reverberators provide reverberation times from 0 to 5 seconds. For speech

the system is operated without any synthetic reverberation. Under these conditions the over-all ratio of the direct to reflected sounds for the sound picked up on the stage and the sound reproduced in the audience area is three. This corresponds to a very low reverberation time of 1/12 second. The level of the speech at the listeners ears is maintained at a level of 78 dB which corresponds to conversational speech at three feet. The high ratio of direct to reflected sound combined with the normal level of the reproduced speech leads to practically perfect intelligibility and naturalness of speech. For every type of musical rendition there is an optimum reverberation time as depicted in Table 8.2. The reverberation time of the ordinary passive architectural enclosure can be correct for only one type of musical rendition. The acousto-electronic system of Fig. 8.10

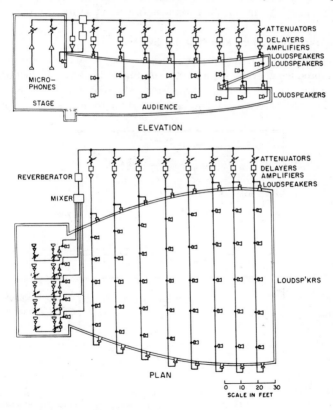

FIG. 8.10. Elevation and plan view of a large auditorium and concert hall equipped with a sophisticated acousto-electronic system consisting of microphones, delayers, reverberators, amplifiers, and loudspeakers.

TABLE 8.2. EFFECTIVE REVERBERATION

Type of Music	Time in Seconds
Organ	2 to 5
Band	2 to 3
Symphony	1.5 to 2
Chorus	1 to 2
Opera	.7 to 1
Popular Orchestra	.5 to 1

provides the correct reverberation time for all types of musical renditions. The complete flexibility of the system makes it possible to vary the reverberation during a rendition. Furthermore, the acoustics are the same for each and every seat location in the entire architectural enclosure.

E. Sound-motion-picture Reproducing System

A typical sound-motion-picture theater and a sound-motion-picture sound-reproducing system are shown in Fig. 8.11. The elements of a sound-motion-picture reproducing system are shown in Figs. 8.11 and 9.59. The photoelectric cell and optical system are housed in the picture

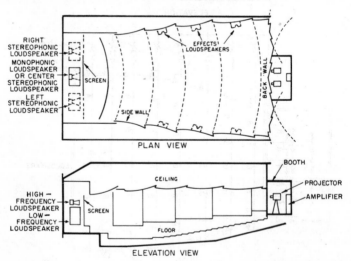

FIG. 8.11. Elevation and plan view of a sound-motion-picture theater showing the elements of the sound-reproducing system. For monophonic sound reproduction, the center loudspeaker is used. For three-channel stereophonic sound reproduction, three loudspeakers are used. The loudspeakers located around the theater are used for sound effects.

projector of Fig. 8.11. For the reproduction of film with magnetic strips as shown in Fig. 9.60B, the magnetic heads are housed in the picture projector of Fig. 8.11. The picture projector, amplifier, and volume controls are located in the projection booth. The loudspeakers are located behind the perforated screen on the stage.

The theater[14] should be designed so that the reverberation time corresponds to the volume of the theater (Sec. 8.2C). In addition, care should be taken to avoid echoes or objectionable concentrations due to focusing of the reflected sound by the use of convex rather than concave walls and ceiling sections.

When a sound wave strikes a wall of a theater, a part is reflected, a part absorbed, and a part transmitted. The reflection, for surfaces large compared to the wavelength, is analogous to specular reflection. The reflected sound, in a poorly designed theater, produces highly concentrated zones of reflected energy. For proper sound-reflection control in an auditorium, the acoustical treatment and shape of the walls and ceiling must be such as thoroughly to diffuse the reflected sound. The reflected sound energy received in any auditorium location should not come from one particular reflecting area but should be contributed by numerous reflecting surfaces. The sound energy from any reflection should be small compared with the total reflected sound energy at any point in the auditorium. This also provides a more uniform decay of the reverberant sound.

In a theater, free of acoustical difficulties, the energy density of the generally reflected sound is practically the same for all parts of the theater. Therefore, the solution to the problem of achieving uniform sound-energy density is to employ directional loudspeakers arranged so that the direct sound-energy density is the same for all parts of the theater.

Stereophonic sound was introduced into sound-motion-picture reproduction several years ago (see Sec. 9.12). The stereophonic sound is reproduced on three channels (see Fig. 9.60). The three loudspeakers located behind the screen in Fig. 8.11 reproduce the sound in auditory perspective. The loudspeakers located around the walls of the theater operate from the fourth channel and are used to reproduce various sound effects which enhance the artistic aspects of the reproduced sound.

A typical sound-motion-picture "drive-in" theater and sound-reproducing system are shown in Fig. 8.12. The elements of a sound-motion-picture reproducing system are shown in Figs. 8.12 and 9.59. The photoelectric cell and optical system are housed in the picture projector of Fig. 8.12. For the reproduction of film with magnetic strips as shown in Fig. 9.60B, the magnetic heads are housed in the picture projector of Fig. 8.12. The

[14] Standardizing Committee, *Jour. Soc. Motion Picture Engrs.*, Vol. 36, No. 3, p. 267, 1941.

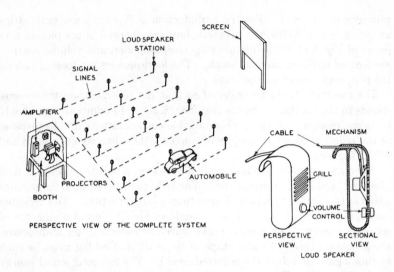

Fig. 8.12. A perspective view showing the elements of a "drive-in" sound-motion-picture theater. Perspective and sectional views of the loudspeaker used in the "drive-in" theater are shown in the lower right.

picture projector, amplifier, and master volume controls are located in the projection booth. In the "drive-in" theater, the customers are seated in their automobiles to view the picture and hear the sound. Individual loud-speakers are provided to supply the sound to each automobile. Perspective and sectional views of the loudspeaker are shown in Fig. 8.12. In use, the loudspeaker is hung inside the motorcar. The volume control makes it possible for the user to adjust the level of the reproduced sound. When not in use, the loudspeakers are hung on posts provided for the purpose. Signal lines connect the individual loudspeakers to the amplifier in the booth.

F. Radio Receiver, Phonograph, and Television Sound-reproducing Systems in the Home[15],[16]

The radio receiver, phonograph, and television sound-reproducing systems represent by far the largest number of complete sound-reproducing systems. For this reason, the operation of these systems in the average living room or den is an extremely important problem. A monophonic sound-reproducing system, as represented by a radio receiver, phonograph, magnetic tape reproducer, and television receiver, is shown in Fig. 8.13.

[15] Olson, H. F., *Jour. Audio Eng. Soc.*, Vol. 6, No. 2, p. 80, 1958.
[16] Olson and Belar, *Jour. Audio Eng. Soc.*, Vol. 8, No. 1, p. 37, 1960.

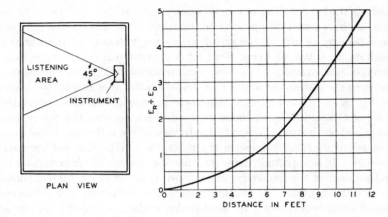

Fig. 8.13.   Ratio of the reflected sound energy E_R to the direct sound energy E_D for a conventional console-type radio receiver, phonograph, or television receiver in the listening area shown above in a typical living room, as a function of the distance between the instrument and the listener.

The reverberation time characteristic of the average living room in the home is shown in Fig. 8.14.   In fact, the reverberation time characteristic of all the rooms in the home have this same general shape.   For each room though, there will be a deviation from the average characteristic of Fig. 8.14; however, the variation is not particularly significant.

The reverberation time in the average residence as shown in Fig. 8.14 is short.   This is a fortunate state of affairs because the over-all effective reverberation of the complete chain from the sound source in the studio to the listener in a room in a residence is determined by the studio.   In the case of the theater it is possible to adjust the loudspeakers so that the direct sound is the same in all parts of the auditorium.   It is not practical to

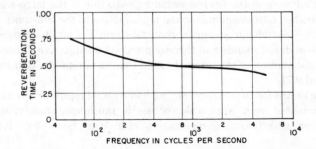

Fig. 8.14.   The reverberation time characteristic of a room in a home.

arrange the loudspeakers in a radio receiver so that there will be no variation of direct sound with position in the room. The ratio of the reflected
sound-energy density to the direct sound-energy density at the listening
position, in a typical living room for a conventional console-type radio
receiver, phonograph, or television receiver, as the function of the distance
between the instrument and the observation point is shown by the graph
of Fig. 8.13. It will be seen that there is considerable variation in the
ratio of reflected to direct sound as the distance between the observation
point and the instrument is varied. In view of the rather small distances
and relatively small volume of the room, the deviation is not very important. It is important, however, that the directional characteristic be
independent of the frequency and sufficiently broad to send direct sound
to all listening areas. In addition, considerable enhancement in the reproduced sound is obtained if the loudspeaker in the instrument is located at
approximately ear level, that is, 42 inches from the floor.

The response-frequency characteristics upon the ears, Fig. 7.10, considered in Sec. 7.4B, show that corresponding to the intensity of a 1,000-
cycle note there is an intensity at another frequency that will sound as
loud. These characteristics show that if the sound is reproduced at a
lower level than that of the original sound it will appear to be deficient in
low-frequency response. In general, the reproduction level of the sound
in the home is much lower than the level of the original sound in the studio.
In order to compensate for the low-frequency deficiency, the volume control
in most radio receivers and phonographs is designed so that the low-frequency response is accentuated in an inverse ratio to the relative sensitivity
of the ear in going from the original level to the lower level of reproduction.
This type of volume control is termed an acoustically compensated volume
control.

The preceding considerations have been concerned with monophonic
sound reproduction (see Sec. 9.6C). The most recent significant and important advance in the field of sound reproduction is the large-scale commercialization of stereophonic sound reproduction in the consumer or mass
market. Stereophonic sound reproduction provides auditory perspective
of the reproduced sounds and thereby preserves a subjective illusion of the
spatial distribution of the original sound sources in reproduction (see Secs.
9.6D and 9.7).

Three means for two-channel stereophonic sound reproduction have been
commercialized on a large scale for use in the home, namely, magnetic
tape, disc and frequency-modulation radio (see Secs. 9.9, 9.10 and
9.11).

A plan view of a typical living room in the home and the location of the

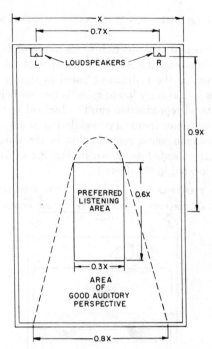

Fig. 8.15. The arrangement of the loudspeakers for a two-channel stereophonic sound-reproducing system showing the optimum relative dimensions, the preferred listening area for excellent auditory perspective, and the listening area for satisfactory auditory perspective.

loudspeakers[17] for the reproduction of stereophonic sound is shown in Fig. 8.15. The arrangement of the loudspeakers, the preferred listening area and the relative dimensions apply to practically any room in the average home. If the dimensional ratios and preferred listening area, depicted in Fig. 8.15 are maintained, the important auditory perspective effects, namely, the angular and depth distribution of the sound sources, will be maintained.

G. Radio Receiver and Magnetic Tape Sound Reproduction in an Automobile[18]

Radio receivers are installed as standard equipment in more than 75 per cent of the automobiles sold today. Therefore, sound reproduction in

[17] Olson, H. F., *Proceedings of the Third International Congress on Acoustics*, Elsevier Publishing Co., Amsterdam, p. 791, 1959.

[18] Olson, H. F., *Proceedings of the Third International Congress on Acoustics*, Elsevier Publishing Co., Amsterdam, p. 791, 1959.

automobiles is a large factor in the mass dissemination of information and entertainment.

In monophonic sound reproduction in an automobile, the loudspeaker is usually placed in a central location in the front of the instrument panel or in the top cover of the instrument panel as shown in Fig. 8.16. In some installations an auxiliary loudspeaker is located behind the rear seat.

Stereophonic sound reproduction can be obtained in the automobile by means of a stereophonic frequency-modulation radio receiver (see Sec. 9.11). In stereophonic sound reproduction in the automobile, the loudspeakers are usually located as shown in Fig. 8.16. In some cases, the loudspeakers are located in the doors.

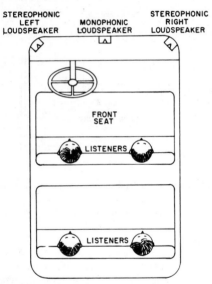

FIG. 8.16. The location of the loudspeakers for monophonic and stereophonic sound reproduction in an automobile.

Recently, stereophonic-magnetic tape reproducers have been commercialized for automobiles. The programs are recorded on endless loop cartridges.

In conclusion, excellent auditory perspective is obtained, in an automobile, from stereophonic sound reproduction which also provides more realistic reverberation characteristics, improved low-frequency response and greater discrimination against ambient noise. Furthermore, good auditory perspective of the reproduced sound is obtained in every seat location in the automobile.

H. General-announce and Paging Systems[19]

General-announce systems and paging systems are useful in factories, department stores, warehouses, railroad stations, airport terminals, schools, offices, hospitals, hotels, etc. Typical installations are depicted in Fig. 8.17. For this type of sound reproduction, intelligibility is more important than quality. The deleterious effect of reverberation upon articulation can be reduced, and a better control of sound distribution can be obtained, by reducing the low-frequency response of the system. Furthermore, the cost of the amplifiers and loudspeakers is also reduced by limiting the frequency range. To find the power required, the sound-intensity level under actual operating conditions should be determined. The system should be designed to produce an intensity level 20 to 40 decibels above the general noise level. Under no conditions should the system be designed to deliver an intensity level of less than 80 decibels. The loudspeakers should be selected and arranged following an analysis similar to that outlined in the preceding sections, so that uniform sound distribution and adequate intensity levels are obtained.

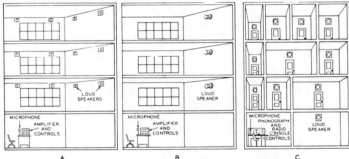

Fig. 8.17. Schematic views of call, general-announce, and sound-distributing systems. *A*. A large number of direct-radiator loudspeakers are used to obtain a moderate sound level over a large floor area, as for example, in department stores, schools, and offices. *B*. High-efficiency high-power horn loudspeakers are used to obtain a high sound level over a large floor area, as for example, in factories, warehouses, and railroad stations. *C*. Small direct-radiator loudspeakers are used to supply small rooms, as for example, in hotels and hospitals.

For large rooms in department stores, schools, offices, etc., where the noise level is relatively low and where it is desirable to employ a low level in the reproduced sound, adequate power and uniform distribution of the reproduced sound can be obtained by the use of a large number of direct-radiator dynamic loudspeakers of the permanent-magnet dynamic type, as shown in Fig. 8.17*A*. In this connection, it should be mentioned that

[19] Olson, H. F., *RCA Rev.*, Vol. 1, No. 1, p. 49, 1936.

these loudspeakers have a very low efficiency, being of the order of 3 per cent as compared with 25 to 50 per cent for horn loudspeakers. However, since the power requirements for the applications in Fig. 8.17A are relatively low, direct radiator loudspeakers may be used. On the other hand, for large rooms in factories, warehouses, railroad stations, airport terminals, etc., where the noise level is relatively high, which in turn requires a high level for the reproduced sound and a correspondingly large acoustical output from the loudspeakers, it is more economical to use a high-efficiency horn-loudspeaker installation, as shown in Fig. 8.17B. The use of a high-efficiency horn loudspeaker makes it possible to reduce the size of the power amplifier, which becomes an important factor in high-power large-scale sound reproduction.

For general-announce, paging, and sound-distributing installations used in schools, hospitals, hotels, etc., the intensity level required is relatively low and the volume of the average room is usually small (Fig. 8.17C). For most installations of this type, save in noisy locations, an intensity level of 70 decibels is more than adequate. Higher intensity levels tend to produce annoyance in adjacent rooms. Therefore, the power require-

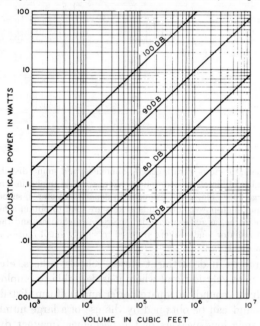

FIG. 8.18. The acoustical power required to produce intensity levels of 70, 80, 90, and 100 decibels in a room as a function of the volume of the auditorium or room.

ments for the loudspeaker will be small. To blend with the furnishings of the room, it is desirable to mount the loudspeakers flush with the wall surface. Therefore, for these applications, a direct-radiator loudspeaker of the permanent-magnet dynamic type is the most suitable.

I. Power Requirements for Sound-reproducing Systems[20]

The power requirement is an important factor in the motion-picture and sound-reinforcing systems. The minimum intensity which these systems should be capable of producing is 80 decibels, where 0 decibels = 0.0002 dyne per square centimeter. The graphs of Fig. 8.18 show the acoustical power required, as a function of the volume, in auditoriums to produce sound levels of 70, 80, 90, and 100 decibels. In large auditoriums where the orchestra is also reinforced, the power available should be greater. For example, to render full artistic appeal in the reinforcement of a symphony orchestra, the system should be capable of producing a level up to 100 decibels.

J. Outdoor Theater Sound-reinforcing Systems[21]

Two types of sound-reinforcing installations for an outdoor theater are shown in Fig. 8.19. The system on the left employs a single loudspeaker station located either above or below the stage, as shown in Fig. 8.19. If

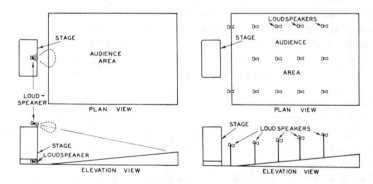

FIG. 8.19. Two arrangements of sound-reinforcing systems for an outdoor theater. The arrangement on the left employs a single loudspeaker located above the stage and having suitable directional characteristics to produce a uniform intensity level over the audience area. An alternate arrangement employs a loudspeaker located below the stage floor. The arrangement on the right employs a number of loudspeakers, each covering a small portion of the audience. (*After Olson, Acoustical Engineering, D. Van Nostrand Company, Inc., Princeton, 1957.*)

[20, 21] Olson, H. F., *RCA Rev.*, Vol. 1, No. 1, p. 49, 1936.

the stage is quite low, the logical position for the loudspeakers is at the top of the stage. The same procedure for obtaining uniform sound coverage and adequate intensity level of the sound from the loudspeakers as used in the preceding considerations is applicable in this case. If the stage is very high, the separation between the loudspeakers and the action on the stage will be very large. As a consequence, the wide difference in the direction of the direct and reinforced sound will be particularly disconcerting to listeners in the front portion of the seating area. Under these conditions, it may be desirable to locate the loudspeaker under the stage, as shown in Fig. 8.19. The system depicted on the right employs a large number of loudspeakers, each one supplying a small portion of the audience. The individual characteristics of the loudspeakers should be selected so that each individual area is adequately supplied. Cognizance must be taken of the energy supplied from adjacent loudspeakers.

There are certain advantages in each system. In the case of the single-loudspeaker system, better illusion is obtained because the augmented sound appears to come from the stage. On the other hand, the intensity level outside the audience area in a backward direction falls off very slowly. At a distance equal to the length of the audience area, the level is only 6 decibels lower than that existing in the audience area. In certain locations the sound levels produced by such systems will cause considerable annoyance to those located in the vicinity of the theater. By dividing the theater area into small plots, each supplied by a loudspeaker, and by directing the loudspeakers downward, the sound-intensity level outside the audience area will be considerably lower than in the case of the single-loudspeaker station and usually eliminates any annoyance difficulties. The short sound-projection distance is another advantage of the multiple loudspeaker system.

The sound-reinforcing system for the outdoor theater shown in the right of Fig. 8.19 may be equipped with a delay similar to that used for the system shown in Fig. 8.9. The intelligibility will be superior to the system without delay. In addition, greater realism will be achieved by the use of a delay system.

The above typical examples of outdoor public-address and sound-reinforcing systems illustrate the principal factors involved in this field of sound reproduction.

K. Orchestra and Stage Shell[22]

When orchestra and stage productions are conducted in outdoor theaters, it is desirable to provide a shell to augment and direct the sound to the audience, to surround the orchestra with reflecting surfaces; and to protect

[22] Olson, H. F., *Acoustical Engineering*, D. Van Nostrand Company, Inc., Princeton 1957.

the performers and instruments against wind, dew, and other undesirable atmospherics. Most of the outdoor orchestra shells have been of the concave type which produce intense and sharp concentrations of reflected sound in both the shell and audience area. These acoustical effects are particularly undesirable when the sound is picked up by microphones on the stage for sound reinforcing and broadcasting. Under these conditions the intensifications and discriminations make it appear that the orchestra is unbalanced with relation to the various instruments. Furthermore, it is impossible for the conductor to obtain a true balance because these undesirable acoustical effects also exist at the conductor's platform on the stage. The undesirable acoustical effects can be overcome by means of an orchestra shell in which the boundaries are polycylindrical surfaces, as shown in Fig. 8.20. These surfaces reflect the sound in a diffuse manner and thereby obviate concentrations of sound energy on the stage and in the audience area. The acoustics of this type of orchestra shell make it possible for the conductor to obtain an improved balance of the orchestra. This type of shell produces a uniform distribution of sound in the audience area. The polycylindrical shell provides good acoustics for microphone pickup for sound reinforcing or broadcasting.

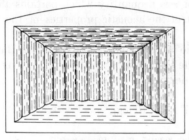 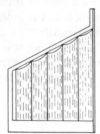

PERSPECTIVE VIEW　ELEVATION VIEW

Fig. 8.20. Front and side views of an orchestra shell with polycylindrical surfaces.

L. Sound Transmission through Partitions[23-30]

The problem of sound transmission through partitions and walls is complicated because of the many factors involved. The problem of the mass-

[23] Rayleigh, *Theory of Sound*, Macmillan & Co., Ltd., London, 1926.
[24] Eckhardt and Chrisler, Bureau of Standards, Paper 526.
[25] Knudsen, *Architectural Acoustics*, John Wiley & Sons, Inc., New York, 1932.
[26] Sabine, *Acoustics and Architecture*, McGraw-Hill Book Company, Inc., New York, 1932.
[27] Watson, *Acoustics of Buildings*, John Wiley & Sons, Inc., New York, 1923.
[28] Morrical, K. C., *Jour. Acoust. Soc. Amer.*, Vol. 11, No. 2, p. 211, 1939.
[29] Harris, *Handbook of Noise Control*, McGraw-Hill Book Co., New York, 1951.
[30] Beranek, *Noise Reduction*, McGraw-Hill Book Co., New York, 1960.

controlled single-wall partition is very simple. The sound insulation of this type of partition is proportional to the mass and frequency. For the usual building materials and walls of ordinary dimensions supported at the edges, the problem is that of the clamped rectangular plate with distributed resistance throughout the plate and lumped damping at the edges. Obviously, the performance of this system depends upon the size, the ratio of the two linear dimensions, the weight of the material, the damping in the material, and the edge supports. This type of problem is not amenable to an analytical solution.

The transmittivity of a partition is defined as the ratio of the intensity in the sound transmitted by the partition to the intensity in the sound incident upon the partition. The transmission loss, in decibels, introduced by the partition is given by

$$T.L. = 10 \log_{10} \frac{I_i}{I_t} = 10 \log_{10} \frac{1}{\tau} \qquad (8.11)$$

where I_i = intensity of the incident sound

I_t = intensity of the transmitted sound

τ = transmittivity or transmission coefficient

The coefficient of transmission τ is a quantity which pertains alone to the partition and is independent of the acoustic properties of the rooms which it separates.

The reduction factor is the ratio of the sound-energy density in the room containing the sound source to the sound energy in the adjoining receiving room. The reduction factor, in decibels, is given by

$$R.F. = T.L. + 10 \log_{10} \frac{A}{S} \qquad (8.12)$$

where A = total absorption in the receiving room, in sabins

S = area of the partition, in square feet

Equation (8.12) shows that the reduction is due to both the loss introduced by the partition and the absorption in the receiving room.

The choice of a partition for insulating a room against sound involves a number of considerations. Some of the factors are the frequency distribution and intensity level of the components of the objectionable sound, the transmission-frequency characteristics of the partition, the ambient noise or sound level in the receiving room which will mask the objectionable sound, and the response-frequency characteristics of the ear.

Measurements have been made by various investigators upon the transmission by single partitions. The results of these measurements are shown in Table 8.3.

TABLE 8.3 NOISE REDUCTION THROUGH VARIOUS STRUCTURES

Material	Weight, pounds per square foot	Frequency					Average transmission loss, decibels	Average τ	Author
		128	256	512	102	2,048			
		Transmission loss, decibels							
Iron, 0.03-inch galvanized..........	1.2	....	25	20	29	35	25	0.0032	B.S.
Lead, ⅛ inch.....................	8.2	....	31	27	37	44	32	0.00063	B.S.
Plywood, ¼ inch..................	0.73	....	21	21	25	26	21	0.008	B.S.
Celotex, standard ½ inch..........	0.66	....	22	17	23	27	20	0.01	B.S.
Hair felt, 1 inch..................	0.75	4.9	4.6	6.0	7.1	6.7	5	0.32	P.E.S.
Hair felt, 4 inches................	....	7.5	12.5	15	19	19	14	0.04	P.E.S.
Wood studs, wood lath, lime plaster.	18	....	29	38	47	43	43	0.00005	P.E.S.
Tile 2-inch gypsum...............	20	....	34	44	51	63	48	0.000016	P.E.S.
Tile clay, 6 by 12 by 12 inches, plastered both sides................	37	....	41	35	45	52	40	0.0001	B.S.
Brick, 8 inches plastered both sides..	87	....	50	48	55	63	50	0.00001	B.S.
Door, light 4 panel................	....	13	16	20	23	22	22	0.0063	P.E.S.
Door, oak........................	....	15	18	23	26	25	25	0.0032	P.E.S.
Door, steel, ¼ inch................	....	25	27	31	36	31	35	0.00032	P.E.S.
Window glass, plate ¼ inch........	3.5	....	33	31	33	35	30	0.001	B.S.
Window glass, small panes 3/16 inch..	....	19	20	24	31	28	29	0.0013	P.E.S.

The abbreviations in the above table are as follows: B.S., Bureau of Standards; P.E.S., P. E. Sabine.

The mass-controlled partition with air between the partition elements is a low-pass filter in which the mass of the wall is the series element and the volume between the partitions is the shunt element. The partitions in this case are mounted in edge supports which allow freedom of motion without cracks which would pass air-borne sound.

8.3 COLLECTION OF SOUND

A. Sound-collecting System[31]

When a source of sound is caused to act in a room, the first sound that strikes a collecting system placed in the room is the sound that comes directly from the source without reflection from the boundaries. Following this comes sound that has been reflected once, twice, and so on, meaning that the energy density of the sound increases with the time, as the number of reflections increase. Ultimately, the absorption of energy by the boundaries equals the output of the source, and the energy density at the collecting system no longer increases; this is called the steady-state condition. Therefore, at a given point in a room there are two distinct sources of sound, namely, the direct and the generally reflected sound. This has been depicted graphically in Fig. 8.1. For rooms that do not

[31] Olson, H. F., *Acoustical Engineering*, D. Van Nostrand Company, Inc., Princeton, 1957.

exhibit abnormal acoustical characteristics, it may be assumed that the ratio of the reflected to the direct sound represents the effective reverberation of the collected sound.

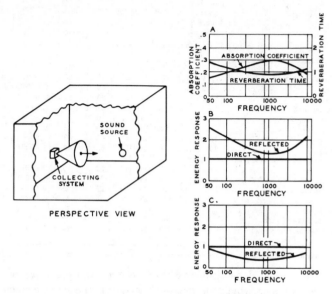

FIG. 8.21. Sound-collecting system in a studio. Graph A shows the reverberation time of the studio and absorption coefficient of the boundaries of the studio. Graph B shows the energy response for the direct and reflected sound for a nondirectional microphone. Graph C shows the energy response for the direct and reflected sound for the bidirectional velocity and the unidirectional microphones.

Consider a sound-collecting system, Fig. 8.21, the efficiency of reception of which may be characterized as a function of the direction with respect to some reference axis of the system. (The nondirectional collecting system is a special case of the directional system in which the efficiency of reception is the same in all directions.) The output of the microphone may be expressed as

$$e = Qpf_1(\psi) \tag{8.13}$$

where e = voltage output of the microphone, in volts

p = sound pressure, in dynes per square centimeter

Q = sensitivity constant of the microphone

ψ = angle between incident pencil of sound and the reference axis of the microphone

If the distance, in centimeters, between the source of the sound and the collecting system is D, the energy density at the microphone due to the direct sound is

$$E_D = \frac{E_0}{D^2 4\pi c} \tag{8.14}$$

where E_0 = power output of the sound source, in ergs per second

c = velocity of sound, in centimeters

To simplify the discussion, assume that the effective response angle of the microphone is the solid angle Ω, in steradians. The direction and phase of the reflected sound are assumed to be random. Therefore, the reflected sounds available for actuating the directional microphone are the pencils of sound within the angle Ω. The energy response of the directional microphone to generally reflected sound will be $\Omega/4\pi$ that of a nondirectional microphone. The generally reflected sound to which the directional microphone is responsive is, therefore, given by

$$E_R = \frac{4E_0\Omega}{caS4\pi}(1 - \epsilon^{\{cS[\log\epsilon(1-a)]t\}/4V})(1 - a) \tag{8.15}$$

where a = absorption per unit area, absorption coefficient

S = area of absorbing material, in square centimeters

V = volume of room, in cubic centimeters

Ω = solid angle of reception, in steradians

t = time, in seconds

The ratio of the generally reflected sound to the direct sound is a measure of the effective reverberation of the collected sound and is given by

$$\frac{E_R}{E_D} = \frac{4D^2\Omega(1 - \epsilon^{\{cS[\log\epsilon(1-a)]t\}/4V})(1 - a)}{aS} \tag{8.16}$$

If the sound continues until the conditions are in a steady state, Eq. (8.16) becomes

$$\frac{E_R}{E_D} = \frac{4D^2}{aS}\Omega(1 - a) \tag{8.17}$$

From Eqs. (8.16) and (8.17), it will be seen that the received reverberation can be reduced by decreasing the distance D, by increasing the absorption aS, or by decreasing Ω.

For a given room employing a directional microphone, the receiving distance can be increased $\sqrt{4\pi/\Omega}$ times that in the nondirectional system with the same collected reverberation in both cases.

The absorption characteristic of a studio is shown in Fig. 8.21A. The direct sound picked up by a nondirectional microphone and two directional microphones is the same because the distance between the sound source and the microphone is assumed to be the same for the two cases (Figs. 8.21*B* and 8.21*C*). The generally reflected sound picked up by a nondirectional microphone is shown in Fig. 8.21*B*. The generally reflected sound picked up by a velocity or unidirectional microphone in which $\Omega = 4\pi/3$ is shown in Fig. 8.21*C*. The effectiveness of a directional sound-collecting system in overcoming reverberation and undesirable sounds is graphically depicted in Fig. 8.21*C*.

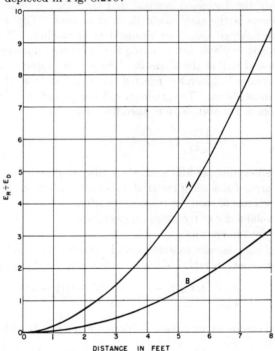

Fig. 8.22. Ratio of the reflected sound-energy response E_R to the direct sound-energy response E_D in a typical studio for nondirectional and directional microphones. *A*. Nondirectional microphone. *B*. Bidirectional velocity (cosine pattern) or unidirectional (cardioid pattern).

The ratios of the reflected sound-energy response to the direct sound-energy response, in a typical studio, for a nondirectional microphone and a bidirectional velocity microphone or a unidirectional microphone with a cardioid pattern, as a function of the distance, are shown in Fig. 8.22. These characteristics show that, for the same effective reverberation, the directional microphones may be operated over a much greater distance than the nondirectional microphone. These characteristics also show how the effective reverberation is increased as the distance between the source and the microphone is increased.

Directional microphones,[32] in addition to discriminating against noise and generally reflected sounds, have been found to be extremely useful in arranging actors in dialogue and for adjusting the relative loudness of the instruments of an orchestra.

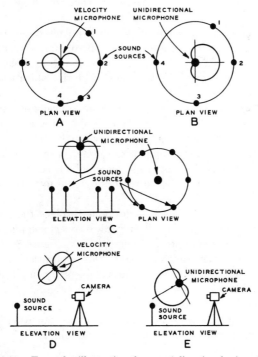

Fig. 8.23.   Examples illustrating the use of directional microphones.

A plan view of a velocity microphone and a number of sound sources is shown in Fig. 8.23A.   Suppose that sources 2 and 5 represent two actors who are carrying on a dialogue.   In view of the fact that this microphone receives with the same efficiency in two directions, it is possible to have the actors face each other, which is an advantage from a dramatic standpoint.   Suppose that the sources of sound 1, 2, 3, and 5 of Fig. 8.23A represent the instruments of an orchestra.   All the sources are located at the same distance.   This means that 1 will produce 0.7 times the voltage output by the microphone as that produced by 2 for the same sound out-

[32] Olson, H. F., Proc. Inst. Radio Engrs., Vol. 21, No. 5, p. 655, 1933.

put by the two instruments.   In the same way, 3 will be 0.5 of 2.   Source 4 is considered as objectionable and is placed in the zero-reception zone. With the microphone the relative intensity of these sources in the reproduced sound can be adjusted by the angular position relative to the microphone axis as well as the distance.   Obviously, this is a great advantage in balancing the instruments of an orchestra.   In the case of a nondirectional microphone, the relative intensity in the reproduced sound can be adjusted only by the distance.

The same procedure for balancing the instruments of an orchestra may be used in connection with a unidirectional microphone (Fig. 8.23B).   The unidirectional microphone is particularly useful when all the instruments are grouped in front of the microphone and the objectionable sounds originate behind the microphone.

The velocity microphone is used in recording music and other sound in sound motion pictures.   In some instances, noises produced by the camera and devices are objectionable and must be reduced.   The directional characteristics of the velocity microphone are useful in overcoming objectionable noises.   It is possible to orient the microphone so that the objectionable noise lies in the plane of zero response of the microphone, as shown in Fig. 8.23D.

In certain types of recording it is desirable to place the microphone at the center of action and directed downward to collect sounds over an angle of 360 degrees with respect to the microphone.   Figure 8.23C illustrates the use of a unidirectional microphone for this application.

The unidirectional microphone is almost universally employed for speech pickup in sound motion pictures and television (Fig. 8.23E).   The broad coverage in the forward direction makes it possible to follow the action. The high discrimination against pickup of sounds originating in the rear is useful in eliminating noises from the camera and lights.

B. Radio and Television Broadcasting Studios[33-42]

In the early days of broadcasting it was customary to make the reverberation time of the studios as low as possible.   This imposed quite a

[33] Morris and Nixon, *Jour. Acoust. Soc. Amer.*, Vol. 8, No. 2, p. 81, 1936.

[34] Chinn and Bradley, *Proc. Inst. Radio Engrs.*, Vol. 27, No. 7, p. 421, 1939.

[35] Potwin and Maxfield, *Jour. Acoust. Soc. Amer.*, Vol. 11, No. 1, Part 1, p. 48, 1939.

[36] Nixon, G. M., *RCA Rev.*, Vol. 6, No. 3, p. 259, 1942.

[37] Volkmann, J. E., *Jour. Acoust. Soc. Amer.*, Vol. 13, No. 3, p. 234, 1942.

[38] Boner, C. P., *Jour. Acoust. Soc. Amer.*, Vol. 13, No. 3, p. 234, 1942.

[39] Content and Green, *Proc. Inst. Radio Engrs.*, Vol. 32, No. 2, p. 72, 1944.

[40] Nygren, A., *FM and Television*, Vol. 6, No. 5, p. 25, 1946.

[41] Rettinger, *Applied Architectural Acoustics*, Chemical Publishing Co., New York, 1947.

[42] Knudsen and Harris, *Acoustical Designing in Architecture*, John Wiley and Sons, New York, 1950.

strain upon the orchestra and singers to keep in tune. The almost universal use of directional microphones during the past few years has eliminated the necessity of extremely dead studios. As a result, the quality and artistic effects of the collected sound are materially enhanced.

The studios in a large radio or television broadcasting station should be graduated in size and in corresponding acoustical condition to accommodate anticipated loading to the best advantage. The control booths should be provided with soundproof windows located so that the studio engineer has an unobstructed view of the studio.

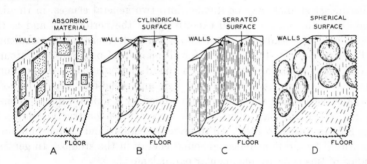

FIG. 8.24. Wall structures for diffusing the reflected sound. *A*. Absorbing material located in spots. *B*. Cylindrical surfaces. *C*. Serrated surfaces. *D*. Spherical surfaces. (*After Olson, Acoustical Engineering, D. Van Nostrand Company, Inc., Princeton, 1957*).

The studios should be insulated against all types of extraneous noises. Cinder composition has been found to give very good insulation. Resilient mounting of the walls, floor, and ceiling reduces mechanical transmission.

Air-borne noises carried in the air-conditioning ducts may be suppressed by lining the ducts with felt, rock wool, etc., to obtain suitable attenuation. Mechanical transmission of sound by the ducts may be reduced by isolating the sections of the duct.

The reflected sound in a studio produces standing-wave systems. These standing-wave systems exhibit variations in sound pressure from point to point in a room. It is desirable to reduce this variation to as small a value as possible. This can, of course, be done by making the walls very absorbing, which leads to the undesirable characteristic of a very low reverberation time. The problem of obtaining a better sound-pressure distribution can be accomplished by the use of wall surfaces which diffuse, distribute, and disperse the sound reflected from the walls. Four typical wall treatments for obtaining a diffuse and uniform sound-pressure-distribution characteristic are shown in Fig. 8.24. In Fig. 8.24*A*, the absorbing material is distributed in discrete spots[43] on the wall surface. This distri-

[43] Potwin and Maxfield, *Jour. Acoust. Soc. Amer.*, Vol. 11, No. 1, Part 1, p. 48, 1939.

bution of material breaks up the reflected wavefront and thereby produces a diffuse sound field. In Fig. 8.24B, C, and D, polycylindrical,[44] serrated,[45] and spherical[46] surfaces are employed to produce a diffused sound field. These surfaces have been used for walls and ceilings in broadcast studios. The polycylindrical and spherical surfaces increase the wavefront of the reflected sound. The convex surfaces also reduce the interference effect between direct and reflected sounds. The treatments shown in Fig. 8.24 are also applied to the ceiling. The use of these expedients produces a more uniform sound-decay curve and reduces echoes and flutters.

Broadcast studios may be divided into two general classes: in the first class the entire studio is used exclusively by the performers, and in the second class the studio is used for both the performers and the audience.

The first class of studio is used for all manner of programs and groups. Under these conditions, it has been found that a studio with uniform acoustics is more useful than the live and dead-end type. Uniform acoustics throughout the studio are obtained by a uniform distribution of the absorbing material. The various types of wall construction shown in Fig. 8.24 are used to break up discrete reflections and thereby obtain a uniform distribution of reflected sound energy in the studio. In general, studios of this type are rectangular parallelepipeds.

Various expedients, in addition to the wall structures of Fig. 8.24, are used to break up flutters and echoes. In some designs the walls are in-

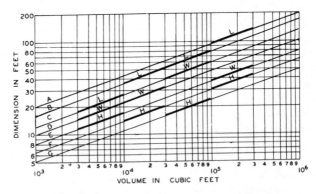

Fig. 8.25. Preferred studio dimensions. In the graph, H = weight, W = width, L = length. Small rooms, $H:W:L = 1:1.25:1.6 = E:D:C$. Average shape rooms, $H:W:L = 1:1.6:2.5 = F:D:B$. Low-ceiling rooms, $H:W:L = 1:2.5:3.2 = G:C:B$. Long rooms, $H:W:L = 1:1.25:3.2 = F:E:A$. (After Volkmann.)

[44] Volkmann, J. E., *Jour. Acoust. Soc. Amer.*, Vol. 13, No. 3, p. 234, 1942.
[45] Nixon, G. M., *RCA Rev.*, Vol. 6, No. 3, p. 259, 1942.
[46] Nygren, A., *FM and Television*, Vol. 6, No. 5, p. 25, 1946.

clined to eliminate parallelism between opposite walls. In other designs the ceiling and/or walls are broken up into two or more nonplanar surfaces.

The ratio of the dimensions[47] of the studio is important in distributing the characteristic resonant frequencies uniformly over the frequency range. The graph of Fig. 8.25 shows the ratio of the dimensions for small, medium, and large studios. The most desirable ratio of the dimensions would be in the ratio of the cube root of 2. This separates the dimensions by one-third octave. This ratio is possible for small studios but is not practical for large studios, in that the ceiling height becomes too great. The dimensions of the small rooms are given by the lines C, D, and E of Fig. 8.25. For medium studios the ratio of the dimensions is near the cube root of 4. The dimensions of medium studios are given by the lines B, D, and F of Fig. 8.25. This is approximately the ratio $2:3:5$ which has been frequently used in the design of radio and television broadcast studios. For very large studios the dimensions are given by the lines B, C, and G of Fig. 8.25. For very long studios the dimensions are given by A, E, and F of Fig. 8.25.

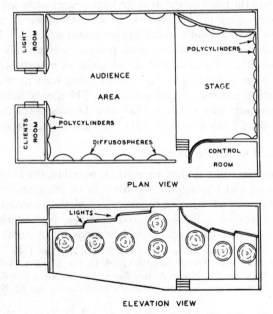

FIG. 8.26. Plan and elevation of an auditorium-type radio and television broadcasting studio. (*After Nixon and Nygren.*)

[47] Volkmann, J. E., *Jour. Acoust. Soc. Amer.*, Vol. 13, No. 3, p. 234, 1942.

In the second class of studio, termed the auditorium type, the performers occupy one end of the room and the audience the other end. A plan and sectional view of an auditorium-type radio and television broadcast studio is shown in Fig. 8.26. The wall and ceiling surfaces of the stage are arranged to provide sound diffusion so that the reflected sounds are properly mixed and the tonal quality of the performer or performing group is enhanced. The stage ceiling is broken in a saw-tooth fashion to provide a sound-diffusing condition and to conceal the border lights and spotlights from the eyes of the audience. The rear wall of the stage is constructed of a series of plaster polycylindrical surfaces to provide diffused reflection of the sound from this boundary. The side walls of the stage are provided with spherical surfaces for diffusing the reflected sound. The acoustical treatment of the stage, save for the diffusospheres on the side walls and polycylinders at the rear of the stage, is rock wool, 2 inches in thickness, covered with perforated sheet metal or perforated sheet asbestos. The diffusospheres throughout the studio and polycylinders at the rear of the stage are made of plaster and backed by rock wool. The ceiling and the side walls in the rear two-thirds of the auditorium section are untreated. The walls and ceiling in the front of the auditorium section are treated with 2 inches of rock wool covered by perforated asbestos. Heavy upholstered chairs in the audience area provide substantially the same acoustical conditions with and without an audience present in the studio. The control room is located so that the occupants have an unobstructed view of the stage and studio seating section. The clients' room is located so that the sponsors may watch and listen to the progress of the program. Lighting booths are also provided in the rear for lighting the stage.

C. Scoring and Recording Studios[48-55]

Scoring and recording studios are used for recording the music in sound motion pictures and phonograph records. In recent years, considerable effort has been expended in the improvement of the acoustics of scoring and recording studios. To obtain good acoustics, particularly for large musical aggregations, the studio should be large. The studio should be

[48] Rettinger, M., *Proc. Inst. Radio Engrs.*, Vol. 28, No. 7, p. 296, 1940.

[49] Volkmann, J. E., *Jour. Acoust. Soc. Amer.*, Vol. 13, No. 3, p. 234, 1942.

[50] Rettinger, M., *Jour. Soc. Motion Picture Engrs.*, Vol. 39, No. 3, p. 186, 1942.

[51] Livadary and Rettinger, *Jour. Soc. Motion Picture Engrs.*, Vol. 42, No. 6, p. 361, 1944.

[52] Slyfield, C. O., *Jour. Soc. Motion Picture Engrs.*, Vol. 42, No. 6, p. 367, 1944.

[53] Ryder, L. L., *Jour. Soc. Motion Picture Engrs.*, Vol. 42, No. 6, p. 379, 1944.

[54] Rettinger, M., *Jour. Audio Eng. Soc.*, Vol. 9, No. 3, p. 178, 1961.

[55] Bolle, Voldner, Pulley, Stevens, and Volkmann, *Jour. Audio Eng. Soc.*, Vol. 11, No. 2, p. 80, 1963.

designed so that the reflected sound is thoroughly diffused. The studio should be well soundproofed. A scoring and recording stage satisfying these requirements is shown in Fig. 8.27. The maximum dimensions for the height, width, and length are respectively 30, 50, and 75 feet. A shell is provided for the orchestra at the live end of the studio. The voluminous part of the studio is made sound-absorbent to simulate an imaginary audience. The reflecting portion of the convex surfaces constituting the side walls of the stage is made of ¼-inch plywood. One-fourth of the convex surface is made absorbent, as shown in Fig. 8.27. The ceiling construction is similar to the wall surface, save that one-fourth of the convex surface is equipped with ventilating grills instead of absorbing material. Wood

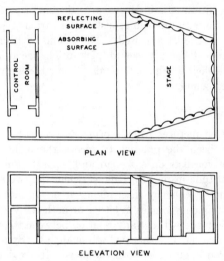

PLAN VIEW

ELEVATION VIEW

FIG. 8.27. Plan and elevation view of a scoring and recording studio. (*After Rettinger.*)

polycylindrical surfaces comprise the rear wall of the shell. The treatment on the side walls and rear wall consists of rock wool packed between 2- by 4-inch vertical studs. Wood strips, 1 by 2 inches, are applied to vertical studs graduated in spacing from 27 inches near the wainscoting to 12 inches near the ceiling. Fiberboard ½ inch thick and plywood ⅜ inch thick are applied to the studs between the stripping to produce a series of horizontal rock wool, fiberboard, and plywood panels. The construction of the ceiling is similar to the walls, save that the plywood panels are omitted and the fiberboard panels made narrower because of the reflective floor parallel to it. Since the wood strips are thicker than the plywood and fiberboard panels, a sheet of muslin is stretched over the walls and ceiling to form a monolithic surface broken only by a narrow decorative molding fastened to the furring strips.　The wall construction provides a uniform absorption and eliminates concentrations of the reflected sound. The live shell with convex surfaces provides an ideal environment for the orchestra as well as a means for eliminating sound concentrations and for directing the flow of sound toward the absorbing part of the studio.

In more recent times the practice is to design a recording studio in the form of a rectangular parallelepiped with uniform distribution of the sound absorbing material which leads to uniform acoustics throughout the studio. This is in contrast to the recording studio of Fig. 8.27 in which the end containing the orchestra shell is relatively live and the remainder of the studio relatively dead.　In order to provide uniform dispersion of the sound in the studio with uniform acoustics the various types of wall treatments shown in Fig. 8.24 are used.　The studio with uniform acoustics provides a more satisfactory sound pickup for recording stereophonic sound.　The live shell of Fig. 8.27 tends to blur the auditory perspective of the reproduced sound.

D. Vocal Studios[56–58]

In sound motion pictures when scoring an orchestra and one or more vocalists, it has been the practice to record the orchestra on one film channel and the vocalists on a second and separate film channel.　This permits great latitude in musical balance when the two sound tracks are dubbed together.　Frequency discrimination or accentuation of various portions of the frequency ranges in either or both the vocal and orchestra recording may be made without any relation between the two.　Compression may be carried out in either or both channels.　Synthetic reverberation may be added in either or both channels.　It is evident that the use of two separate channels permits a wide range of artistic effects which would be impossible if a single original record were made.

The vocal studio should be located adjacent to the orchestra studio.　A window between the two studios should be placed so that the vocalists or vocal group can see the conductor.　The vocalists hear the orchestra by means of telephone receivers which reproduce the orchestra.　In general, the number in the vocal studio will not exceed 30.

The acoustics of the vocal room should be similar to that of a small standard broadcast studio (see Sec. 8.3B).　One of the most important

[56] Mounce, Portman, and Rettinger, *Jour. Soc. Motion Picture Eng.*, Vol. 42, No. 6 p. 375, 1944.

[57] Ryder, L. L., *Jour. Soc. Motion Picture Eng.*, Vol. 42, No. 6, p. 379, 1944.

[58] Rettinger, *Applied Architectural Acoustics*, Chemical Publishing Co., New York, 1947.

considerations in the design of a vocal studio is the sound isolation between the vocal studio and the orchestra. The sound level of the orchestra in the vocal studio must be sufficiently low so that no sound from the orchestra will be recorded in the output of the microphones in the vocal studio. The type of wall and window construction for the vocal studio to obtain the desired value of sound isolation can be determined as outlined in Sec. 8.2L.

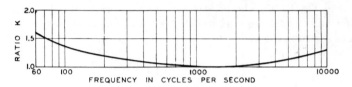

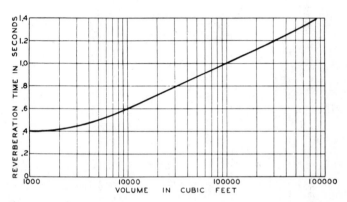

FIG. 8.28. Lower graph shows the reverberation time for a recording or broadcasting studio as a function of the volume for 1,000 cycles. Upper graph shows the relation between the reverberation time and the frequency; that is, the reverberation time at other frequencies is obtained by multiplying by K. (*After Morris and Nixon.*)

E. Reverberation Time of Broadcasting, Recording, and Scoring Studios[59-62]

The optimum reverberation time of broadcasting, recording, and scoring studios as a function of the volume of the studio, for a frequency of 1,000 cycles, is shown in the lower graph of Fig. 8.28. The reverberation for other frequencies can be obtained by multiplying by the factor K, obtained

[59] Morris and Nixon, *Jour. Acoust. Soc. Amer.*, Vol. 8, No. 2, p. 81, 1936.

[60] Knudsen and Harris, *Acoustical Designing in Architecture*, John Wiley and Sons, New York, 1950.

[61] Rettinger, *Applied Architectural Acoustics*, Chemical Publishing Co., New York, 1947.

[62] Rettinger, M., *Jour. Audio Eng. Soc.*, Vol. 9, No. 3, p. 178, 1961.

from the upper graph of Fig. 8.28. The reverberation time is greater at the lower and higher frequencies so that the aural rate of decay of pure tones will be approximately the same for all frequencies.

F. Sound-pickup Arrangements for Orchestras

The sound-pickup arrangement for recording a symphony orchestra is shown in Fig. 8.29. The orchestra is usually located on the stage of a concert hall or in a recording studio. A large number of microphones are used as shown in Fig. 8.29.

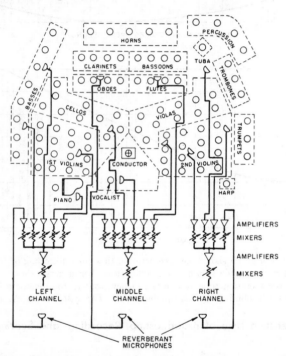

FIG. 8.29. Arrangement of the instruments and microphones for three-channel stereophonic recording of a symphony orchestra.

There are many different ways of employing multiple microphones for stereophonic sound reproduction. In some cases, a major microphone is used in each channel to pick up the left, center and right sections of the orchestra. Additional microphones are used to provide the required definition of particular instruments or sections of the orchestra. The use of multiple microphones makes it possible to obtain any desired balance of

he various sections of the orchestra. Accurate control of the auditory
perspective is also obtained by the use of multiple microphones with appropriate mixing. The use of multiple microphones with relatively short
pickup distances leads to a high ratio of direct to reflected sound and a
resultant high definition. Reverberation is, of course, necessary. The
reverberant sound is supplied by three microphones, one for each channel,
located at a suitable distance from the orchestra. The sound picked up
by the reverberation microphones contains very little direct sound. The
direct and reverberant sounds are mixed to provide the proper over-all
blending which the reverberant sound supplies. The microphones are

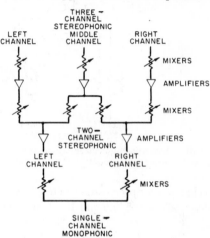

Fig. 8.30. Schematic arrangement
of the equipment for the conversion
of three-channel stereophonic sound
to two-channel stereophonic sound
or to one-channel monophonic
sound.

mixed into three channels for master stereophonic recording for tape and
disc. The three channels are converted to two channels for prerecorded
stereophonic magnetic tape or stereophonic disc records as described in
Sec. 9.9 and Fig. 9.36. The schematic arrangement of the equipment for
conversion to two-channel stereophonic sound for radio broadcasting of a
symphony orchestra in stereophonic sound is shown in Fig. 8.30. A further step, the conversion to monophonic sound for radio and television
broadcasting of a symphony orchestra in monophonic sound is also depicted
in Fig. 8.30.

The sound-pickup arrangement for the recording of a small concert
orchestra in three-channel stereophonic sound is shown in Fig. 8.31. The
general arrangement of the multiple microphones and the mixing procedures are similar to that of the symphony orchestra. Although reverberation microphones may be used, as in the case of the symphony orchestra,
the general procedure for smaller orchestras is to use synthetic reverberation (see Sec. 8.3K and Fig. 8.39). The three channels are converted to

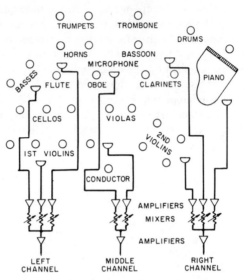

Fig. 8.31. Arrangement of the instruments and microphones for stereophonic recording of a small concert orchestra.

two channels for prerecorded stereophonic tape or stereophonic disc records as described in Sec. 9.9 and Fig. 9.36.   The schematic arrangement of the

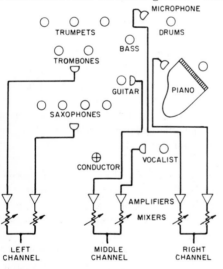

Fig. 8.32. Arrangement of the instruments and microphones for recording a dance band or popular music orchestra.

quipment for conversion to two-channel stereophonic sound for broadcast-
ng a concert orchestra in stereophonic sound is shown in Fig. 8.30.   A
urther step, the conversion to monophonic sound for radio and television
roadcasting of a concert orchestra in monophonic sound, is also depicted
n Fig. 8.30.

The sound pickup of a dance band or popular music orchestra for either
roadcasting or recording employs a large number of microphones as shown
n Fig. 8.32.   Each microphone covers one instrument or a group of instru-
nents with very intimate coupling between the microphone and the sound
ource.   In the case of popular music a high order of definition is desirable.
Because of the fast tempo of the music, the ratio of direct to reflected
ound must be kept large or the reproduced music will be blurred.   In
order to accomplish this objective, it is necessary to keep the distance very
mall between the microphone and instruments.   When dance bands are
oicked up for broadcasting from very noisy surroundings, a large number
of microphones must be placed close to the instruments.   This type of
pickup insures that a larger ratio of desired sound to noise will be obtained.
The conversion to two-channel stereophonic sound and one-channel mono-
ohonic sound can be carried out as described for the symphony and concert
orchestras.

G. Sound-pickup Arrangements for a Sound Broadcast

Most of the material for sound broadcasting today is in the form of
records and prerecorded material; nevertheless, there are still a large num-

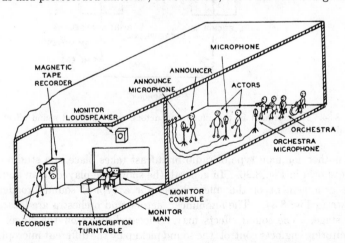

Fig. 8.33.   Perspective view of the studio showing the personnel and the arrangement
of the elements used in radio broadcasting.

ber of live radio broadcasts involving a small number of actors and musicians.

A broadcast studio with a setup for the sound pickup of a sound broadcast is shown in Fig. 8.33.   The monitor room is located next to the studio.
A large window between the monitor room and the studio makes it possible
for the monitor man to view the action in the studio.   Separate microphones
are used for the orchestra, actors and announcer.   A transcription turntable supplies the recorded program material.   The program may be also
recorded on a magnetic tape recorder for broadcast at a later time.

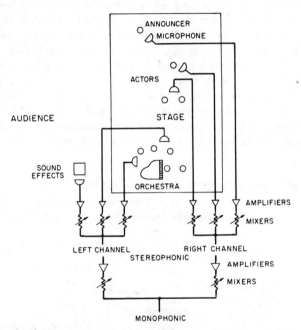

FIG. 8.34.   Arrangement of the announcer, actors, orchestra, and sound effects for a
sound-broadcast show.

Another common type of radio broadcast takes place in a studio of the
type shown in Fig. 8.26.   In general, the shows are played to an audience.
The arrangement of the microphones for a typical sound broadcast is
shown in Fig. 8.34.   The announcer, actors, and orchestra are located on
the stage.   The sound effects may be located on or off the stage.   The
monitoring engineer controls the sound pickup by the different microphones.

The system shown in Fig. 8.34 provides for either stereophonic or monophonic radio broadcasting.

1. Sound-pickup Arrangements for a Television Show

A common type of television show combines action on the stage with off-stage music. A plan view of the arrangement is shown in Fig. 8.35. The pickup usually takes place in a studio of the type shown in Fig. 8.26. Both picture and sound are picked up on the stage, whereas only sound is picked up from the orchestra. Titles and other effects are picked up on a special camera. The announcer uses a separate microphone. Velocity microphones are used for the orchestra and announcer pickup. A unidirectional microphone is used on the boom to pick up the sound on the stage. In general, shows of this kind are played to an audience.

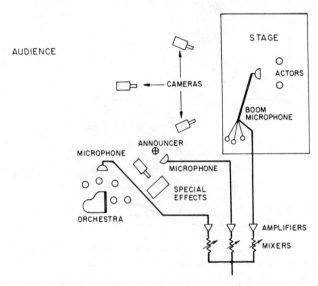

FIG. 8.35. Arrangement of the announcer, actors, special effects, orchestra, cameras, and microphones for a television show.

Some of the television shows played to an audience are recorded on video magnetic tape for a television broadcast at a later time. Other television shows are produced in a studio as described in Sec. 8.3I. These shows are also recorded on video magnetic tape. Still other television shows are recorded in motion picture studios by means of sound motion picture recording techniques.

I. Sound Stages for Motion Pictures and Television[63,64]

A sound stage is a large acoustically treated room used to house a stage

[63] Ringel, A. S., *Jour. Soc. Motion Picture Engrs.*, Vol. 15, No. 3, p. 352, 1930.

[64] Rettinger, *Applied Architectural Acoustics*, Chemical Publishing Co., New York, 1947.

setting in sound-motion-picture recording or television broadcasting. Th
sound stage is equipped with catwalks, power outlets, air conditioning, an
other facilities required for the production of sound motion pictures o
television. A sound stage with a setup for sound-motion-picture recordin
is shown in Fig. 8.36. In this case the initial monitoring is done at
console located on the sound stage. This permits the monitor man t
have an excellent view of the action. Most sound stages are equippe
with a recording room located next to the studio. This room may also b
used as a monitoring room. In the case of very large studios, portabl
monitoring and recording booths are located on the stage.

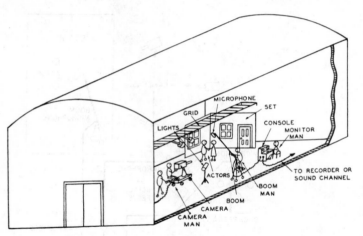

Fig. 8.36. Perspective view of a sound stage showing the personnel and the arrange
ment of the elements used in recording sound motion pictures.

A large number of motion pictures are now recorded in stereophoni
sound. In general, three channels are used for the auditory perspectiv
as shown in the reproducing system of Sec. 9.3E and Fig. 8.11. Th
recording of the action for stereophonic sound motion pictures is carrie
out with three channels similar to the procedures outlined in Secs. 8.3C
and H. For the stereophonic sound pickup of dramatic action, three space
microphones are usually mounted on a single boom and connected to thre
channels.

A sound stage with a setup for television broadcast pickup is shown i
Fig. 8.37. In this case the monitoring room is located next to the stage
A large window between the monitoring room and the stage makes it pos
sible for the audio and video monitor men to view all the action on th
stage. The actors, orchestra, announcer, and sound-effects man are a

located on the stage. The stage cameras pick up the action of the actors. The title camera picks up the titles. Separate microphones are used for the different sounds at the appropriate time. The video men monitor and select the appropriate camera.

The television show depicted in Fig. 8.37 may be broadcast as it is being played. The show may also be recorded on video magnetic tape for television broadcast at a later time.

The technique of the pickup of sound in motion pictures and television differs from that of radio broadcasting and phonograph recording in that the microphone must be kept out of the picture. This is done by suspend-

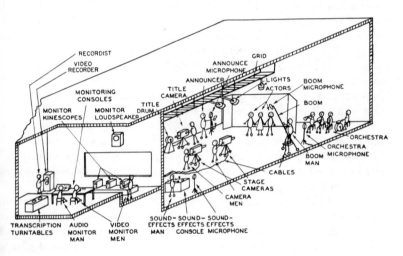

FIG. 8.37. Perspective view of a sound stage and monitor room showing the personnel and the arrangement of the elements used in television broadcasting.

ing the microphone from a boom so that it can be raised and lowered by the boom man (Figs. 8.36 and 8.37). The boom is also equipped with a suitable manually operated swivel arrangement so that the boom man can keep the directional microphone pointed at the action. Since the microphone must be kept out of the picture, the pickup distance is usually quite large. In the case of radio broadcasting and phonograph recording, the microphone can be placed in a position which yields the best sound pickup. For the broadcasting of speech, the distance from the speaker to the microphone can be made very small so that the received reverberation is negligible. However, for sound stages with large pickup distances, the received reverberation is kept low by making the reverberation time of the sound stage as low as possible and by the use of directional microphones.

When the reverberation time of the stage is low, the setting determines the acoustics of the sound picked up by the microphone. In the case of sets consisting of small rooms, the acoustics of the set mask the acoustics of the sound stage. In the early days of sound-motion-picture recording it was customary to make the sets of acoustical materials having good transmission at the low frequencies and high absorption at the high frequencies. In this way it was possible to keep the reverberation time of the set very low. With the advent of directional microphones it has been possible to use conventional materials for the construction of sets.

Typical over-all dimensions for a large studio are as follows: height, 45 feet; width, 100 feet; and length, 140 feet. A reverberation time of about one-half second is possible for stages with a volume of about 500,000 cubic feet. In the case of smaller stages a lower reverberation time may be obtained. It is usually standard practice to erect several sets on a single stage. This procedure may render some of the absorbing material of the stage ineffective and thereby increase the reverberation time. These undesirable effects may be overcome by the use of heavy sound-absorbing curtains which shield the different sets from each other.

The floors of the sound stage should be rigid and massive to prevent transmission of sound along the floor due to impacts, as, for example, in the case of large dancing groups. An improvement in the case of the floor can be effected by dividing the floor into sections and isolating each section mechanically.

The sound stage should be soundproof and isolated against vibrations coming through the ground and from adjacent rooms and buildings. For average conditions a relatively light double-wall construction may be used. Under these conditions, the outer wall consists of 1-inch fiberboard sheathing nailed to the vertical studs. On the outer face of this fiberboard a layer of building paper and stucco wire netting is applied. Stucco 1 inch thick is applied to the wire netting. The inner wall consists of vertical 2 by 4-inch studs spaced from the outer wall by at least 2 inches. A layer of ½-inch plasterboard is applied to the outside face of the studs. The space between the 2- by 4-inch studs is filled with rock-wool battens 4 inches thick. In the case of very noisy locations, massive double-wall construction will be required as, for example, a concrete outer wall. The only noise in which the roof and ceiling are involved is that of airplanes. With the ever-increasing number of airplanes, particular consideration must be given to the roof and ceiling. In the past, 4-inch rock-wool battens have been applied directly to the underside of the roof. This, in general, does not provide adequate shielding, and a ceiling of fiberboard and rockwool separated from the roof is required. In severe noise conditions, the addition of a concrete outer roof may be necessary.

Audio Delayers[65]

An audio delayer is a means which provides a time delay between the input and output audio signals of the delay system.

Audio delayers are used in sound-reinforcing systems to enhance the naturalness and improve the intelligibility of the reproduced sound in sound-reinforcing systems. See Sec. 8.2D and Figs. 8.9 and 8.10.

Audio delayers are used in synthetic reverberation systems to provide the time delay between the pencils of sound (see Sec. 8.3K).

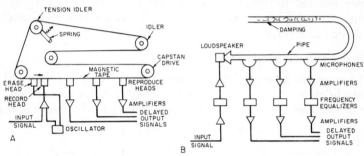

FIG. 8.38. Schematic arrangements for introducing time delay of an audio signal. A. Magnetic tape delayer. B. Electroacoustic delayer.

The most common delayer in use today is a magnetic tape recorder and reproducer with a magnetic record head and several magnetic reproduce heads spaced along the tape from the magnetic record head as shown in Fig. 8.38A. The delay is given by

$$t = \frac{d}{v} \tag{8.18}$$

where t = time delay, in seconds

d = spacing between the magnetic record head and the magnetic reproduce head, in inches

v = tape speed, in inches per second

The appropriate number of reproduce heads are included to provide the required audio delays as depicted in Fig. 9.38A. The customary design of delayers employs an endless loop of magnetic tape as shown in Fig. 9.38A.

Another type of audio delayer consists of a long pipe with a loudspeaker at the input end and microphones spaced along the pipe as shown in Fig. 8.38B. The delay is given by

$$t = \frac{d}{c} \tag{8.19}$$

[65] Olson, H. F., *Jour. Acoust. Soc. Amer.*, Vol. 31, No. 7, p. 872, 1959.

where t = time delay, in seconds

d = spacing between the loudspeaker and microphone, in feet

c = velocity of sound, in feet per second

There is progressive attenuation in the pipe with distance along the pipe and with increase of the frequency. Therefore, some amplitude equalization with respect to frequency is required in both the loudspeaker and microphone electrical lines in order to provide a uniform transfer-frequency response characteristic. The end of the pipe is damped with sound absorbing material to thereby reduce deleterious reflections to a negligible amount. The appropriate number of microphones are included to provide the desired time delays.

K. Synthetic Reverberation[66,67]

The reverberation time of studios may be changed and controlled within certain limits by varying the absorption. The amount of control that may be obtained by varying the amount of absorption by means of hard

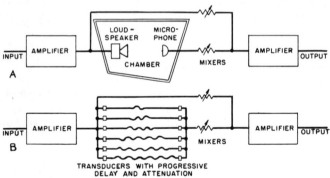

Fig. 8.39. Schematic arrangements for introducing synthetic reverberation in reproduced sound. *A*. Loudspeaker, microphone, and reverberant chamber combination. *B*. System consisting of transducers with progressive delay and attenuation.

panels which cover the absorbing material or other similar systems is limited. Furthermore, the reproducing conditions may also require additional reverberation. Where the reverberation time of reproduced sound is far below the optimum value, the reproduction may be enhanced by artificially adding reverberation.

Artificial reverberation may be added to a sound signal by means of the loudspeaker, reverberant chamber, and microphone combination shown in Fig. 8.39*A*. The reverberant chamber consists of an enclosure with highly reflecting, nonparallel walls, ceiling, and floor. The cubical content varies

[66] Olson and Bleazey, *Jour. Audio Eng. Soc.*, Vol. 8, No. 1, p. 37, 1960.

[67] Schroeder and Logan, *Jour. Audio Eng. Soc.*, Vol. 9, No. 3, p. 192, 1961.

rom 1,000 to 10,000 cubic feet. If a reduction in the reverberation time s desired, flats of absorbing material may be brought into the chamber. . modification of the single-chamber system is the addition of a second hamber coupled to the first by means of a door. The use of two rooms aakes it possible to obtain a wide variety of reverberant effects by varying oth the reverberation time of the chambers and the coupling between the hambers. The loudspeakers, microphones, and amplifiers used for these ystems should be of the highest quality. Mixers are provided so that any atio of the original sound to reverberant sound may be obtained.

Reverberation, in the chamber described above, consists of the multiple eflection of a large number of pencils of sound. Each pencil of sound uffers a decrease in intensity with each reflection (see Sec. 8.2). These onditions may be simulated by a system shown in Fig. 8.39B. The amlified sound signal is passed through a number of transducers with proressive delay and attenuation. These transducers may be a series of ipes with loudspeakers and microphones at the input and output loads r a series of springs with electromechanical transducers at the input and utput ends or a magnetic tape recorder with endless belt of tape and a ingle recording magnetic head and a series of spaced-reproducing magetic heads (see Sec. 8.3J). Feedback may be used to provide a larger umber of components in the reverberant sound as shown in Fig. 8.39B.

The velocity of transmission of longitudinal waves along a spring is elatively slow. This characteristic of a spring has been employed for roviding artificial reverberation. The sending transducer is located at ne end of the spring and the receiving transducer is located at the other nd. The resultant standing wave pattern on the spring simulates reerberation in one dimension.

The velocity of transmission of transverse waves in a stretched membrane r plate is relatively slow. This characteristic of a stretched membrane or late has been employed for providing artificial reverberation. The sendng and receiving transducers are arranged for the production and reception f transverse vibrations of the membrane or plate. The resultant standing vave pattern simulates reverberation in two dimensions. The system is quipped with a variable and adjustable damping means for the membrane r plate so that reverberation time can be varied.

.. Volume Limiters and Compressors[68-71]

A volume compressor is a system that reduces the amplification of an mplifier when the signal being amplified is large and increases the ampli-

[68] Sinnett, C. M., *Electronics*, Vol. 8, No. 11, p. 14, 1935.
[69] Norman, N. C., *Bell Labs. Record*, Vol. 13, No. 4, p. 98, 1934.
[70] Mathes and Wright, *Bell System Tech. Jour.*, Vol. 13, No. 3, p. 315, 1934.
[71] Steinberg, J. C., *Jour. Acoust. Soc. Amer.*, Vol. 13, No. 2, p. 107, 1941.

fication when the signal is small. Compressors are used to reduce the volume range in sound-motion-picture and phonograph recording, sound broadcasting, public-address and sound-reinforcing systems, etc.

Volume compressors and limiters are amplifiers in which the amplification varies as a function of the general level of the signal. The elements of a compressor or limiter are shown in Fig. 8.40. The input signal is amplified and rectified. The rectified signal is applied to a resistance capacitor network. The direct-current voltage across the capacitor is used to vary the bias and, as a consequence, the amplification of a push-pull amplifier employing tubes with variable transconductance.

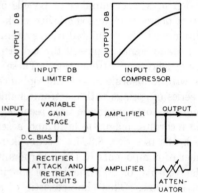

FIG. 8.40. The input versus output characteristic of a limiter and a compressor and a schematic diagram of the elements of a limiter or compressor.

Transistors and similar electronic solid-state devices have been developed for use in compressor and limiting systems. The performance is essentially the same as that for the vacuum tube systems. The constants of the electronic systems can be developed to provide either compression or limitation as well as different characteristics for these two functions.

In the limiter characteristic, shown in Fig. 8.40, the relation between the output and input is linear up to a certain level, beyond this point the output remains constant regardless of the input. The limiter type is useful for protection against a sudden overload, as, for example, in the input to a broadcast transmitter.

In the case of the compressor characteristic, shown in Fig. 8.40, there is a gradual reduction in the gain with increase of the input. A reduction in the volume range in radio and phonograph reproduction makes it possible to reproduce the wide range of orchestra music in the home without excessive top levels. It also improves the signal to noise ratio. It improves the intelligibility of speech and enhances music reproduction when the ambient noise is high, as, for example, in sound-motion-picture theater reproduction.

The attack time for a gain reduction of 10 decibels, in compressors and imiters, is of the order of a millisecond. The retreat to normal is of the order of 1 second.

8.4 NOISE[72-74]

Noise is any undesired sound. Noise occurs in many different ways. There is the noise generated in sound-reproducing systems. There is the ambient noise in studios, rooms, and theaters. The ease with which speech may be heard or understood depends upon the noise conditions as well as upon the acoustical characteristics of the room and the sound-reproducing systems. The full artistic effects of musical reproduction can be obtained only with a wide volume range. This volume range depends upon the noise level at the listening point. The tolerable level of the noises generated in any reproducing system depends upon the noise level at the reproducing point.

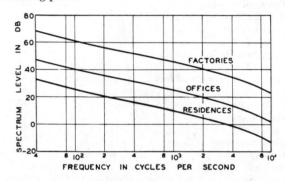

Fig. 8.41. Average noise spectrum for residences, offices, and factories. 0 db = 0.000204 dyne per square centimeter. (*After Hoth.*)

The noise level in the average residence is about 43 decibels. The average noise level in the average business office is 58 decibels. The average noise level in the average factory is 78 decibels. The average noise level in empty theaters is 25 decibels. With an audience, the noise level may rise to 48 decibels and may go as low as 32 decibels during a quiet dramatic passage.

The average noise spectrum may be obtained from the measurements of room noise. In general, the noise spectrum is the same for all types of rooms. The difference in rooms exists in the absolute noise level. The total noise level for the average living room in the home is 43 decibels.

[72] Olson, H. F., *Acoustical Engineering*, D. Van Nostrand Company, Inc., Princeton, 1957.

[73] Beranek, *Noise Reduction*, McGraw-Hill Book Co., New York, 1960.

[74] Hoth, D. F., *Jour. Acoust. Soc. Amer.*, Vol. 12, No. 4, p. 499, 1941.

The spectrum for the average room noise having a total level of 43 is de
picted by the lower curve in Fig. 8.41. The ordinates, depicting the
spectrum level, are given by

$$B = 10 \log \frac{I}{WI_0} \tag{8.20}$$

where B = spectrum level in a bandwidth of one cycle per second

     I = sound intensity in a bandwidth W cycles per second

    I_0 = zero reference level of 10^{-16} watts per square centimeter

The noise level in various locations is shown in Table 8.4.

TABLE 8.4 NOISE LEVELS FOR VARIOUS SOURCES AND OCCU-
PIED LOCATIONS

Source or Description of Noise	Noise Level, Decibels
Threshold of pain	130
Hammer blows on steel plate, 2 feet	114
Riveter, 35 feet	97
Factory	78
Busy street traffic	68
Large office	65
Ordinary conversation, 3 feet	65
Large store	63
Factory office	63
Medium store	62
Restaurant	60
Residential street	58
Medium office	58
Garage	55
Small store	52
Theater	43
Hotel	43
Apartment	43
House, large city	41
House, country	35
Average whisper, 4 feet	20
Quiet whisper, 5 feet	10
Rustle of leaves in gentle breeze	10
Threshold of hearing	0

Owing to the complexity of the human hearing mechanism and to the
various types of sounds and noises, it is impractical, at the present time, to
build a noise meter which will show the true loudness level. The dis
crepancies can be determined by actual use and suitable weighting factors
applied to the results. Objective measurements are almost indispensable
in any scientific investigation. The noise meter or sound-level meter
provides a system for measuring the absolute sound level of a sound.

A schematic diagram of a sound-level meter or noise meter is shown in

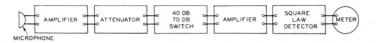

FIG. 8.42. Schematic arrangement of the components of a noise meter. (*After Olson, Elements of Acoustical Engineering, D. Van Nostrand Company, Inc., New York, 1957.*)

Fig. 8.42. The microphone should be calibrated in terms of a free wave. The attenuator and meter should be calibrated in decibels. A sound level of 60 decibels means a sound level of 60 decibels above a reference level. The reference point of the decibel scale incorporated in the noise meter shall be the reference intensity at 1,000 cycles in a free progressive wave, namely, 10^{-16} watt per square centimeter. The response-frequency characteristic of the human ear shows less sensitivity above and below 3,000 cycles (Fig. 7.10). The over-all response-frequency characteristic of an ideal noise meter should be the reciprocal of the ear response-frequency characteristics. This would make the noise meter unduly complicated. As a consequence, a compromise system has been established. The response-frequency characteristics recommended for noise meters by the American Standards Association[75,76] are shown in Fig. 8.43. Curve A is

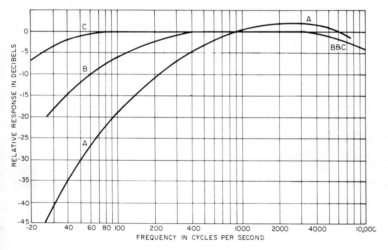

FIG. 8.43. Recommended characteristics for sound-level meters. (*American Standards Association.*)

[75] American Standard Specification for General-Purpose Sound Level Meters, S1.4-1961. American Standards Association, New York, 1961.

[76] Proposed American Standard Procedure for the Computation of Noise, S3.4, American Standards Association, New York, 1963.

recommended for measurements at the lower levels, and curve B for measurements around 70 decibels above the threshold. For very loud sounds (80 to 100 decibels), the flat characteristic of curve C should be used.

The noise meter may be used for noise analysis in offices, factories, restaurants, homes, etc. In these measurements, a large number of observations should be made in various positions. The noise meter may also be used to measure the transmission or attenuation by walls, floors, ceilings, and doors.

The highest noise level which can be tolerated within an enclosure over long periods of time without annoyance, undue fatigue, or discomfort due to the noise is termed the tolerable noise level. A list of tolerable noise levels for different enclosures are shown in Table 8.5. This table gives

TABLE 8.5. TOLERABLE AVERAGE NOISE LEVELS
UNOCCUPIED STUDIOS, THEATERS, AND ROOMS

Type of Enclosure	Noise Level, Decibels
Phonograph recording studio	20 to 25
Radio studio	20 to 30
Television studio	25 to 35
Legitimate theater	25 to 35
Auditoriums	25 to 40
Motion picture theaters	25 to 35
Homes	35 to 45
Hotels	40 to 50
Private and small offices	40 to 45
Large offices	45 to 50
Restaurants	45 to 55

the average noise level as measured on the standard noise meter described above. The values in this table assume that there are no periodic or discrete sounds of a certain frequency, which would of course stand out in the operation of recording, radio, and television studios, and still be low enough so that the average limits of Table 8.5 would be satisfied. It should be mentioned in passing that the lower limit in this table is the most desirable, and attempts should be made to achieve this noise level.

CHAPTER NINE

Sound-reproducing Systems

9.1 INTRODUCTION

Reproduction of sound[1] involves the picking up of sound at one location
and reproducing it at the same location or at another location either at the
same time or at some subsequent time.   In the reproduction of sound,
sound waves are converted into the corresponding waves in some other
medium.   This medium may be the undulations in a phonograph record,
the magnetic variations in a magnetic tape, the density or area variations
in a photographic film, the electrical waves in a wire, or the electromagnetic
waves in the ether.   The variations in these different media correspond to
the variations in the sound pressure in the original sound wave.   To complete the process of the reproduction of sound, the variations in these
various media are transformed back into sound waves.   These sound
waves correspond to the variations in the different media and hence correspond to the original sound waves.   This process is termed the reproduction of sound.   The reproduction of sound requires a number of different
transducers for accomplishing the various energy conversions required in
these systems.   There are two electroacoustic transducers which are almost
universally used in sound reproduction today, namely, the microphone for
converting sound waves into electrical variations and the loudspeaker for
converting electrical variations into sound waves.   In addition to these
two there are the various transducers for further conversion into the various
media.   It is the purpose of this chapter to consider microphones, loudspeakers, and complete sound-reproducing systems.

9.2 MICROPHONES

A microphone is an electroacoustic transducer actuated by energy in an
acoustical system and delivering energy to an electrical system, the waveform in the electrical system being substantially equivalent to that in the

[1] For a more complete discussion on microphones, loudspeakers, and sound-reproducing
systems and references, see Olson, H. F., *Acoustical Engineering*, D. Van Nostrand Company, Inc., Princeton, 1957.   Also Olson, H. F., *Advances in Sound Reproduction*,
Proceedings 5th Congress International D'Acoustique, Liege, 1965.

acoustical system.   A pressure microphone is a microphone in which the
electrical response corresponds to the variations in pressure in the actuating
sound wave.   A velocity microphone is a microphone in which the electrical
response corresponds to the particle velocity resulting from the propagation
of a sound wave through an acoustical medium.   All microphones in use
today may be classified as follows: pressure, velocity, or combination pressure and velocity.   For the conversion of the acoustical variations into the
corresponding electrical variations the following transducers may be used:
carbon, magnetic, dynamic, condenser, crystal, magnetostrictive, electronic,
and hot-wire.

It is beyond the scope of this book to consider in detail all types of microphones.   Therefore, the considerations of microphones will be confined to
the following, namely, the pressure-responsive carbon, crystal, dynamic,
magnetic, and condenser microphones; the velocity-responsive dynamic
ribbon microphone; and various unidirectional microphones.

A. Pressure Microphones

1. Carbon Microphone.   A carbon microphone is a microphone which
depends for its operation on the variation in resistance of carbon contacts.
The high sensitivity of this microphone is due to the relay action of the
carbon contacts.   The carbon microphone is almost universally employed
in telephonic communications, where the prime requisite is sensitivity
rather than low distortion and uniform response over a wide frequency
range.

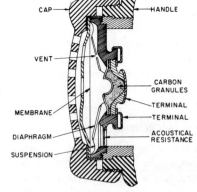

Fig. 9.1.   Sectional view of a carbon-
pressure microphone.

A sectional view of a carbon microphone is shown in Fig. 9.1.   The
carbon button consists of a cylindrical cavity filled with carbon granules.
The carbon granules make contact with the central portion of the dia-

phragm and base of the carbon granule cavity. A displacement of the diaphragm produces a change in the electrical resistance between the two terminals of the carbon chamber. The resistance decreases as the motion of the diaphragm compresses the carbon granules and increases as the motion of the diaphragm expands the volume occupied by the carbon granules. If the carbon microphone is connected in a series electrical circuit consisting of a fixed electrical resistance and battery, the current will be an inverse function of the resistance of the carbon microphone. The voltage developed across the fixed electrical resistor is proportional to the current. Under these conditions, since the motion of the diaphragm is proportional to the impinging sound pressure, the alternating voltage developed across the fixed electrical resistor will correspond to the pressure undulations in the sound wave.

2. Crystal Microphone. A crystal or piezoelectric microphone is a microphone which depends upon the generation of a voltage by the deformation of a crystal having piezoelectric properties. A piezoelectric crystal generates a voltage when it is deformed. Rochelle salt exhibits the greatest piezoelectric activity of all the known crystals. A barium titanate ceramic element may be used in place of the Rochelle salt crystal element. The action and the general performance is the same except that a microphone with a ceramic element exhibits slightly lower sensitivity. The barium titanate element will operate at higher temperatures than the Rochelle salt element without suffering permanent damage.

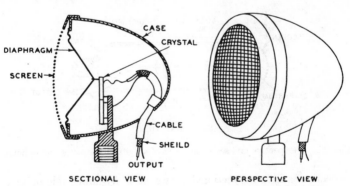

Fig. 9.2. Sectional and perspective views of a crystal-pressure microphone.

Perspective and cross-sectional views of a crystal microphone are shown in Fig. 9.2. The apex of the diaphragm is coupled to one corner of the crystal. The remaining three corners are attached to the case of the microphone. The motion of the diaphragm due to an impinging sound wave deforms the crystal which produces an alternating voltage which corresponds to the pressure undulations in the sound wave. The crystal microphone has a high electrical impedance and is therefore coupled directly to the grid of the first vacuum tube in the amplifier or a high-impedance field-effect transistor.

3. Dynamic Microphone. A dynamic microphone is a microphone in which the output results from the motion of a conductor in a magnetic field. When a conductor moves in a magnetic field, a voltage is generated which is proportional to the number of magnetic flux lines which are cut per second.

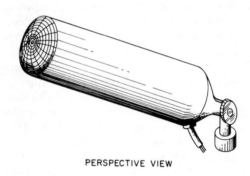

PERSPECTIVE VIEW

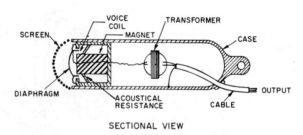

SECTIONAL VIEW

Fig. 9.3.   Perspective and sectional views of a dynamic-pressure microphone.

Perspective and cross-sectional views of a dynamic microphone are shown in Fig. 9.3. The diaphragm is coupled to a coil located in a magnetic field. The motion of the diaphragm due to an impinging sound wave is transmitted to the coil. The motion of the coil generates an alternating voltage which corresponds to the pressure variations in the sound wave. The electrical impedance of the dynamic microphone is relatively low, being of the order of 10 to 50 ohms. Therefore, a transformer is used to couple the output of the microphone to the grid of a vacuum tube. The low electrical impedance of the dynamic microphone makes it suitable for direct coupling to a transistor amplifier. Very small dynamic microphones of the type shown in Fig. 9.3 have been built for personal use and worn in the manner of a lavaliere.

4. Magnetic Microphone. A magnetic microphone is a microphone which depends for its operation on variations in the reluctance of a magnetic circuit. Perspective and sectional views of a magnetic microphone are shown in Fig. 9.4. The variation in reluctance in the magnetic circuit

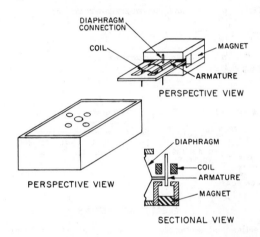

Fig. 9.4. Perspective and sectional views of a magnetic-pressure microphone.

produces a variation in magnetic flux and leads to the induction of a voltage in the coil surrounding the magnetic circuit. In the magnetic microphone shown in Fig. 9.4 the diaphragm is coupled to the armature. Motion of the armature produces variation in the magnetic flux in the armature and leads to the induction of a voltage which corresponds to the motion of the diaphragm. The electrical impedance can be of almost any value up to 5,000 ohms. The electrical impedance can be selected for coupling directly to a transistor. One of the main uses for the magnetic microphone is in hearing aids.

5. Condenser Microphone (Electrostatic Microphone). A condenser or electrostatic microphone is a microphone which depends for its operation on the variations in electrical capacitance. Perspective and sectional views of a condenser microphone are shown in Fig. 9.5. The elements of the condenser microphone consist of the diaphragm and fixed back plate. The impinging sound wave produces motion of the diaphragm and a corre-

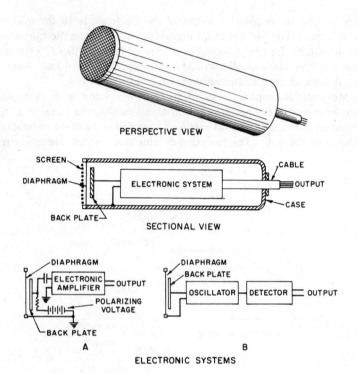

FIG. 9.5. Perspective and sectional views of a condenser or electrostatic-pressure microphone. *A.* Polarized condenser and electronic amplifier. *B.* Condenser as an element of an oscillator with a detector.

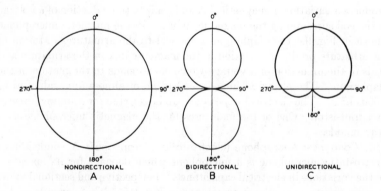

FIG. 9.6. Directional patterns of microphones.

sponding variation in capacitance. In Fig. 9.5A, a polarizing voltage is applied between the diaphragm and the back plate. Under these conditions the electrical output of the diaphragm corresponds to the motion of the diaphragm. The output is amplified by a vacuum tube or transistor amplifier. In Fig. 9.5B, the capacitance of the diaphragm and back plate is a part of the resonant circuit of vacuum tube or transistor oscillator. Motion of the diaphragm produces a variation in the frequency of the oscillator. The output of the oscillator is detected by means of a frequency-modulation detector. The resultant electrical output of the detector corresponds to the motion of the diaphragm.

The response of the carbon, crystal, magnetic, condenser and dynamic microphones described in the preceding sections corresponds to the pressure component in a sound wave. The directivity pattern of these microphones is a nondirectional characteristic, as shown in Fig. 9.6A. This means that the efficiency of pickup is the same in all directions. There is some deviation from this at the high frequencies, where the dimensions of the microphone become comparable to the wavelength of the impinging sound wave. That is to say, the diffraction of the sound wave by the case of the microphone makes the unit somewhat directional in the high-frequency portion of the audio-frequency range.

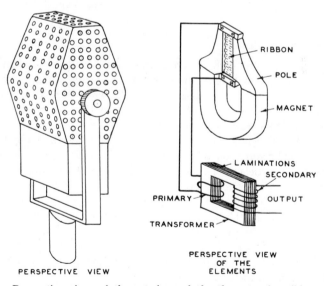

PERSPECTIVE VIEW

RIBBON

POLE

MAGNET

LAMINATIONS
SECONDARY

PRIMARY

OUTPUT

TRANSFORMER

PERSPECTIVE VIEW
OF THE
ELEMENTS

FIG. 9.7. Perspective views of the exterior and the elements of a ribbon-velocity microphone.

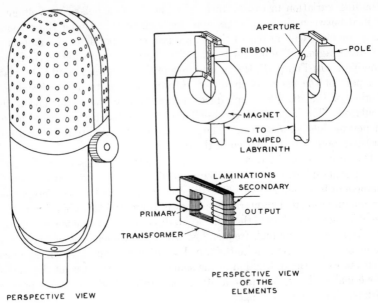

FIG. 9.8. Perspective views of the exterior and the elements of a ribbon-unidirectional microphone.

B. Velocity Microphone

The response of the carbon, crystal, magnetic, condenser and dynamic microphones corresponds to the pressure variations in a sound wave. The response of a velocity microphone corresponds to the particle velocity in a sound wave. The most common type of velocity microphone consists of a ribbon suspended in a magnetic field with both sides of the ribbon freely accessible to the surrounding air (Fig. 9.7). The ribbon is driven by the difference in sound pressure on the two sides. This difference in pressure corresponds to the particle velocity in the sound wave. Therefore, the motion of the ribbon is proportional to the particle velocity in the sound wave. The resultant motion of the ribbon in a magnetic field produces a voltage which corresponds to the particle velocity in the sound wave. The electrical impedance of the ribbon is very low, being about one-fourth ohm. Therefore, a transformer is located in the microphone case to step up the electrical impedance to a value suitable for transmission over a line to the vacuum-tube or transistor amplifier.

The directivity pattern of the velocity microphone is a bidirectional characteristic, as shown in Fig. 9.6B. The directional pattern of the velocity microphone is useful in discriminating against undesirable sounds, in balancing the instruments of an orchestra, and in monologue and dialogue pickup.

C. Unidirectional Microphones

A unidirectional microphone is a microphone with a substantially unidirectional pattern over the response-frequency range. A unidirectional microphone may be constructed by combining a bidirectional microphone and a nondirectional microphone or by combining a single-element microphone with an appropriate acoustical-delay system.

The elements of a single-unit unidirectional microphone are shown in Fig. 9.8. The back of the ribbon is coupled to an acoustical resistance in the form of a damped pipe which is folded to fit in the lower part of the case. The addition of the aperture in the portion of the pipe covering the back of the ribbon introduces an acoustical element in the form of an inertance which shifts the phase so that the response from the back is very low. The directional pattern under these conditions is a cardioid, as shown in Fig. 9.6*C*. This characteristic is useful in picking up sounds from the front and discriminating against sounds originating from the rear. The unidirectional microphone is almost universally used in sound-motion-

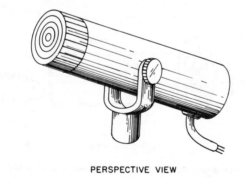

PERSPECTIVE VIEW

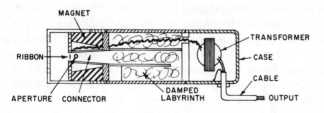

SECTIONAL VIEW

FIG. 9.9. Perspective and sectional views of an axial-ribbon unidirectional microphone.

picture recording, television, stage-sound pickup, and sound-reinforcing systems.

Other directional patterns may be obtained from the microphone shown in Fig. 9.8. For example, if the aperture is closed, a pressure-responsive microphone is obtained with a nondirectional characteristic, as shown in Fig. 9.6A. If the aperture is made very large, a velocity microphone is obtained with a bidirectional characteristic, as shown in Fig. 9.6B.

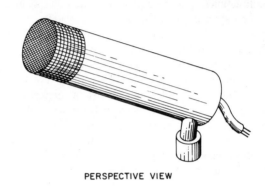

PERSPECTIVE VIEW

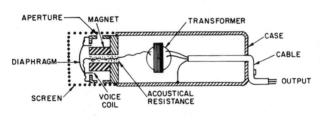

SECTIONAL VIEW

Fig. 9.10.   Perspective and sectional views of a dynamic-unidirectional microphone.

The perspective and sectional views of Fig. 9.9 depicts another version of the single-unit unidirectional microphone employing a ribbon coupled to an acoustical resistance in the form of a damped folded pipe. The directivity pattern is the cardioid of Fig. 9.6C. The maximum sensitivity of the unidirectional microphone of Fig. 9.9 is along the axis of the cylinder.

A unidirectional microphone employing a dynamic transducer is illustrated by the perspective and sectional views shown in Fig. 9.10. The back of the diaphragm is coupled to the inertance of the ports in the side of the case and to an acoustical resistance and acoustical capacitance of the air volume of the case. The phase shift introduced by the acoustical network consisting of the inertance, acoustical resistance and acoustical capacitance is the same as the distance from the ports to the front of the diaphragm. As a result, for sound arriving from the back, the forces on the front and back of the diaphragm are practically equal in phase and amplitude and the response is low. The phase shift is at a maximum for sound arriving in a forward direction with a resultant maximum sensitivity. The directivity pattern of the dynamic-unidirectional microphone of Fig. 9.10 is the cardioid of Fig. 9.6C.

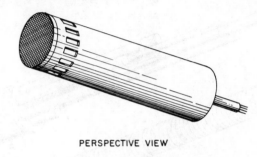

PERSPECTIVE VIEW

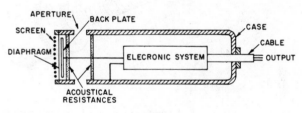

SECTIONAL VIEW

Fig. 9.11. Perspective and sectional views of a condenser or electrostatic-unidirectional microphone.

A unidirectional microphone employing a condenser or electrostatic transducer is depicted by the perspective and sectional views of Fig. 9.11. An acoustical resistance is located behind the diaphragm to provide resistance control of the vibrating system. The back of the diaphragm is coupled to the inertance of the ports in the side of the case and an acoustical resistance and acoustical capacitance of the air volume in the case. The acoustical network is similar to that of the dynamic-unidirectional microphone. The directional pattern is the cardioid of Fig. 9.6C. The maximum sensitivity is along the cylindrical axis of the microphone. The electronic systems which may be used with this microphone are the same as those for the condenser microphone of Fig. 9.5.

A line microphone is a microphone consisting of a large number of equally spaced pickup points arranged on a line. The arrangement may consist of a series of tubes graduated in length and connected to a diaphragm of a condenser microphone as shown in Fig. 9.12A or a pipe with openings along the length with one end closed and the other connected to a diaphragm of a dynamic microphone as shown in Fig. 9.12B. To obtain a high order of directivity the line must be at least a wavelength in length.

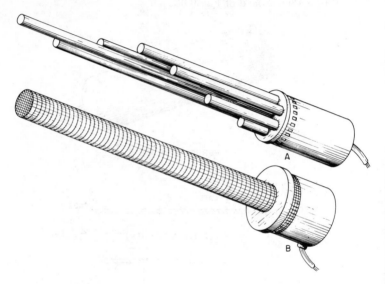

FIG. 9.12.   Line microphones.   A. A bundle of pipes graduated in length connected to a condenser-unidirectional microphone.   B. A pipe with openings along the length and connected to a dynamic-unidirectional microphone.

Therefore, the line becomes long and cumbersome in the low-frequency range. The microphones shown in Fig. 9.12 are cardioid microphones of Figs. 9.11 and 9.10 with a line connected to the diaphragm. The directivity pattern is a cardioid in the low-frequency range. In the mid- and high-frequency ranges the directivity is greater than that of a cardioid microphone.

9.3 LOUDSPEAKERS

A loudspeaker is an electroacoustic transducer actuated by energy in an electrical system and delivering energy to an acoustical system, the waveform in the acoustical system being substantially equivalent to that in the electrical system. There are two general classifications of loudspeakers in use today, namely, the direct-radiator type and the horn type. In the direct-radiator type a large diaphragm is coupled directly to the air. In the horn type a small diaphragm is coupled to the throat of a horn. Through the years, many types of systems have been used for driving the diaphragm. However, today the dynamic type of drive is almost universally used.

A. Direct-radiator Dynamic Loudspeakers

The almost universal use of the dynamic direct-radiator loudspeaker in radio receivers, phonographs, announce and intercommunicating systems is due to the simplicity of construction, small space requirements, and the relatively uniform response-frequency characteristic. The vibrating system of the dynamic direct-radiator loudspeaker consists of a conical paper diaphragm coupled to a voice coil located in a magnetic field (Fig. 9.13).

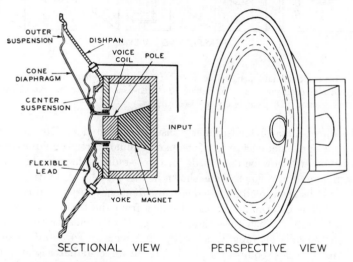

SECTIONAL VIEW   PERSPECTIVE VIEW

Fig. 9.13. Sectional and perspective views of a dynamic direct-radiator loudspeaker mechanism.

When a current is caused to flow in the voice coil by the application of a voltage, a force is produced by the interaction of the current and the magnetic field which corresponds to the applied voltage. This force is transmitted to the diaphragm and, as result, produces motion of the coil and diaphragm. The motion of the diaphragm produces sound waves. Since the motion of the diaphragm corresponds to the applied voltage, the resultant undulations in the sound wave produced by the motion of the diaphragm corresponds to the variations in the applied alternating voltage.

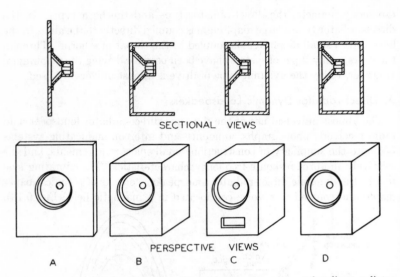

SECTIONAL VIEWS

PERSPECTIVE VIEWS

A   B   C   D

Fig. 9.14. Sectional and perspective views of mounting systems for direct-radiator loudspeaker mechanisms. *A.* Flat baffle. *B.* Open-back cabinet. *C.* Phase inverter or ported cabinet. *D.* Completely enclosed cabinet.

The direct-radiator loudspeaker is usually mounted in a flat baffle or some type of cabinet enclosure. The most common direct-radiator mounting systems will now be described. A direct-radiator loudspeaker mechanism mounted in a flat baffle is shown in Fig. 9.14*A*. A relatively large baffle must be used to obtain adequate low-frequency response because the baffle cutoff is determined by the distance from the front to the rear of the mechanism. When this distance is less than a quarter wavelength, the response falls off rapidly below this frequency. To obtain good response to 70 cycles requires a 4-foot baffle. A direct-radiator loudspeaker mechanism housed in a completely enclosed cabinet is shown in Fig. 9.14*D*. This system will exhibit uniform response to about one-half octave below the fundamental resonant frequency of the combination of the mechanism and cabinet. A direct-radiator loudspeaker mechanism mounted in a completely enclosed cabinet save for a port is shown in Fig. 9.14*C*. The addition of the port accentuates the low-frequency response but raises the cutoff frequency as compared with a completely enclosed cabinet. A direct-radiator loudspeaker mechanism housed in an open-back cabinet is shown in Fig. 9.14*B*. This system is used in practically all radio receivers and phonographs both of the table-model and console types. The fundamental resonance of the cabinet as a pipe coupled to the mechanism produces considerable accentuation of the low-frequency response and is termed cabinet resonance. This accentuation of the low-frequency response has received public acceptance because of the lack of adequate low-frequency components in most music, both original and reproduced.

A single direct-radiator dynamic loudspeaker mechanism is shown in the baffle and cabinets in Fig. 9.14. For wide-frequency range systems, multiple loudspeakers in various arrangements are employed as shown in Fig. 9.15. In Fig. 9.15A, a large direct-radiator dynamic loudspeaker is used to cover the low-frequency range and a small direct-radiator dynamic loudspeaker is used to cover the high-frequency range. In Fig. 9.15B, there are three direct-radiator loudspeakers covering the low-, mid- and high-frequency ranges respectively. In Fig. 9.15C, the direct-radiator

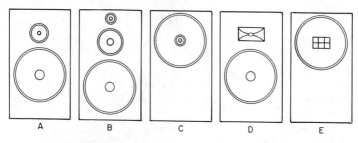

FIG. 9.15. Multiple loudspeaker arrangements. A. Low- and high-frequency direct-radiator loudspeaker mechanisms. B. Low-, mid- and high-frequency direct-radiator loudspeaker mechanisms. C. Low- and high-frequency direct-radiator loudspeaker mechanisms mounted coaxially. D. Direct-radiator low-frequency loudspeaker mechanism and horn high-frequency mechanism. E. Direct-radiator low-frequency mechanism and cellular-horn loudspeaker mechanism mounted coaxially.

high-frequency loudspeaker is coaxially located in the low-frequency direct-radiator loudspeaker. The axial location of the high frequency is such that the surface of the large cone is a continuation of the surface of the small cone. This reduces interference effects in the overlap region and thereby provides a more uniform response and directivity. In Fig. 9.15D, a horn loudspeaker is used to cover the high-frequency range and a direct-radiator loudspeaker is used to cover the low-frequency range. In Fig. 9.15E, a cellular horn loudspeaker, located on the axis of the low-frequency direct-radiator loudspeaker, is used to cover the high-frequency range. In the loudspeaker systems shown in Fig. 9.15, suitable electrical networks are used to divide and allocate the power input to the loudspeaker mechanisms appropriate to the frequency-response ranges of the loudspeaker mechanisms.

B. Horn Loudspeakers

A horn loudspeaker consists of an electrically or mechanically driven diaphragm coupled to a horn. The principal virtue of a horn resides in the possibility of presenting practically any value of acoustical resistance to the driving system. The horn provides a matching means between the large acoustical impedance of the relatively heavy diaphragm and the small acoustical impedance of the relatively light air. The horns used are of the flared type, because this shape provides a more uniform coupling between the diaphragm and the air. The efficiency of horn loudspeakers ranges from 25 to 50 per cent.

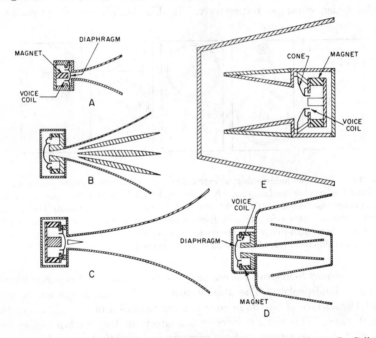

Fig. 9.16. Horn loudspeakers. *A*. High-frequency horn loudspeaker. *B*. Cellular high-frequency horn loudspeaker. *C*. Full-range horn loudspeaker. *D*. Full-range folded horn loudspeaker. *E*. Low-frequency folded horn loudspeaker.

Five typical horn loudspeakers are shown in Fig. 9.16. All of the horns are coupled to dynamic-loudspeaker mechanisms. The small horn loudspeaker shown in Fig. 9.16*A* is used to cover the high-frequency range. An example is the system of Fig. 9.15*D*. Fig. 9.16*B* shows a cellular-type horn. The use of a cellular horn maintains essentially uniform directivity over the response range. The horn loudspeaker of Fig. 9.16B is also used to cover the high-frequency range. In the horn loudspeaker of Fig. 9.16C, a large straight-axis horn is coupled to the dynamic-loudspeaker mechanism. This long and large horn covers the entire frequency range. The horn loudspeaker of Fig. 9.16D is similar to that of 9.16C except that the horn is folded to conserve space in the axial direction. A low-frequency horn loudspeaker is shown in Fig. 9.16E. A large folded horn is coupled to a dynamic-loudspeaker mechanism similar to that of the direct-radiator type.

A horn loudspeaker system consisting of a straight-axis horn loudspeaker for reproducing the high-frequency range and a large folded horn for reproducing the low-frequency range is shown in Fig. 9.17. The loudspeaker system of Fig. 9.17 is used for large-scale applications in sound motion pictures and sound-reinforcing systems.

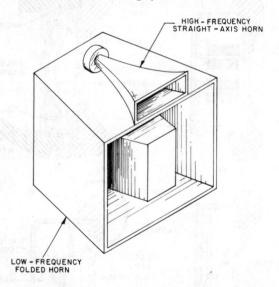

HIGH – FREQUENCY
STRAIGHT – AXIS HORN

LOW – FREQUENCY
FOLDED HORN

Fig. 9.17. A high-power horn loudspeaker consisting of a low-frequency folded horn loudspeaker and a high-frequency straight-axis horn loudspeaker.

The efficiency of a loudspeaker is the ratio of acoustical-power output to the electrical-power input. The efficiency of the direct-radiator loudspeaker is relatively low, being about 3 to 5 per cent. Therefore, the direct-radiator loudspeaker is not particularly suitable for installations requiring large amounts of sound power as, for example, in sound-motion-picture and large-scale high-power sound-reinforcing and public-address systems because the power amplifiers required to drive the loudspeaker become too large. Under these conditions, it is more economical to employ loudspeakers with higher efficiency. With a horn loudspeaker it is possible to obtain an efficiency of 25 to 50 per cent.

9.4 EARPHONES

An earphone is an electroacoustic transducer actuated by energy in an electrical system and supplying energy to an acoustical system, the waveform in the acoustical system being substantially the same as in the electrical system. The bipolar earphone consists of a steel diaphragm spaced a small distance from the pole pieces which are wound with insulated wire

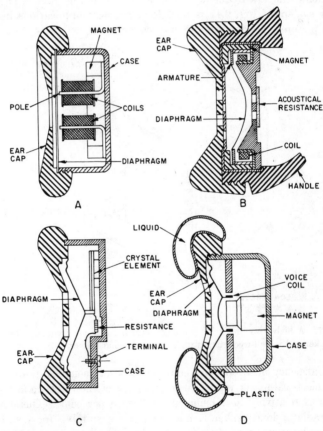

Fig. 9.18. Earphones. *A*. Bipolar earphone. *B*. Ring armature earphone. *C*. Crystal earphone. *D*. Dynamic earphone.

(Fig. 9.18A). A permanent magnet supplies the steady flux. In the bipolar earphone, the alternating force due to the alternating current in the electromagnet operates directly upon a steel diaphragm. The alternating force is superposed upon the steady force supplied by the permanent magnet. Under these conditions, the motion of the diaphragm and the resultant sound pressure delivered to the ear correspond to the electrical current in the coils.

The frequency response of the bipolar earphone is limited in frequency range and somewhat nonuniform in response with respect to frequency. Therefore, the bipolar earphone is not suitable for high quality monitoring. The use of the bipolar earphone is confined to limited frequency range communication and test applications.

Earphones of the type shown in Fig. 9.18A have been used in telephone systems. However, a more recent development employs an armature of a highly permeable alloy coupled to a diaphragm of a light aluminum alloy as shown in Fig. 9.18B. The diaphragm is damped by means of an acoustical resistance. Relatively smooth response is obtained over a frequency range of 200 to 4,000 cycles. The earphone shown in Fig. 9.18B is used in the modern telephone handset.

In the crystal or ceramic earphone shown in Fig. 9.18C, a small paper cone is coupled to a Rochelle salt or barium titanate element. One of the features of the crystal or ceramic earphone is excellent high-frequency response and light weight.

The vibrating system of the dynamic earphone shown in Fig. 9.18D is similar to that of a small dynamic loudspeaker. Response in the low-frequency part of the audio-frequency range can be maintained by employing a vibrating system with a low-resonant frequency and soft ear caps to eliminate leakage between the cap and the external ear in the low-frequency range. The use of a light-weight vibrating system insures that response will be maintained in the high-frequency part of the audio-frequency range. A properly designed dynamic earphone will reproduce the entire audio-frequency range with good fidelity.

9.5 AMPLIFIERS

An audio amplifier is an electronic transducer in which the output voltage current or power corresponds to and usually exceeds the input voltage current or power. External power is supplied for the transduction process. The most common amplifiers employ vacuum tubes and transistors. The purpose of this section is to describe the action of simple vacuum tube and transistor amplifiers.

The vacuum-tube amplifier revolutionized the reproduction of sound. The advent of the vacuum tube made radio broadcasting, sound-reinforcing systems, public-address systems, sound motion pictures, long-distance telephony, and other sound-reproducing systems possible. This was due to the fact that the vacuum tube made many electroacoustic transducers practical, whereas before the use had been precluded because of the low efficiency of transduction. The use as an amplifier is the most common application of the vacuum tube. Another application is the oscillator described in Sec. 5.6C.

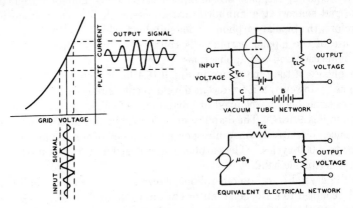

Fig. 9.19. Triode vacuum-tube network, the equivalent electrical network, and the grid-voltage versus plate-current characteristic. e_g = variable grid voltage, μ = amplification factor of the tube, r_{EG} = plate electrical resistance, r_{EL} = load electrical resistance, r_{EC} = grid-bias electrical resistance.

The simplest form of amplifying vacuum tube, termed the triode, consists of the following elements: the cathode, the grid, and the plate, as shown in Fig. 9.19. Electrons are emitted by the heated cathode. The grid, located between the cathode and the plate, controls the electron flow from the cathode to the plate. The number of electrons attracted to the plate depends upon the combined effects of the grid and plate polarities. When the tube is used as an amplifier, a steady negative voltage, termed the bias voltage, is applied to the grid. The input signal is applied to the grid. The variations in the resultant output across the load electrical resistance are essentially of the same form as the input signal voltage, but of an increased amplitude. Since the voltage variation obtained in the plate circuit is much larger than that required to swing the grid, amplification of the signal is obtained. A graphical illustration of the amplification process is depicted in Fig. 9.19. The grid-voltage plate-current characteristic gives the relation between the applied grid signal and the resultant plate-current variation. In general, the voltage developed across the plate electrical resistance load is greater than the voltage applied to the grid.

The plate current flowing through the load resistance r_{EL}, of Fig. 9.19, causes a voltage drop which varies directly with the plate current. The ratio of this voltage variation produced in the load resistance to the input-signal voltage is the voltage amplification, or gain, provided by the tube. The voltage amplification, $V.G.$, due to the tube is expressed by the following equation:

$$V.G. = \frac{\mu r_{EL}}{r_{EL} + r_{EG}} = \frac{g r_{EG} r_{EL}}{(r_{EL} + r_{EG})10^6} \tag{9.1}$$

where μ = amplification factor

r_{EG} = plate electrical resistance of the tube, in ohms

r_{EL} = load electrical resistance of the load, in ohms

g = electrical transconductance, in micromhos

From Eq. (9.1) it can be seen that the gain actually obtainable from the tube is less than the tube's amplification factor but that the gain approaches the amplification factor when the load resistance is large compared with the plate resistance of the tube.

The equivalent generator and electrical circuit of a vacuum-tube amplifier are shown in Fig. 9.19.

The transistor is a three-element solid-state device which may be employed for a wide variety of control functions, including amplification, oscillation and frequency conversion.

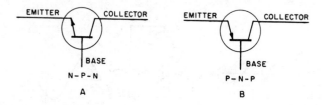

FIG. 9.20. The elements of a transistor. *A.* n-p-n type. *B.* p-n-p type.

The transistor is rapidly replacing the vacuum tube in audio amplifiers and oscillators. There are three elements in the common transistor as depicted in Fig. 9.20, namely, the emitter, the base and the collector. There are two varieties of transistors, namely, the n-p-n and the p-n-p as shown in Fig. 9.20.

In the n-p-n transistor shown in Fig. 9.20, electrons flow from the emitter to the collector. In the p-n-p transistor shown in Fig. 9.20, electrons flow from the collector to the emitter. In other words, the direction of dc electron current is always opposite to that of the arrow on the emitter head. The "conventional current flow" is in the direction of the arrow

The first two letters of the n-p-n and p-n-p designations indicate the respective polarities of the voltages applied to the emitter and the collector in normal operation. In an n-p-n transistor, the emitter is made negative with respect to both the collector and the base, and the collector is made positive with respect to both the emitter and the base. In a p-n-p transistor, the emitter is made positive with respect to both the collector and the base, and the collector is made negative with respect to both the emitter and the base.

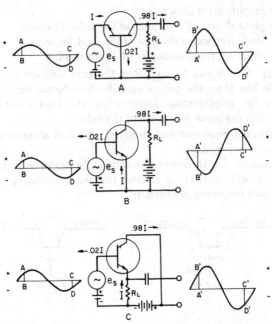

FIG. 9.21. Transistor circuit configurations with input and output waveforms A. Common base. B. Common emitter. C. Common collector.

There are three basic ways of connecting transistors in a circuit: common-base, common-emitter, and common-collector. In the common-base (or grounded-base) connection shown in Fig. 9.21A the signal is introduced into the emitter-base circuit and extracted from the collector-base circuit (Thus the base element of the transistor is common to both the input and output circuits.) Because the input or emitter-base circuit has a low impedance, (resistance plus reactance) in the order of 0.5 to 50 ohms, and the output or collector-base circuit has a high impedance in the order of 1,000 ohms to 1 megohm, the voltage gain is large. The power gain in this type of configuration may be in the order of 1,500.

The direction of the arrows in Fig. 9.21A indicates electron current flow. As stated previously, most of the current from the emitter flows to the collector; the remainder flows through the base. In practical transistors, from 95 to 99.5 per cent of the emitter current reaches the collector. The current gain of this configuration, therefore, is always less than unity, usually in the order of 0.95 to 0.995.

The waveforms in Fig. 9.21A represent the input voltage produced by the signal generator e_s and the output voltage developed across the load resistor R_L. When the input voltage is positive, as shown at AB, it opposes the forward bias produced by the base-emitter battery, and thus reduces current flow through the n-p-n transistor. The reduced electron-current flow through R_L then causes the top point of the resistor to become less negative (or more positive) with respect to the lower point, as shown at $A'B'$ on the output waveform. Conversely, when the input signal is negative, as at CD, the output signal is also negative, as at $C'D'$. Thus, the phase of the signal remains unchanged in this circuit, that is, there is no voltage phase reversal between the input and the output of a common-base transistor amplifier.

In the common-emitter (or grounded-emitter) connection shown in Fig. 9.21B, the signal is introduced into the base-emitter circuit and extracted from the collector-emitter circuit. This configuration has more moderate input and output impedances than the common-base circuit. The input (base-emitter) impedance is in the range of 20 to 5,000 ohms, and the output (collector-emitter) impedance is about 50 to 50,000 ohms. Power gains in the order of 10,000 (or approximately 40 decibels) can be realized with this circuit because it provides both current gain and voltage gain.

Current gain in the common-emitter configuration is measured between the base and the collector, rather than between the emitter and the collector as in the common-base circuit. Because a very small change in base current produces a relatively large change in collector current, the current gain is always greater than unity in a common-emitter circuit; a typical value is about 50.

The input-signal voltage undergoes a phase reversal of 180 degrees in a common-emitter amplifier, as shown by the waveforms in Fig. 9.21B. When the input voltage is positive, as shown at AB, it increases the forward bias across the base-emitter junction, and thus increases the total current flow through the transistor. The increased electron flow through R_L then causes the output voltage to become negative, as shown at $A'B'$. During the second half-cycle of the waveform, the process is reversed, i.e., when the input signal is negative, the output signal is positive (as shown at CD and $C'D'$).

The third type of connection, shown in Fig. 9.21C, is the common-collector (or grounded-collector) circuit. In this configuration, the signal is introduced into the base-collector circuit and extracted from the emitter-collector circuit. Because the input impedance of the transistor is high and the output impedance low in this connection, the voltage gain is less than unity and the power gain is usually lower than that obtained in either a common-base or a common-emitter circuit. The common-collector circuit is used primarily as an impedance-matching device. As in the case of the common-base circuit, there is no phase reversal of the signal between the input and the output.

The circuits shown in Figs. 9.21A, B and C are biased toward n-p-n transistors. When p-n-p transistors are used, the polarities of the batteries must be reversed. The voltage phase relationships, however, remain the same.

9.6 MONAURAL, BINAURAL, MONOPHONIC AND STEREOPHONIC SOUND-REPRODUCING SYSTEMS[2]

The reproduction of sound is the process of picking up sound at one point and reproducing it either at the same point or some other point or at the same time or some subsequent time. There are many different types of systems employed for the reproduction of sound. In this connection, sound reproducing systems in use today may be classified as follows: monaural, binaural, monophonic, and stereophonic. The purpose of this section is to define and describe the salient features of monaural, binaural, monophonic and stereophonic sound-reproducing systems.

A. Monaural Sound-reproducing System

A monaural sound-reproducing system is a closed circuit type of sound-reproducing system in which one or more microphones are connected to a single transducing channel which in turn is coupled to one or two earphones worn by the listener, as in Fig. 9.22. The most common example of a monaural sound-reproducing system is the telephone in which there is, in general, a single source of sound, one microphone, a transducer, and one earphone coupled to one ear of the listener. In most local applications, the carbon microphone is coupled directly to the earphone. In long-distance telephony vacuum-tube and transistor amplifiers may be used between the microphone and earphone. For other more limited applications, as for example, monitoring purposes, the transducer may be a radio transmitter and receiver, a television sound transmitter and receiver, a disc

[2] Olson, H. F., *Audio*, Vol. 42, No. 9, p. 28, 1958.

phonograph recorder and reproducer, a sound-motion-picture recorder and reproducer and/or a magnetic tape recorder and reproducer. In some applications, there may be more than one sound source. One or more microphones may be used. In some applications, two earphones may be used transmitting the same program to each of the ears of the listener.

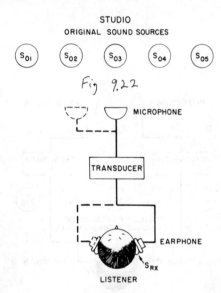

Fig 9.22

$$S_{RX} = S_{R1} + S_{R2} + S_{R3} + S_{R4} + S_{R5}$$
REPRODUCED SOUND SOURCES

FIG. 9.22. Monaural sound-reproducing system.

The monaural sound-reproducing system is of the closed-circuit type in which the ear of the listener is transferred to a microphone location by means of the microphone, transducer, and earphone combination. The acoustics of a single room are involved in the reproduction of the sound, namely, the studio or room in which the microphone is located.

B. Binaural Sound-reproducing System

A binaural sound-reproducing system is a closed-circuit type of sound-reproducing system in which two microphones, used to pick up the original sound, are each connected to two independent corresponding transducing channels which in turn are coupled to two independent corresponding earphones worn by the listener, as in Fig. 9.23. There is no widespread use of the binaural sound-reproducing system. The use is limited to

specific applications.   The binaural sound-reproducing system consists of
two separate channels.   Each channel consists of a microphone, transducer, and earphone.   The microphones are mounted in a dummy simulating the human head in shape and dimensions and at the locations corresponding to the ears of the human head.   The transducer may be an

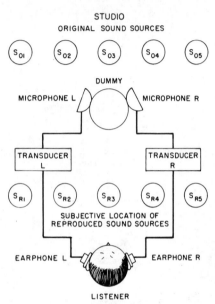

FIG. 9.23.   Binaural sound-reproducing system.

amplifier, a radio transmitter and receiver, a phonograph recorder and
reproducer, a motion picture recorder and reproducer, or a magnetic tape
recorder and reproducer.   The binaural sound-reproducing system is of
the closed-circuit type.   The listener is transferred to the location of the
dummy by means of a two-channel sound-reproducing system.

C. Monophonic Sound-reproducing System

A monophonic sound-reproducing system is a field-type sound-reproducing system in which one or more microphones, used to pick up the
original sound, are coupled to a single transducing channel which in turn
is coupled to one or more loudspeakers in reproduction (Fig. 9.24).   The
monophonic sound-reproducing system is the most widely employed
of all sound-reproducing systems.   Examples are the disc phonograph,
radio, sound motion picture, television, magnetic tape reproducer and

sound systems. The monophonic sound-reproducing system is of the field type, in which the sound is picked up by a microphone in a field and reproduced by means of a loudspeaker into a field. The sound at the

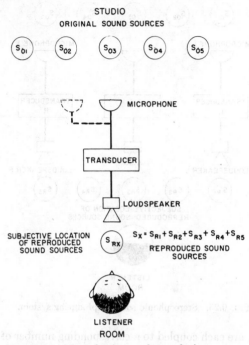

FIG. 9.24. Monophonic sound-reproducing system

microphone is reproduced at the loudspeaker. The transducer may be an amplifier, radio transmitter and receiver, a phonograph recorder and reproducer, a sound-motion-picture recorder and reproducer, a television transmitter and receiver, or a magnetic tape recorder and reproducer. The acoustics of two rooms are involved in monophonic sound reproduction, namely, the studio in which the sound is picked up and the room in which the sound is reproduced.

D. Stereophonic Sound-reproducing System[3,4,5]

A stereophonic sound-reproducing system is a field-type sound-reproducing system in which two or more microphones, used to pick up the

[3] Fletcher, H., *Elec. Eng.*, Vol. 53, No. 1, p. 9, 1934.
[4] Olson, H. F., *Jour. Audio Eng. Soc.*, Vol. 6, No. 2, p. 80, 1958.
[5] Olson and Belar, *Jour. Audio Eng. Soc.*, Vol. 8, No. 1, p. 7, 1960.

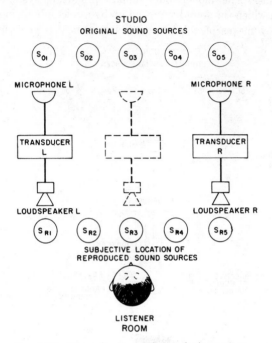

FIG. 9.25. Stereophonic sound-reproducing system.

original sound, are each coupled to a corresponding number of independent transducing channels which in turn are each coupled to a corresponding number of loudspeakers arranged in substantial geometrical correspondence to that of the microphones (Fig. 9.25).   The transducer may be an amplifier, radio transmitter and receiver, a phonograph recorder and reproducer, a sound-motion-picture recorder and reproducer, a television transmitter and receiver, or a magnetic tape recorder and reproducer.   Two channels are used in the disc phonograph and frequency-modulation radio. Two and three channels are used in the magnetic tape reproducer.   Two, three and more channels are used in motion picture reproducers.   The acoustics of two rooms are involved in stereophonic sound reproduction, namely, the studio in which the sound is picked up and the room in which the sound is reproduced.

9.7   AUDITORY PERSPECTIVE

Stereophonic sound reproduction as depicted in Fig. 9.25 provides auditory perspective of the reproduced sound.   Reproduction of sound in auditory perspective provides the illusion of the distribution of the reproduced sound in lateral directions as well as in depth in a geometrical configuration and correspondence which approximate the disposition of the original sound sources. This can be illustrated by means of a two-channel stereophonic reproducing system as depicted in Fig. 9.25. The five original sources of sound are S_{O1}, S_{O2}, S_{O3}, S_{O4} and S_{O5}. Under favorable conditions in reproduction, the corresponding reproduced sources of sound appear to originate at S_{R1}, S_{R2}, S_{R3}, S_{R4} and S_{R5}. The two-channel stereophonic sound-reproducing system does provide in actual practice very good reproduction of sound in auditory perspective.

The microphone placement with respect to the sound sources plays an important part in the auditory perspective aspects of the reproduced sound in the living room. The sound sources and the microphone place-

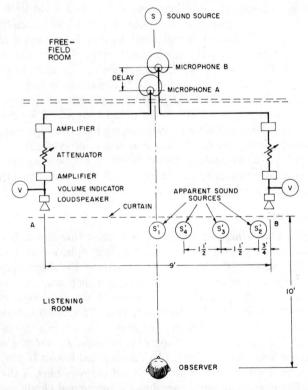

Fig. 9.26. Schematic arrangement of the apparatus which illustrates the effect of the relative delay and amplitude of two spaced sound sources in determining the apparent lateral location of the reproduced sound source.

ment in the studio have been studied by means of listening tests in a typical living room. In this work the most important result has been the observation that the subjective location of a sound source in stereophonic sound reproduction is determined by the phase and amplitude. This can be illustrated by the experiment depicted in Fig. 9.26. In this experiment the difference in phase is obtained by the separation between the microphones. The intensities in the two channels are adjusted by means of the attenuators. When the intensity and phase of the sound emanating from the two loudspeakers are the same, the sound appears to originate from a point midway between the two loudspeakers designated as S_1'. If the intensity of the sound emanating from loudspeaker B is higher than loudspeaker A the source of sound can be made to appear to originate at S_4', S_3' and S_2' as the difference in level of the sound emanating from the two loudspeakers is increased. If microphone B is placed closer to the source of sound S than microphone A, the phase of the sound emanating from the loudspeaker in channel B will lead the sound emanating from the loudspeaker in channel A and if the intensity of the sound in the two channels is adjusted to be the same, then the sound can be made to appear at S_4', S_3' and S_2' as the phase between the two channels is increased. The experiment of Fig. 9.26 can be extended by employing a difference in both the intensity and phase to shift the location of the sound source as perceived by the listener. This experiment shows that the apparent position of the sound source can be shifted in a lateral direction from a point directly in front of either loudspeaker to any point between by varying the relative phases and/or relative intensities of the sound emanating from the loudspeakers of channels A and B.

9.8   TELEPHONE

The telephone is a sound-reproducing system consisting of a carbon microphone (sometimes termed the transmitter), a telephone receiver, and a battery (Fig. 9.27). When sound impinges upon the diaphragm of the carbon microphone, as shown in Fig. 9.27, the diaphragm is moved by the alternating sound pressure. The motion of the diaphragm produces a change in the resistance of the carbon button. The resistance decreases as the motion of the diaphragm compresses the carbon granules and increases as the motion of the diaphragm expands the volume occupied by the carbon granules. Referring to the electrical circuit it will be seen that the current will be an inverse function of the resistance of the carbon button. The change in current produces a corresponding change in the force applied to the diaphragm of the earphone (sometimes termed the receiver) by the electromagnets. This force causes the diaphragm of the earphone to move and thereby produces compressions and rarefactions in

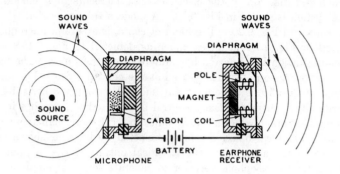

FIG. 9.27. Schematic arrangement of a simple telephone system.

the air which are, of course, the elements of a sound wave. Since the motion of the diaphragm of the microphone, the change in resistance of the carbon button, the current in the electrical circuit, the force applied to the earphone diaphragm, and the motion of the earphone diaphragm all correspond to the undulations in pressure in the impinging sound wave, the sound wave produced by the motion of the earphone diaphragm corresponds to the original sound wave.

The system shown in Fig. 9.27 depicts the action of a telephone system. Much more equipment is needed for a complete telephone system, so that the different telephone subscribers can be connected to each other at will.

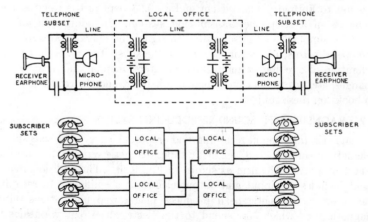

FIG. 9.28. Schematic arrangement of the apparatus in a telephone system. The upper diagram depicts a simple local office. The lower diagram depicts four offices with interconnecting lines.

A schematic diagram of the electroacoustic elements of a modern telephone system is shown in Fig. 9.28. A modern subscriber dial telephone set is shown in Fig. 9.29. The handset contains a carbon microphone as depicted in Fig. 9.1 and an earphone as depicted in Fig. 9.18B. Each telephone station is connected by a line to a central office. The battery supply or other equipment is located at the central office. The function of the central office is to connect any subscriber to any other subscriber. In large cities, there are many central or local offices, because it is not economical or practical for a central office to serve more than about 10,000 stations. The local offices are interconnected by lines, as shown in Fig. 9.28. In local transmission, the electrical output of the microphone is sufficient for the telephone receiver to generate sound of ample loudness

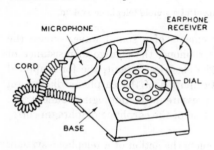

FIG. 9.29. A perspective view of a subscriber dial telephone set.

for intelligent transmission of speech. In long-distance telephony, electronic repeaters are used at regular intervals to restore the level of transmission to normal. The system of Fig. 9.28 depicts the electroacoustic elements of a telephone transmission system. In addition, equipment must be supplied for the subscriber to signal the operator and to permit the operator to send ringing currents to the subscriber. The further consideration of circuits, switchboards, repeaters, manual and automatic exchanges, etc., are outside the scope of this book, and the reader is referred to books on these subjects.

9.9 MAGNETIC TAPE SOUND-REPRODUCING SYSTEM

A magnetic tape sound-recording and -reproducing system consists of a magnetic recording head, magnetic tape, magnetic reproducing head and tape transport mechanism as shown in Fig. 9.30. The additional equipment includes the input and output electronic amplifiers. In recording, the recording magnetic head produces magnetized undulations on the magnetic tape which correspond to the electrical current variations impressed upon the coil in the magnetic recording head. In reproducing, as the tape moves past the reproducing magnetic head, the magnetized elements of the tape produce corresponding magnetic flux variations in the

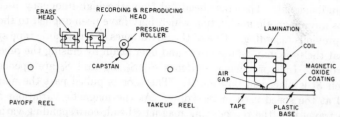

FIG. 9.30. The elements of a magnetic tape sound-recording and -reproducing system showing the reels, magnetic tape recording, reproducing and erase magnetic heads and the capstan drive. The details of the magnetic head and magnetic tape are shown on the right.

magnetic head which lead to an induction of the corresponding electrical voltage in the coil of the reproducing magnetic head.

A complete monophonic-magnetic tape sound-recording and -reproducing system is shown in Fig. 9.31. In recording, switches S_1, S_2, S_4, and S_5 are closed and switch S_3 is placed in position B. The output of the microphone is amplified by an amplifier equipped with suitable equalization to compensate for the response-frequency characteristics of the magnetic tape and the recording and reproducing magnetic heads. A volume control in the form of an attenuator is used to control the recording level. The power amplifier drives the magnetic recording head and loudspeaker. A volume indicator is used in recording so that the appropriate magnetic amplitude will be applied to the magnetic tape. Details of the magnetic-recording system are shown in Fig. 9.30. The recording magnetic head located between the two reels magnetizes the tape in a series of magnetizations which correspond to the variations in the original sound wave. To overcome magnetic nonlinearity in the tape, a high-frequency electrical

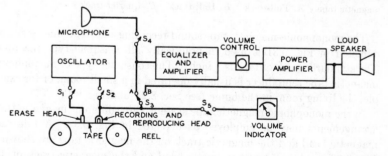

FIG. 9.31. The elements of a complete monophonic-magnetic tape sound-recording and -reproducing system.

signal of 30,000 cycles or higher is fed to the recording head together with
the audio signal.   The erase head fed from the high-frequency oscillator
erases any magnetic modulation which may have been applied to the tape
in a previous recording.   One of the advantages of a magnetic tape system
is that the tape may be used over and over again by erasing the preceding
recording.   In reproducing, switches S_1, S_2, S_4, and S_5 are opened and
switch S_3 is placed in position A.   The tape is pulled past the magnetic
head at the speed used in recording.   As the magnetic variations in the
tape travel past the reproducing magnetic head, corresponding variations
in flux are induced in the head.   These magnetic variations in the magnetic head induce voltages in the coil of the magnetic head which correspond
to the magnetic flux variations.   The output of the magnetic reproducing
head is fed to the equalized amplifier.   The output of the amplifier is fed
to a volume control which controls the intensity of the reproduced sound
in the room.   The volume control is followed by the power amplifier which
drives the loudspeaker.

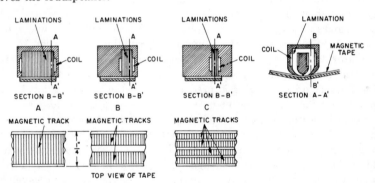

Fig. 9.32.   Sectional views of monophonic-magnetic heads and magnetic tracks on the
magnetic tape. A. Full-track. B. Half-track. C. Quarter-track.

The monophonic-magnetic tape sound-recording and -reproducing system
depicted in Fig. 9.31 is used in practically all consumer-type recorders.
The standard width of tape is $\frac{1}{4}$ inch.   The consumer type of monophonic-
magnetic tape reproducer is usually operated in a small room, or for example, the living room in the home (see Sec. 8.2F and Fig. 8.13).

In the monophonic-magnetic tape system, three different magnetic-track
arrangements may be employed as depicted in Fig. 9.32.   A full-track
magnetic head and the magnetic track on the magnetic tape are shown in
Fig. 9.32A.   A half-track magnetic head and the magnetic track on the
magnetic tape are shown in Fig. 9.32B.   A quarter-track magnetic head
and the magnetic track on the magnetic tape are shown in Fig. 9.32C.

The tape speeds used in audio-magnetic tape recorders and reproducers are 30, 15, $7\frac{1}{2}$, $3\frac{3}{4}$, $1\frac{7}{8}$, and $\frac{15}{16}$ inches per second.

The magnetic tape is stored on reels of various diameters, as follows, 3, 5, 7 and 10 inches. Magnetic tape is also stored on cartridges of three different designs, namely, two-reel, single-reel and continuous-loop cartridges.

A two-channel stereophonic-magnetic tape recording system is depicted in Fig. 9.33. In addition to the essential elements shown in Fig. 9.33, there may be additional microphones, mixers, compressors, limiters and other electronic compensating elements in each channel. The details of the magnetic head and the magnetic tracks on the tape are shown in Fig. 9.34.

The tape speeds, reels and cartridges are the same as those for the monophonic system described above.

A two-channel stereophonic-magnetic tape reproducing system is shown in Fig. 9.35. In addition to the essential elements shown in Fig. 9.35, there may be various

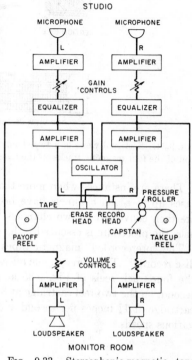

FIG. 9.33. Stereophonic-magnetic tape recording system.

electronic compensating elements. This type of system is usually operated in a room in the home as described in Sec. 8.2F and depicted in Fig. 8.15.

In the case of the master recording, the most common procedure is to record with a three-channel stereophonic system on $\frac{1}{2}$-inch magnetic tape. The system is similar to Fig. 9.33 except that there are three channels instead of two channels. There may be more than one microphone in each channel with appropriate mixers. The system usually includes compressors and other electronic-compensating elements. The three channels of stereophonic sound are converted to two channels of stereophonic sound as shown in Fig. 9.36. The resultant two-channel master tape is used for the master in the recording of "prerecorded" magnetic tape and the recording of disc records.

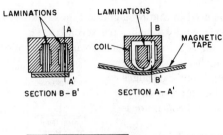

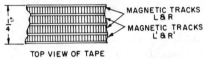

Fig. 9.34. Sectional views of a two-channel stereophonic magnetic head and the location of the four magnetic tracks on the magnetic tape.

There are instances when more than three channels are used. In some cases up to seven channels have been used in master recording. These are then converted to two channels. Regardless of the number of channels, the procedure is essentially the same as for three channels.

The "prerecorded" magnetic tapes contain programs similar to those on disc records. The stereophonic releases are made in various forms, namely, the four-track reels of Fig. 9.35 at a tape speed of $3\frac{3}{4}$ or $7\frac{1}{2}$ inches per second, on the two-reel cartridge at $3\frac{3}{4}$ inches per second, on the single-reel cartridge at $1\frac{7}{8}$ inches per second and on the continuous-reel cartridges at various speeds.

Fig. 9.35. Stereophonic-magnetic tape reproducing system.

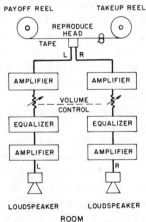

The magnetic tape system is also used for recording the master sound record in sound motion pictures.
The recording is usually carried out on 35-millimeter magnetic tape (see Fig. 9.60*A*). The tape is perforated with sprocket holes along the edge for synchronizing the magnetic tape with the picture film in the recording process. The number of magnetic tracks recorded on the 35-millimeter tape may be from two to seven. In general, four magnetic tracks are used.

9.10 DISC PHONOGRAPH

The phonograph was the first sound-reproducing system which made it possible for all the people of the world to hear statesmen, artists, operas, symphonies, bands, and all manner of musical aggregations where only a relatively few had been able to hear these first hand. During the first two decades after the commercialization of the phonograph, the recording and reproducing were made by mechanical means. After the advent of the vacuum tube, recording and reproducing have been made by a combination of electrical and mechanical means.

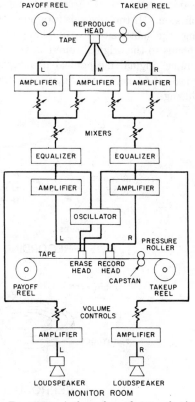

Fig. 9.36. A three-channel stereophonic-magnetic tape reproducer and a two-channel stereophonic-magnetic tape recorder with auxiliary equipment for converting from three-channel stereophonic sound to two-channel stereophonic sound.

The mechanical-disc phonograph sound-reproducing system consists of the cutter, the disc record, and the pickup reproducer (Fig. 9.37). The sound wave generated by the sound source travels down the horn and causes the diaphragm located at the throat of the horn to vibrate with a motion corresponding to the variation in the impinging sound wave. This motion in turn is transmitted to the cutting stylus which cuts a wavy groove in the record. Since the motion of the diaphragm and the cutting stylus correspond to the undulations in the pressure of the impinging sound wave, the shape of the wavy groove also corresponds to these variations in sound pressure. In reproduction, when the record is rotated, the wavy groove drives the reproducing stylus, which in turn drives the diaphragm. If the record is rotated at the same speed as that used in recording, the motion of the diaphragm produces a sound wave which corresponds to the original sound wave. The system shown in Fig. 9.37 is the simple original mechanical phonograph. Today, a combination of electrical and mechanical means is used for both recording and reproducing.

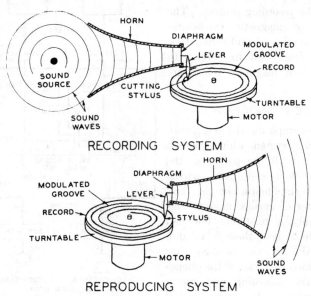

Fig. 9.37. Schematic views of a mechanical-disc phonograph sound-recording and reproducing system.

A complete monophonic-disc phonograph recording system is shown in Fig. 9.38. The sound wave is converted into the corresponding electrical wave by the microphones. The microphones and the source of sound to be recorded are located in the studio (Sec. 8.3C and Fig. 8.27). The output of the microphones is amplified by vacuum-tube amplifiers and applied to the cutter. The mixers make it possible to adjust the outputs from the various microphones and thereby obtain a suitable balance in the recorded sound when several microphones are used to pick up several sound sources. The monitoring system is located in the monitoring room. Compressors, equalizers, and filters are employed to enhance the recorded sound and to compensate for limitations in the system. It will be noted that the output is also fed to a monitor system which enables the recording engineer to determine the quality of the recording as applied to the cutter.

The cutter used for recording monophonic-lateral phonograph records is shown in Fig. 9.39. The driving system is of the dynamic type with two wire coils located in a magnetic field. One coil, used as the driving coil, is connected to the output of the power amplifier. The other coil, used as a sensing coil, is fed in an out of phase relationship to the input of the amplifier as shown in Fig. 9.38. The use of a feedback system provides uniform response and low nonlinear distortion. A current in the driving coil causes motion of the cutting stylus in a lateral direction which corresponds to the electrical input. The motion is transferred to the stylus which cuts a wavy groove in the lacquer disc. The undulations in the wavy groove correspond to the variations in the original sound wave.

The general practice is to record the master on magnetic tape instead of on the lacquer original. The lacquer original is cut from the magnetic tape master as shown in Fig. 9.40. The process of cutting the lacquer disc is the same as that described for Figs. 9.38 and 9.39.

The steps for the mass production of discs are shown in Fig. 9.41. In the record plant the lacquer disc recording termed the "original" of Fig.

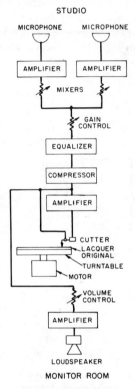

FIG. 9.38. Monophonic-disc recording system.

9.41A is metalized and then electroplated. The plating is separated from the lacquer and reinforced by backing with a solid metal plate. The assembly is termed the "master" (Fig. 9.41B). The master is electro-

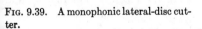

FIG. 9.39. A monophonic lateral-disc cutter.

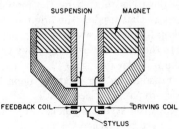

plated. This plating is separated from the master and reinforced by backing with a solid metal plate. The assembly is termed the "mother" (Fig. 9.41C). Several mothers may be made from the master. The mother is electroplated. This plating is separated from the mother and reinforced by a solid metal plate. The assembly is termed the "stamper" (Fig. 9.41D). Several stampers may be made from each mother. One stamper containing a sound selection to be placed on one side of the final record is mounted in the upper jaw, and another stamper containing a sound selection to be placed on the other side of the record is placed in the lower jaw of a hydraulic press equipped with means for heating and cooling the stampers (Fig. 9.41E). A preform or biscuit of thermoplastic material such as a shellac compound or vinylite is placed between the two stampers. The stampers are heated, and the jaws of the press are closed which presses the two stampers against the thermoplastic material. When an impression of the stampers has been obtained in the thermoplastic material, the stampers are cooled which cools and sets the plastic record. The jaws of the hydraulic press are opened, and the record is removed from the press. The modulated grooves in the record correspond to those in the original lacquer disc (Fig. 9.41F). The stamping procedure is repeated again and again until sufficient records are obtained. The above process constitutes the "mass-production system" for the production of phonograph records.

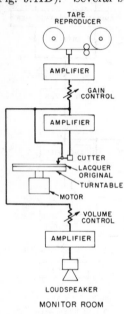

FIG. 9.40. A monophonic-magnetic tape reproducer and a monophonic disc recorder and auxiliary equipment for the production of disc masters from the master magnetic tape.

The monophonic system for reproducing the disc phonograph record is shown in Fig. 9.42. The record is rotated by the reproducing turntable at the same angular speed as that used in the original recording. The stylus or needle of the pickup follows the wavy spiral groove and generates a voltage corresponding to the undulations in the wavy groove of the record. The phonograph pickup shown in Fig. 9.42 is of the crystal type. Many other methods of transduction are used for phonograph pickups as, for example, magnetic, dynamic, electrostatic, ceramic, and electronic. In the crystal pickup, the stylus is connected to a Rochelle salt crystal by means of a lever system. When the crystal is twisted by the motion of the stylus, a voltage is developed which corresponds to the deformation.

A barium titanate ceramic element may be used instead of the Rochelle salt element. The action and general performance is the same except that a phonograph pickup with a ceramic element exhibits slightly lower sensi-

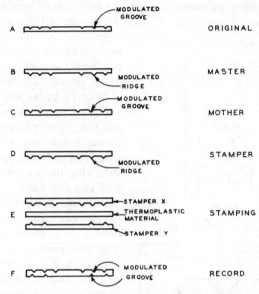

FIG. 9.41. The steps in the process for the mass production of disc phonograph records.

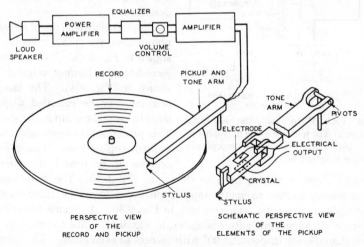

FIG. 9.42. A monophonic-disc reproducing system. Perspective view, schematic diagram of the elements and a schematic perspective view of the elements of a monophonic phonograph pickup.

tivity. The barium titanate element will operate at higher temperatures than the Rochelle salt element without suffering permanent damage.

The output of the pickup is amplified by means of an electronic amplifier. Suitable electrical equalizers are used to compensate for the response of the pickup and the recording characteristics. A volume control in the form of an attenuator is employed to control the intensity of the reproduced sound. The volume control is followed by an electronic power amplifier which drives the loudspeaker. The phonograph reproducing instrument is usually operated in a small room as, for example, the living room in the home (Sec. 8.2F and Fig. 8.13).

Dynamic and magnetic pickups are also employed for the reproduction of monophonic disc records. The construction of stereophonic dynamic and magnetic pickups will be described later in the text. Therefore, monophonic dynamic and magnetic pickups will not be described.

The essential elements of a stereophonic-disc recording system[6] are shown in Fig. 9.43. Two means of recording the lacquer original are shown in Fig. 9.43. The lacquer original may be recorded directly from the sound pickup in the studio. This procedure is employed in very limited applications. The general practice is to record the lacquer

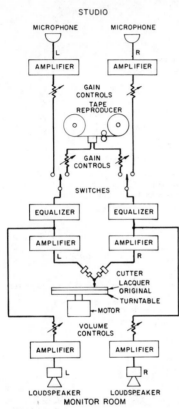

FIG. 9.43. Stereophonic-disc recording system. The source may be microphones in a studio or a two-channel stereophonic master magnetic tape.

original from a master tape as shown in Fig. 9.43. The two-channel stereophonic-master magnetic tape is produced from the three-channel stereophonic magnetic tape as shown in Fig. 9.36. A schematic-sectional view of a stereophonic cutter is shown in Fig. 9.44. The elements in the two channels are disposed at 90° with respect to each other. The driving

[6] Frayne and Davis, *Jour. Audio Eng. Soc.*, Vol. 7, No. 4, p. 146, 1959.

coil is connected to the output of the amplifier.   The feedback coil is connected to the input of the amplifier.

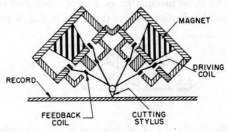

FIG. 9.44.   Sectional view of a stereophonic disc cutter.

Four modulation possibilities of the 45°–45° stereophonic-disc recording system are shown in Fig. 9.45 as follows: signal on the left channel only,

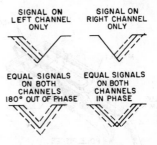

FIG. 9.45.   Groove modulations for various signal conditions in the 45°-45° stereophonic disc.

signal on the right channel only, equal signals on both channels 180° out of phase and equal signals on both channels in phase.   The signal recorded

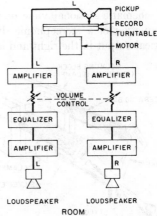

FIG. 9.46.   Stereophonic-disc reproducing system.

in the right and left channels are independent in both recording and reproducing.

The mass production of stereophonic disc records is the same as for the monophonic disc records depicted in Fig. 9.41.

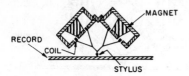

FIG. 9.47. Sectional view of a stereophonic-dynamic pickup:

The elements of stereophonic-disc reproducing systems are shown in Fig. 9.46. The stereophonic-disc reproducing system shown in Fig. 9.46 is usually operated in a room in the home as described in Sec. 8.2F and depicted in Fig. 8.15.

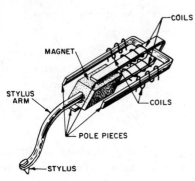

FIG. 9.48. Perspective view depicting the elements of a stereophonic-magnetic pickup.

A schematic-sectional view of a stereophonic-dynamic pickup for the reproduction of a stereophonic disc record is shown in Fig. 9.47. The electrical output in the right and left channels of the pickup correspond to

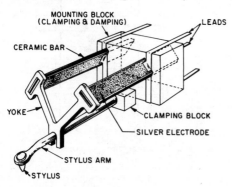

FIG. 9.49. Perspective view depicting the elements of a stereophonic-ceramic pickup.

the modulation in the right and left walls of the record which in turn correspond to the inputs to the right and left channels of the cutter.

A large number of magnetic and stereophonic pickups have been developed for the reproduction of stereophonic disc records. The magnetic pickup depicted in Fig. 9.48 exhibits superior performance from the standpoint of uniform response with respect to frequency and a low mechanical impedance presented to record groove by the stylus. Considerable electrical compensation is required to obtain uniform output from the magnetic pickup. However, the ceramic pickup is most widely used because the output is high and practically no compensation with respect to frequency is required. The essential elements of a ceramic pickup are shown in Fig. 9.49. In general, the frequency response of the ceramic pickup is not as smooth as the magnetic pickup. Furthermore, a relatively high mechanical impedance is presented to the record at certain portions of the frequency range. The electrical impedance of a magnetic pickup is relatively low, whereas the electrical impedance of the ceramic pickup is relatively high.

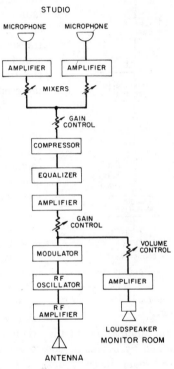

Fig. 9.50. Monophonic amplitude-modulation radio transmitter.

9.11 RADIO SOUND-REPRODUCING SYSTEM

A monophonic amplitude-modulation radio transmitter is shown in Fig. 9.50. In the radio transmitter of Fig. 9.50, the sound wave is converted into corresponding electrical waves by the microphone. The electrical outputs of the microphones are amplified by electronic amplifiers and fed to mixers. The microphones and source of sound to be picked up are located in the studio (Sec. 8.3B and Fig. 8.26). The mixers, volume control, amplifiers, and monitoring system are located in the control room. A soundproof glass wall partition which separates the studio and control room gives the engineer full view of the action in the studio (Sec. 8.3B). The output of the compressor amplifier is coupled to an equalizer followed by an amplifier. The output of the amplifier is fed to the monitoring system and the modulator. The monitoring system is located in the control room. In the standard broadcast band, the frequency of the oscillator lies somewhere between 550 to 1,700 kilocycles. The modulator varies the amplitude of the oscillator. The amplitude of the variations corresponds to the variations in the original sound wave. The output of the oscillator is coupled to the radio-frequency power amplifier. The radio-frequency power amplifier is coupled to the transmitting antenna. The modulated radio-frequency wave is transmitted in the ether in all directions. The input to the AM transmitter shown in Fig. 9.50 is provided by sound pickup by microphones. The input to the transmitter may also be disc records or prerecorded magnetic tape.

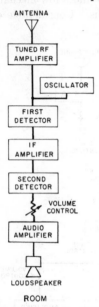

FIG. 9.51. Monophonic amplitude-modulation radio receiver.

In the monophonic amplitude-modulation radio-receiving system of Fig. 9.51 a very small portion of the radio frequency radiated by the transmitter antenna is picked up by the receiving antenna. The output of the receiving antenna is amplified by a radio-frequency vacuum-tube amplifier. The output is combined with that of a higher radio-frequency oscillator and fed to the first detector. The resultant intermediate-frequency carrier, with the original modulation, is amplified by the intermediate-frequency amplifier. The second detector converts the modulated intermediate-frequency carrier to an audio-frequency electrical wave in which the variations correspond to the undulations in the original sound wave. The second detector is followed by a volume control which controls the intensity of the reproduced sound in the room. The volume control is followed by a power amplifier which drives the loudspeaker. The loudspeaker converts the electrical variations into the corresponding sound vibrations. The latter corresponds to the original variations in the studio. The radio receiver is usually operated in a small room, as for example, the living room in the home (Sec. 8.2F and Fig. 8.13).

The radio sound-reproducing system shown in Figs. 9.50 and 9.51 employs amplitude modulation. Frequency modulation is also used for standard sound broadcasting and for television sound broadcasting. The difference between amplitude-modulation (AM) and frequency-modulation (FM) systems resides in the modulator in the transmitter and in the detector in the receiver. In the amplitude-modulation system shown in

Fig. 9.50 the amplitude of the oscillator is varied to correspond to the audio-frequency signal. In the frequency-modulation system, the frequency of the oscillator is varied to correspond to the audio-frequency signal. Some form of frequency discriminator is used in the receiver to convert the frequency modulations into the corresponding audio variations.

Fig. 9.52. Spectrum of the signals of stereophonic frequency-modulation radio.

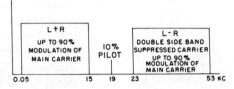

Standards for stereophonic radio transmission in the frequency-modulation band in the United States was established by the U. S. Federal Communications Commission in 1962. The standards provide for the transmission of the sum of the left and right stereophonic audio signals as the direct-frequency modulation of the radiated carrier as depicted by the spectrum diagram of Fig. 9.52. The difference between the left and right stereophonic audio signals is used to amplitude-modulate a 38-kilocycle subcarrier in suppressed-carrier fashion, with the result impressed as frequency modulation on the radiated carrier as shown in Fig. 9.52. In addition, a pilot subcarrier of 19 kilocycles, shown in Fig. 9.52, is transmitted for synchronizing the FM receiver.

A typical FM stereophonic radio transmitter[7] is shown in Fig. 9.53. The outputs of the right, R, and

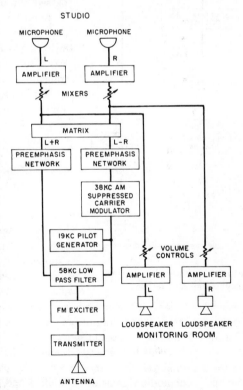

Fig. 9.53. Stereophonic frequency-modulation radio transmitter.

[7] Csicatka and Lima, *Jour. Audio Eng. Soc.*, Vol. 10, No. 1, p. 2, 1962.

left, L, microphones are amplified and fed to a matrix. In general, more than one microphone is used in each channel. The output of the matrix contains two audio signals, the L + R and the L − R. Both are passed through preemphasis networks which accentuate the high-frequency re-

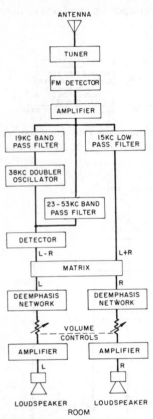

sponse with respect to frequency. The L + R preemphasized audio signal is fed directly to the FM exciter. The L − R preemphasized audio signal is applied to a 38-kilocycle AM double-sideband modulator in which the 38-kilocycle carrier is suppressed. A 19-kilocycle signal is produced by a pilot generator. The suppressed-carrier modulation signal of L − R and pilot signal are combined and passed through a 58-kilocycle filter and on to an FM exciter. The output of the FM exciter is coupled to the FM transmitter, which in turn is coupled to the FM antenna. The input to the FM radio transmitter shown in Fig. 9.53 is provided by sound pickup by microphones. The input to the FM radio transmitter may also be disc records or prerecorded magnetic tape.

A typical FM stereophonic radio receiver is shown in Fig. 9.54. The antenna, tuner, and detector are standard FM units. The output of the detector is amplified and fed to the following: A low-pass filter, a 23- to 54-kilocycle bandpass filter, and a 19-kilocycle narrow-bandpass filter. The output of the 19-

FIG. 9.54. Stereophonic frequency-modulation radio receiver.

kilocycle filter is coupled to a 38-kilocycle doubler oscillator which generates the 38-kilocycle carrier. The carrier is combined with the two sideband signals in the detector. The resultant detector output is the L − R audio signal. The L + R and L − R audio signals are combined in the matrix, the output of which yields the separate L and R audio signals. The deemphasis network normalizes the response with respect to frequency of the audio signals. The audio signals are amplified and converted to left and right sound signals in air by the loudspeakers. The stereophonic frequency-modulation radio receiver shown in Fig. 9.54 is usually operated in a room in the home as described in Sec. 8.2F and depicted in Fig. 8.15.

The FM stereophonic system provides high-quality stereophonic sound reproduction. The essential characteristics are a frequency range of 30 to 15,000 cycles, a separation between left and right signals of 30 decibels and nonlinear distortion in accordance with existing FCC requirements.

9.12 SOUND-MOTION-PICTURE REPRODUCING SYSTEM

A complete monophonic sound-motion-picture recording system is shown in Fig. 9.55. The outputs of the microphones are amplified and fed to attenuators termed mixers. The microphones and source of sound to be recorded are located on the sound stage (Sec. 8.3I and Fig. 8.36). If more than one microphone is used as, for example, a soloist accompanying an

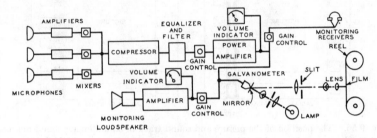

FIG. 9.55. The elements of a monophonic sound-motion-picture film recording system.

orchestra, one microphone for the soloist and one for the orchestra, the output of the two may be adjusted for the proper balance by the mixer system. The mixer system is followed by a compressor which reduces the volume range. A low-pass filter is usually used to reduce ground noise above the upper limits of reproduction. A high-pass filter is used on speech with the lower limit placed below the speech range. This latter expediency reduces low-frequency noises without impairing the speech quality. An equalizer is used to accentuate the high frequencies and thereby compensate for the film-transfer loss in the high-frequency range. A gain control in the form of an attenuator is used to control the over-all volume. The output of the power amplifier is coupled to the light modulator and the monitoring system. By means of the optical system and light modulator the electrical variations are recorded on the film into the corresponding variations in density (termed variable-density recording) or in area (termed variable-area recording). The recording system depicted in Fig. 9.55 is of the variable-density type. The optical and modulating system for both variable-density and variable-area recording will be subsequently described in greater detail in this section. The monitoring system is also connected to the output of the recording amplifier. In general, the monitoring system is equalized so that the reproduced sound heard in the monitoring system simulates that of the ultimate reproduced sound heard in the theater. An amplifier and loudspeaker are used if the mixing and monitoring are carried out in a booth on the sound stage or in a monitoring room next to the sound stage. Headphones are used if the monitoring is carried out on the stage (Sec. 8.3*I* and Fig. 8.36). Volume indicators are provided to ensure that the proper amplitude is obtained in the modulation system.

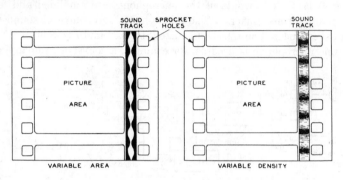

FIG. 9.56. The position of the picture and sound track in 35-millimeter sound-motion-picture film. Two types of track are shown, namely, variable area and variable density. (*After Olson, Acoustical Engineering, D. Van Nostrand Company, Inc., Princeton, 1957.*)

The sound track on 35-millimeter film occupies a space about 0.1 inch wide. There are two types of sound track in general use today, namely, variable area and variable density (Fig. 9.56). The type of variable-area sound track shown in Fig. 9.56 is termed bilateral variable area.

In the variable-area system the transmitted light amplitude is a function of the amount of unexposed area in the positive print. This type of sound track is produced by means of a mirror galvanometer which varies the width of the light slit under which the film passes. The elements of a variable-area recording system are shown in Fig. 9.57. The triangular aperture is uniformly illuminated by means of a lamp and lens system. The image of the triangular aperture is reflected by the galvanometer mirror focused on the mechanical slit. The mechanical slit in turn is focused on the film. The galvanometer mirror swings about an axis parallel to the plane of the paper. The triangular light image on the mechanical slit moves up and down on the mechanical slit. The result is that the width of the exposed portion of the negative sound track corresponds to the rotational vibrations of the galvanometer. In the positive record the width of the unexposed portion corresponds to the signal.

The amount of ground noise produced is proportional to the exposed portion of the positive sound track. For this reason it is desirable to make the unexposed portion of the record just wide enough to accommodate the modulation. This is accomplished by applying a bias signal to the galvanometer. In the absence of a signal a very narrow exposed portion is produced on the negative record, which means a correspondingly narrow

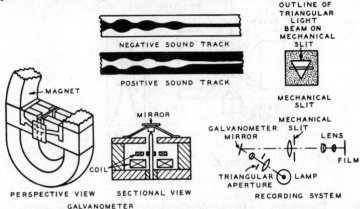

FIG. 9.57. The elements of a variable-area sound-motion-picture film recording system. The negative and positive sound tracks. Perspective and sectional views of the galvanometer.

unexposed portion on the positive record. When a signal appears, the triangular spot on the mechanical slit moves down just enough to accommodate the signal. The initial bias is accomplished within a millisecond. However, the return to normal bias after a large signal followed by a small signal is about 1 second. Faster return action produces thumping in the reproduced record.

A film sound-reproducing system is an amplitude system, that is, the voltage output is proportional to the amplitude on the film. Therefore, in order to obtain a uniform response-frequency characteristic, neglecting the frequency discrimination due to finite recording and reproducing slits, the amplitude of the galvanometer should be independent of the frequency.

In the variable-density system the transmitted light amplitude is an inverse function of the amount of exposure in the positive print. This type of sound track is produced by means of a light valve which varies the amount of light which falls on the moving film. The elements of a variable-density recording system are shown in Fig. 9.58. The ribbons of the valve are illuminated by means of a lamp and lens system. The image of the illuminated slit produced by the ribbons of the light valve is focused on the film. The amount of exposure on the negative film varies with the

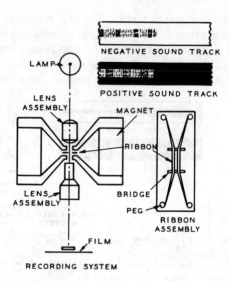

FIG. 9.58. The elements of a variable-density sound-motion-picture film recording system. The negative and positive sound tracks. Sectional and ribbon assembly views.

aperture at the ribbons. In the positive record the amount of exposure is an inverse function of the input to the light valve. Ground-noise reduction can also be obtained with a light valve. In the absence of a signal the light valve is biased so that the aperture between the ribbons is almost closed. When a signal appears, the ribbons open just enough to accommodate the signal. The noise-reducing action is similar to that in the variable-area system.

The elements of a moving-picture-film monophonic sound-reproducing system are shown in Fig. 9.59. The reproducing system save for the loudspeakers is located in the projection booth (Sec. 8.2E and Fig. 8.11). The light source, in the form of an incandescent lamp, is focused upon a

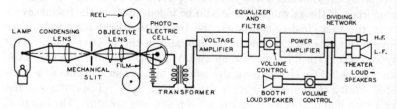

FIG. 9.59. The elements of a monophonic sound-motion-picture film reproducing system.

mechanical slit by means of a condensing lens. The mechanical slit in turn is focused on the negative film. The height of the image on the film is usually about 0.00075 inch. Under these conditions the amount of light which impinges upon the photocell is proportional to the unexposed portion of the sound track in variable-area recording or to the inverse function of the density in variable-density recording. When the film is in motion, the photographic variations in the film produce corresponding undulations in the light which pass through the film and fall upon the photocell. Electrons are emitted when light falls upon the cathode of a photocell. The number of electrons which are emitted is proportional to the intensity of the light. The electrons constitute an electrical current. Thus it will be seen that the undulations in light which fall upon the photocell produce corresponding variations in the current output of the photocell. The voltage output developed by the combination of the voltage polarizing supply and transformer corresponds to the current developed by the photocell and polarizing supply. Therefore, the voltage output at the transformer corresponds to the undulations in density in the film, which in turn correspond to the variations in the original sound which impinged upon the microphone. The lamp, optical system, mechanical slit, photocell, and transformer are located in the motion picture projector. The output of the photocell is fed to a voltage amplifier. The voltage amplifier is followed by an equalizer which compensates for some of the recording characteristics. In some cases a low-pass filter is used to reduce film ground noise. A volume control follows the voltage amplifier, equalizers, and filters. The volume control is followed by the power amplifier which feeds the loudspeakers. The voltage amplifier, equalizers, filters, volume control, and power amplifier are mounted in a rack located in the projection booth. The theater loudspeakers are located behind the screen (Sec. 8.2E and Fig. 8.11). Theater loudspeakers are of the dual type, one set for covering the low-frequency range and another set for covering the high-frequency range (see Sec. 9.3B and Fig. 9.17). A dividing network allocates the power with respect to frequency in the theater loudspeaker system. A monitoring loudspeaker is located in the projection booth.

Some recording of sound motion pictures is still carried out by the direct recording process depicted in Fig. 9.55. However, the general practice is to record the sound on magnetic tape. In this case, the magnetic tape recorder is substituted for the film recorder of Fig. 9.55. The magnetic tape used for recording the sound is in general 35 millimeters in width with sprocket holes on each side as shown in Fig. 9.60. The magnetic tape with sprocket holes makes it possible to synchronize the picture and sound in recording. The magnetic tape master is converted to the film by feeding the output of the tape reproducer to the sound film recorder of Fig. 9.55.

Stereophonic sound was introduced in sound motion pictures several years ago. The original master sound record is recorded on the sprocket hole 35-millimeter magnetic tape shown in Fig. 9.60A. Any number of magnetic tracks up to seven have been used. However, in general, four magnetic tracks are employed. Three channels of stereophonic sound are recorded. The fourth channel may be used for control, sound effects, etc.

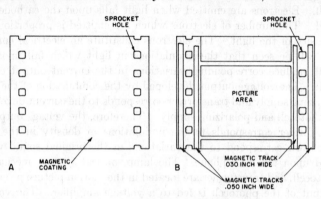

FIG. 9.60. A. Magnetic tape for original sound recording in sound motion pictures. B. Magnetic strips on a motion picture positive film.

The 35-millimeter release print with magnetic tracks for use in projection in the theater is shown in Fig. 9.60B. The three tracks carry the stereophonic sound for reproduction in audio perspective by means of the system shown in Fig. 8.11. The fourth track is used for the reproduction of sound effects, or for control, etc.

9.13   TELEVISION SOUND-REPRODUCING SYSTEM

The elements of a sound channel for a television broadcasting system are essentially the same as those of a monophonic frequency-modulation radio sound-broadcasting system described in Sec. 9.11. The picture and sound electrical variations are transmitted and reproduced simultaneously. In the case of plays, dramas, operas, the sound pickup techniques are a combination of those employed in radio and sound motion pictures (see Sec. 8.3I). In the audience type show, the technique is described in Sec. 8.3H. In some cases, the television show is recorded on sound motion picture film or video magnetic tape for broadcast at a later time. The sound portion of the receiving system in a television receiver is the same as that of a monophonic frequency-modulation radio receiver described in

Sec. 9.11. The reproduction of sound by the television receiver in a room is similar to that of a radio receiver or phonograph (see Sec. 8.2F and Fig. 8.13).

9.14 SOUND-REPRODUCING SYSTEM

A typical sound-reproducing system is shown in Fig. 9.61. The system consists of the following sources: one or more microphones with preamplifiers, a phonograph reproducer, a magnetic tape reproducer, and a radio receiver. Mixers are provided so that any of the sources may be coupled to the reproducing system. The mixers are followed by an amplifier. A volume control in the form of an attenuator is used to control the intensity of the reproduced sound. A volume indicator is used so that the appropriate sound level will be obtained at the reproducing points. The system or parts of the system shown in Fig. 9.61 are used for sound reinforcement, public-address, announce, and paging systems, music reproduction, and

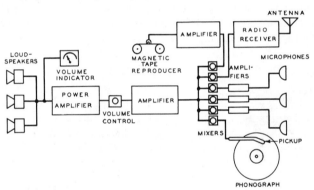

FIG. 9.61. The arrangement of the elements of a sound-reproducing system.

numerous other applications in theaters, churches, auditoriums, outdoor theaters, mass meetings, athletic events, factories, offices, railroad stations, airports, hotels, hospitals, etc. (Chap. 8).

9.15 HEARING AIDS

Tests made upon representative cross sections of the people in this country show a very large percentage to be hard of hearing. Practically all these people may obtain satisfaction from the use of a hearing aid. A hearing aid is a complete sound-reproducing system which increases the sound pressure over that normally received by the ear.

During the past few years, hearing aids employing transistor amplifiers have been developed with a high order of quality combined with small size.

The advent of transistors as described in Sec. 9.5 and miniature batteries have made this possible. A schematic diagram of the elements of a transistor hearing aid is shown in Fig. 9.62. The microphone employed in the transistor hearing aid is the magnetic type shown in Fig. 9.4.

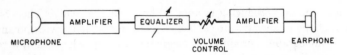

FIG. 9.62.   Elements of a transistor hearing aid.

Two types of earphones are used, namely, the air-conduction type and the bone-conduction type. A perspective and sectional view of an air-conduction insert-type hearing-aid earphone is shown in Fig. 9.63A. A molded plug fits the ear cavity and holds the receiver in place. Under these conditions, the leak at the ear is very small. Therefore, good response is obtained at the low frequencies. The action of the system is essentially the same as the bipolar telephone receiver described in Sec. 9.4.

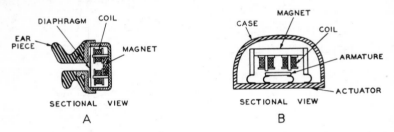

FIG. 9.63.   Sectional views of hearing aid earphones.   A. Insert-type earphone. B. Bone-conduction earphone.

In certain types of deafness, the middle ear, which consists of a series of bones that conduct sound to the inner ear, is damaged while the inner ear, which consists of the cochlea, frequency discriminating means, and nerves, is normal. Under these conditions, sound may be transmitted through the bones of the head to the inner ear by means of the bone-conduction earphone shown in Fig. 9.63B. The term bone-conduction earphone is not universally accepted to designate the electroacoustic transducer shown in Fig. 9.63B. Other designations are bone-conduction receiver, bone-conduction oscillator and bone-conduction vibrator. The face of the bone-conduction earphone is placed against the mastoid back of the external ear. The multiple-resonant system of the bone-conduction earphone delivers a large signal to the mastoid bone.

Complete hearing aids are shown in Fig. 9.64. The most common hearing aid consists of a case housing the microphone, amplifier, battery and controls and insert earphone, shown in Fig. 9.64A. The case is worn at a convenient location on the clothing. A cord connects the output of the amplifier to the earphone. A bone-conduction earphone may also be used with the hearing aid shown in Fig. 9.64A. For this condition, the bone-conduction receiver may be held against the skin and flesh covering the mastoid bone with a headband or by means of double-sided pressure adhesive tape with one side cemented to the earphone and the other side cemented to the skin covering the mastoid bone. An insert-type earphone

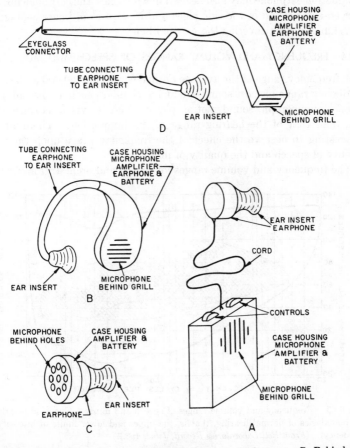

FIG. 9.64. Perspective views of typical hearing aids. *A.* Body-type. *B.* Behind the ear-type. *C.* In the ear-type. *D.* Eyeglass-type.

is shown in Fig. 9.64*B*. The case is worn behind the ear. The tube connecting the earphone in the case to the ear insert is carried over the top of the ear and serves as a support. The microphone, amplifier, controls and batteries are housed in the case behind the ear. A very compact hearing aid is shown in Fig. 9.64*C*. The entire system consisting of the microphone, amplifier, controls, battery and earphone is housed in a small case which fits in the ear. In another form of hearing aid, the entire system consisting of the microphone, amplifier, controls, battery and earphone is housed in one side of the side pieces of the eye spectacles as shown in Fig. 9.64*D*. The air-type earphone is depicted in Fig. 9.64*D*. The bone-conduction earphone may also be used in which case the earphone forms the curved-end portion of the side piece and is worn in intimate contact with the skin and flesh covering the mastoid bone.

9.16   FREQUENCY AND VOLUME RANGES OF SPEECH AND MUSIC

The frequency range of the average normal ear is 20 to 20,000 cycles. The frequency range of most sound-producing channels such as the radio, telephone, phonograph, sound motion picture, and television is considerably less than that of the hearing range of the human ear. Therefore, it is interesting to observe the effect of the frequency range upon the intelligibility of speech and the quality of music.

The frequency and volume ranges[8] of speech and orchestral music for no

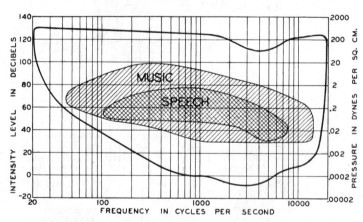

FIG. 9.65.   Frequency and volume ranges of speech and music. The solid line depicts the boundaries of normal hearing, that is, the upper and lower limits of intensity and frequency. (*From Bell Laboratories Record, June, 1934.*)

[8] *Bell Labs. Record*, Vol. 12, No. 10, p. 314, 1934.

perceptible change in quality are shown in Fig. 9.65. The reproduction of speech with subjectively perfect fidelity requires a frequency range of 100 to 8,000 cycles and a volume range of 40 decibels. The reproduction of orchestral music with subjectively perfect fidelity requires a frequency range of 40 to 15,000 cycles and a volume range of 70 decibels.

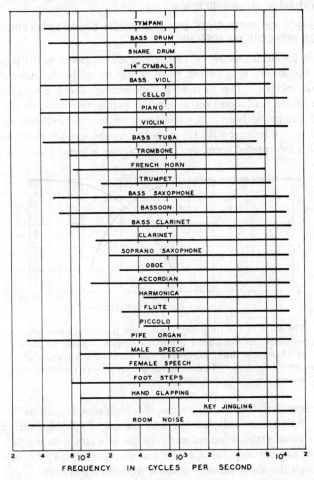

FIG. 9.66. The frequency ranges for the reproduction of speech, musical instruments, and noises without any noticeable distortion. (*After Snow.*)

The frequency ranges required for the reproduction of speech, musical instruments, and noises without any noticeable frequency discrimination or distortion[9] are shown in Fig. 9.66. It will be seen that, owing to the extensive high-frequency range of the overtones of musical instruments, the upper limit of sound reproduction must extend to 15,000 cycles in order to obviate any perceptible change in quality. In order to reproduce the fundamentals of musical instruments, the lower limit of reproduction should extend to at least 40 cycles.

9.17 EFFECT OF FREQUENCY DISCRIMINATION UPON THE ARTICULATION OF REPRODUCED SPEECH[10]

The most important aspect of the perception of speech is the process involving the ability of a person to recognize correctly the speech sounds which are spoken. The method for measuring speech perception is to have a speaker read aloud certain speech sounds to a listener who writes what he thinks he hears. A comparison of sounds, syllables or words recorded by the listener with those spoken by the speaker shows the fraction that is interpreted correctly. This fraction is termed articulation.

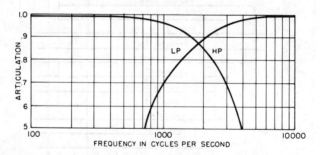

Fig. 9.67. The effect of the frequency range upon the sound articulation of speech. *HP.* High-pass filter: all frequencies below the frequency given by the abscissa are removed. *LP.* Low-pass filter: all frequencies above the frequency given by the abscissa are removed. (*After Fletcher, Speech and Hearing in Communication, D. Van Nostrand Company, Inc., Princeton, 1953.*)

Sound articulation refers to the use of speech sounds such as "p," "a," "t," etc. Syllable articulation refers to the use of syllables such as pat, run, etc. Word articulation refers to the use of a complete word.

The medium between the speaker and the listener may be the direct path through the air or through a sound-reproducing system.

The effect of reducing the high- and low-frequency range upon the sound articulation of speech at a normal conversational level is shown in Fig.

[9] Snow, W. B., *Jour. Acoust. Soc. Amer.*, Vol. 3, No. 1, Part 1, p. 155, 1931.
[10] Fletcher, *Speech and Hearing in Communication*, D. Van Nostrand Company, Inc., Princeton, 1953.

9.67. It will be seen that a relatively high articulation can be obtained with a very narrow transmission band. However, the quality of the reproduced speech is very much impaired by transmission over a narrow frequency band. From the standpoint of articulation, a limited frequency range may be actually superior to a wider frequency band, owing to the introduction of additional noises and distortions in a wider band, unless particular precautions are observed. In the case of speeches, plays, and songs, a limited frequency range impairs the quality and artistic value of the reproduced sound.

9.18 EFFECT OF FREQUENCY DISCRIMINATION UPON THE QUALITY OF REPRODUCED MUSIC

The effect of the frequency range[11] upon the quality of reproduction of orchestral music is shown in Fig. 9.68. It will be seen that the frequency

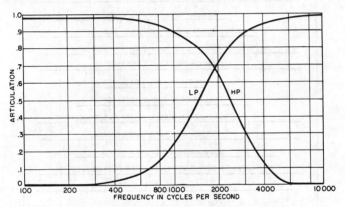

FIG. 9.68. The effect of the frequency range upon the quality of orchestral music. *HP.* High-pass filter: all frequencies below the frequency given by the abscissa are removed. *LP.* Low-pass filter: all frequencies above the frequency given by the abscissa are removed. (*After Snow.*)

range required for no appreciable loss in quality is 40 to 15,000. As the next section will show, the frequency ranges of high-quality magnetic tape recorders and reproducers, disc recorders and reproducers, radio and television transmitters and receivers, sound-motion-picture recorders and reproducers and sound systems cover the frequency range depicted in Fig. 9.68. In some cases, the volume range of the music exceeds the capabilities of the reproducing equipment. Under these conditions, some form of volume compression must be used to cover the amplitude range (see Sec. 8.3L).

[11] Snow, W. B., *Jour. Acoust. Soc. Amer.*, Vol. 3, No. 1, Part 1, p. 155, 1931.

9.19 FREQUENCY RANGES OF SOUND-REPRODUCING SYSTEMS

The frequency ranges of the most common sound-reproducing systems are shown in Fig. 9.69.   The frequency ranges shown are averages of existing systems.   In specific cases the frequency ranges may be greater or less than those shown in Fig. 9.69.

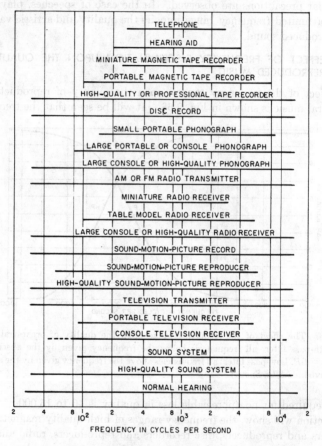

FIG. 9.69.   The frequency ranges of sound-reproducing systems.

The frequency ranges of telephones vary over wide limits, depending upon the type of instrument, the central offices, and the interconnecting lines.   The frequency range depicted is for instruments made in the last decade.   Extending the frequency range would probably result in reduced articulation due to ambient room noise and noises produced by electrical interferences.

The frequency range of the hearing aid shown in Fig. 9.69 represents the average response of high-quality transistor hearing aids in use today. The low-frequency range may be somewhat greater but in general, this added range cannot be used because of "rumble" and other low-frequency noises.

The frequency ranges of magnetic tape recorders and reproducers are shown in Fig. 9.69. The low tape speeds used in miniature and portable tape recorders limits the high-frequency range. The small size of the loudspeaker and cabinet limits the low-frequency response of the miniature and portable magnetic tape recorders. The frequency range of high-quality and professional tape recorders covers the entire audio-frequency range.

The commercial disc record covers the entire audio-frequency range as shown in Fig. 9.69.

The frequency ranges of phonographs are shown in Fig. 9.69. The size of the loudspeaker and cabinet determines the low-frequency response of the portable and small console phonographs. The high-frequency range is limited to provide proper balance. The high-quality phonograph covers the entire audio-frequency range.

AM and FM radio transmitters cover the entire audio-frequency range. The frequency ranges of radio receivers are shown in Fig. 9.69. As in all sound-reproducing systems the low-frequency range is determined by the size of the loudspeaker and cabinet. The frequency ranges increase with the size of the radio receiver as shown in Fig. 9.69. The high-quality AM receiver covers the entire frequency range if there is no cochannel interference; the high-quality FM receiver covers the entire audio-frequency range.

The frequency range of a commercial sound-film record covers the entire audio-frequency range as shown in Fig. 9.69. The frequency range of a sound-motion-picture system shown in Fig. 9.69 refers to the average of systems used in theaters. A small number of high-quality sound-motion-picture systems are in use with a frequency range varying from the commercial system up to the high-quality system shown in Fig. 9.69. The frequency ranges of 16-millimeter and 8-millimeter film record and reproducers are restricted as compared to the commercial 35-millimeter film record and reproducers.

The frequency range of a television sound transmitter is shown in Fig. 9.69. In general, the low-frequency range is limited in transmission due to the high-ambient low-frequency noise level in the studio. The high-frequency range is also limited by electrical noise and the frequency range of the audio-transmission networks.

The low-frequency range of the table model television receiver is limited by the size of the loudspeaker and cabinet. The frequency range of the large console television receiver is limited by the signal input.

The frequency range of the common low-cost sound system is shown in Fig. 9.69. For theaters, auditoriums, and other high-quality installations the frequency range of the sound system covers the entire audio-frequency range. In some cases, the frequency range must be tailored to compensate for the poor acoustics of the auditorium or theater.

The hearing frequency range of a person with excellent hearing is shown in Fig. 9.69.

An examination of the characteristics noted in Fig. 8.69 shows that the consumer can obtain radio receivers, magnetic tape records and reproducers and disc records and phonographs that cover the entire audio-frequency range without discrimination.

9.20 EFFECT OF NONLINEAR DISTORTION UPON THE QUALITY OF REPRODUCED SPEECH AND MUSIC[12]

The various elements in an ideal sound-reproducing system are absolutely invariant with respect to time. However, in actual sound-reproducing systems in use today, all the elements exhibit nonlinear characteristics.

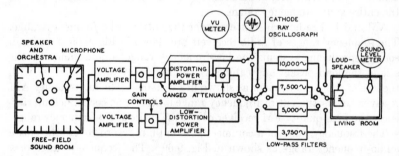

Fig. 9.70. Schematic arrangement of the apparatus for the subjective determination of the relation between nonlinear distortion and the frequency range of the reproduced sound. (*After Olson, Acoustical Engineering, D. Van Nostrand Company, Inc., Princeton, 1957.*)

A nonlinear element introduces nonlinear distortion. The effect of nonlinear distortion is the introduction of harmonic components in the reproduced sound which were not present in the original sound. In order to obtain a better understanding of some of the effects of nonlinear distortion upon the reproduction of sound, a subjective test has been performed to determine the perceptible, tolerable, and objectionable nonlinear distortion

[12] Olson, H. F., *Acoustical Engineering*, D. Van Nostrand Company, Inc., Princeton, 1957.

in a sound-reproducing system as a function of the frequency for different types of distortion.

The effect of various types of nonlinear distortion upon the reproduction of speech and music has been determined by means of the sound-reproducing system shown in Fig. 9.70.   The basic system is a high-quality sound-reproducing system.   The over-all response frequency of the com-

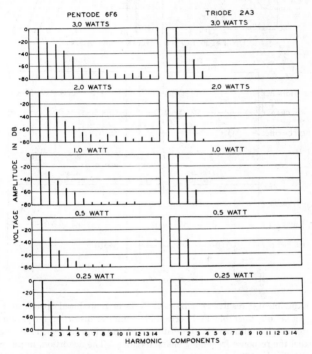

Fig. 9.71.   Distortion characteristics of single-ended pentode and triode amplifiers. (*After Olson, Acoustical Engineering, D. Van Nostrand Company, Inc., Princeton, 1957.*)

bination of the microphone, amplifier, and loudspeaker is flat to within ±2 decibels from 40 to 15,000 cycles.   The over-all nonlinear distortion in the system is less than 0.25 per cent.   Means are provided so that nonlinear distortion of a predetermined amount could be introduced.   Two types of distorting amplifiers were used to introduce distortion of a predetermined amount, namely, a single-ended triode amplifier and a single-ended pentode amplifier.   The distortion power outputs for various power levels are shown in Fig. 9.71.   Referring to Fig. 9.70, it will be seen that by means of the gain control preceding and the attenuator following the distorting

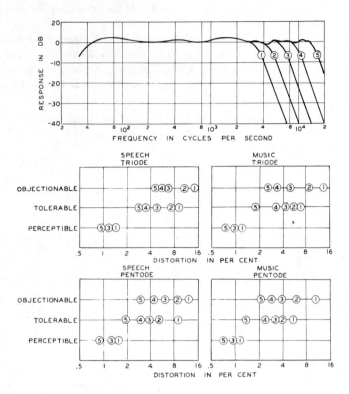

Fig. 9.72. Experimental results of subjective tests of reproduced speech and music depicting objectionable, tolerable, and perceptible nonlinear distortion for various high-frequency cutoffs. The numbers in the distortion data points correspond to the numbers which label the response frequency characteristics. The distortion, in per cent, is the ratio, multiplied by 100, of the total root mean square of the second, third, fourth, etc., components of the distortion to the root mean square of the fundamental. (*After Olson, Acoustical Engineering, D. Van Nostrand Company, Inc., Princeton, 1957.*)

amplifier it is possible to operate the distorting amplifier at practically any distorting level and still retain the same level in the reproduced sound. The pickup studio for these tests was a free field or anechoic sound room, that is, a room devoid of echoes and reverberations as well as having a low noise level. The sound was reproduced in a room with acoustics similar to that of a typical living room (Sec. 8.2*F*). The noise level at the pickup point was 0 decibels in the absence of any performers. The noise level in the listening room was about 20 decibels. The reproducing level was about 70 decibels.

These tests were limited to three subjective gradations of nonlinear distortion, namely, perceptible, tolerable, and objectionable. Perceptible is the amount of distortion in the distorting system required to be just discernible when compared with the reference system. Tolerable and objectionable are not as definite and are a matter of opinion. By tolerable distortion is meant the amount of distortion which could be allowed in low-grade commercial sound reproduction. By objectionable distortion is meant the amount of distortion which would be definitely unsatisfactory for the reproduction of sound in phonograph and radio systems.

Both speech and music were used in making these tests. In the case of music, a six-piece orchestra was employed.

The average results of a few of these tests, with a limited number of critical observers, are shown in Fig. 9.72. As would be expected from the frequency ranges of speech and music together with the masking curves, a distorting system with high-order components is more objectionable than one with low-order components. The amount of tolerable distortion is greater for speech than for music. Referring to Fig. 9.72, it will be seen that the amount of perceptible, tolerable, or objectionable distortion decreases as the high-frequency cutoff increases. This means that as the high-frequency range is decreased the amount of distortion which can be tolerated is increased. For example, these tests show that the tolerable nonlinear distortion in a sound-reproducing system with a 5,000-cycle high-frequency cutoff is of the order of 6 to 10 per cent, whereas the tolerable nonlinear distortion in a sound-reproducing system with a 15,000-cycle high-frequency cutoff is of the order of 1 to 2 per cent. These tests show that unless the nonlinear distortion is a wide-frequency-range system is kept to this relatively low value it will not be acceptable. While it is a comparatively simple matter to design and relatively inexpensive to build a sound-reproducing system with 6 to 10 per cent nonlinear distortion and a limited frequency range, it requires painstaking designs coupled with expensive components to achieve the relatively low order of nonlinear distortion of 1 to 2 per cent in a system of wide frequency range.

9.21 FREQUENCY-RANGE PREFERENCE FOR REPRODUCED SPEECH AND MUSIC[13]

The effect of frequency discrimination upon the quality of reproduced sound has been considered in Sec. 9.18. The effect of frequency discrimination upon the articulation of reproduced speech has been considered in Sec. 9.17. The effect of nonlinear distortion upon the quality of reproduced speech and music has been considered in Sec. 9.20. The frequency and volume ranges of speech and music have been considered in Sec. 9.16.

[13] Chinn and Eisenberg, *Proc. Inst. Radio Engrs.*, Vol. 33, No. 9, p. 571, 1945.

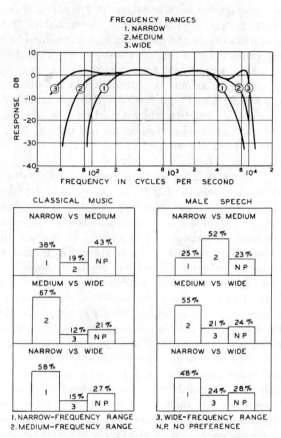

Fig. 9.73. The frequency-range preferences of a cross section of listeners for classical music and speech. The graph depicts the response-frequency characteristics of the following: 1 = narrow-frequency range, 2 = medium-frequency range, 3 = wide-frequency range. The block diagram depicts the preference. *NP* means no preference. When the narrow range was compared with the medium range, there was a preference for narrow for music and medium for speech. However, in the comparison between medium and wide, the preference was markedly for the narrow bands for both classical music and male speech. (*After Chinn and Eisenberg.*)

The frequency ranges of sound-reproducing systems have been considered in Sec. 9.19. A consideration of all the above data indicates that the subject of sound reproduction is exceedingly complex. In order to add to the general understanding, a study has been made of the frequency-range preference of a representative cross section of broadcast listeners.

As contrasted with the other data presented, the purpose of this investigation was the determination of the frequency range of reproduced speech and music that is most pleasant to the average listener. The investigation was made with a variety of musical and voice passages. The tests were made in a room with acoustics similar to those of a large living room. Both high-quality records and direct-wire transmission from the studio were used with very little difference in the results.

The frequency ranges employed for the tests are shown in Fig. 9.73 and were designated as wide, medium, and narrow frequency ranges.

The results of the tests are shown in Fig. 9.73. The general conclusion of these tests is that listeners prefer either a narrow or medium frequency range to a wide one. However, the exact choice of band width varies to some extent within these limits, for different types of program content. Listeners prefer a narrow to a wide tonal range even when informed that one condition is low fidelity and the other high fidelity. Listeners prefer a slightly wider band for female speech, piano, and popular orchestra selections than for male speech, mixed dramatic speech, and classical orchestra selections.

The reasons for the variance of these subjective tests with other subjective tests will be delineated in the sections which follow.

9.22 FREQUENCY-RANGE PREFERENCE FOR SPEECH AND MUSIC[14]

The frequency-range preference for reproduced speech and music has been considered in the preceding section. These tests indicate that listeners prefer a restricted frequency range in monaural reproduced speech and music. There are three possible reasons for the results of these tests, as follows: (1) The average listener, after years of listening to the radio and the phonograph, has become conditioned to a restricted frequency range and feels that this is the natural state of affairs. (2) Musical instruments are not properly designed and would be more pleasing and acceptable if the production of fundamentals and overtones in the high-frequency range were suppressed. (3) The distortions and deviations from true reproduction of the original sound are less objectionable with a restricted frequency range. The distortions and deviations from true reproduction of the original sound are as follows:

1. Frequency discrimination.
2. Nonlinear distortion.
3. Spatial distribution.
   a. Relatively small source.
   b. Separated sources in two-way loudspeaker systems.

14 Olson, H. F., *Jour. Acoust. Soc. Amer.*, Vol. 19, No. 4, p. 549, 1947.

    c. Nonuniform directional pattern with respect to frequency.

4. Single-channel system.
5. Phase distortion.
6. Transient distortion.
7. Microphone placement and balance.
8. Acoustics of two rooms, the pickup studio and the listening room.
9. Limited dynamic range.
10. Difference in level of the original and reproduced sound.
11. Noise.

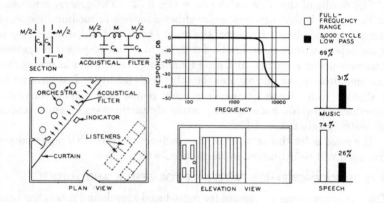

Fig. 9.74. Plan and elevation views of the schematic arrangement of the apparatus for direct testing of frequency-range preference for speech and music. A sectional view, acoustical network, and response-frequency characteristic of the acoustical filter used in the test. The results for speech and music are depicted on the right. (*After Olson, Acoustical Engineering, D. Van Nostrand Company, Inc., Princeton, 1957.*)

In order to obtain a better understanding of the reason for the preference of a restricted frequency range in reproduced sound, a fundamental all-acoustic test of frequency-range preference was made. The general arrangement of the test is shown in Fig. 9.74. An acoustical filter is placed between the orchestra and the listeners and is arranged so that it can be turned in or out. It is composed of three sheets of perforated metal to form a two-section acoustical filter, as shown in Fig. 9.74. The response-frequency characteristic of the acoustical filter shown in Fig. 9.74 approximated commercial good radio or phonograph reproduction in the high-frequency range at the time of the test in 1947. The acoustical filter is composed of 10 units with each unit pivoted at the top and bottom. The 10 units are coupled together and rotated by means of a lever. In this way the acoustical filters can be put in or out by merely turning the units through 90 degrees. The acoustical filters are shown in the full frequency-range position in Fig. 9.74. A sheer cloth curtain which transmits sound with no appreciable attenuation over the frequency range up to 10,000 cycles and less than 2-decibel attenuation from 10,000 to 15,000 cycles is placed between the acoustical filter and the listeners. The curtain is illuminated so that the listeners cannot see what transpires behind the curtain. The particular condition, that is, the full frequency range or 5,000 cycles low-pass transmission, is shown on an AB indicator.

The tests made up to the present time have been conducted in a small room which simulates an average living room in dimensions and acoustics. The orchestra was a six-piece dance band playing popular music. The average sound level in a room was about 70 decibels. The changes from wide open to low pass to wide open, etc., were made every 30 seconds. Two selections were played, and the listeners were asked to indicate a preference. The results of these tests, as shown in Fig. 9.74, indicated a preference for full-frequency range. Similar tests have been made for speech. The preference in the case of speech is also for the full-frequency range. There is a distinct lack of presence in speech with the limited-frequency range.

The results of the all-acoustic frequency-range preference are at variance with similar tests employing reproduced sound, as described in Sec. 9.21. The reason for the difference between the results of the two tests is, without doubt, due to the distortions listed in the first paragraph of this section. The subjective tests of nonlinear distortion, described in Sec. 9.20, indicated that the amount of tolerable distortion decreases as the frequency range is increased. These tests also indicated that a very small amount of nonlinear distortion can be detected when employing the full-frequency range.

9.23 FREQUENCY-RANGE PREFERENCE FOR STEREOPHONICALLY REPRODUCED SPEECH AND MUSIC[15]

Subjective tests of frequency-range preference of live speech and music comparing a restricted frequency range with the full frequency range have shown that the average listener prefers the full frequency range (see Sec. 9.22). These tests were all acoustic. The sound was not reproduced. Therefore, there were no electroacoustic transducers in the form of microphones, amplifiers, modulators, transmitters, records, receivers, pickups, demodulations, loudspeakers, etc. used in these tests. The frequency discrimination was accomplished by means of acoustical filters. The question arises whether similar results can be obtained with reproduced sound. The only reason that the same results could not be obtained with reproduced

[15] Olson, Preston, Woodward, May, Morgan, and Bleazey, Unpublished Report, 1955.

sound would be due to distortions and deviations from true reproduction of the original sound. The principal distortions and deviations from true reproduction of the original sound have been described in Sec. 9.22.

Following the all-acoustic frequency-range preference tests, it appeared logical to follow these tests with frequency-range preference tests employing reproduced sound. In the tests involving reproduced sound it was felt desirable to simulate the conditions of the all-acoustic tests as closely as possible. To attain this objective the orchestra was reproduced in perspective. The acoustics of the studio were eliminated by using a free-field room for the studio housing the orchestra. These expedients eliminated some of the distortions above. The remaining distortions listed, referred to above, were reduced to the lowest possible level.

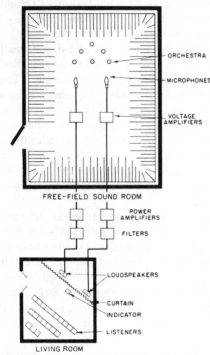

The floor plans of the free-field room used as a studio and listening room and the general arrangement of the tests are shown in Fig. 9.75. The idea of the test is to "transfer" the orchestra to the listening room by means of microphones, amplifiers, and loudspeakers. In order to simulate the all-acoustic tests in this transfer of the orchestra, it is obviously necessary that the studio be devoid of acoustics, that is, reverberation. In order to obtain these conditions the anechoic or free-field sound room is used as the studio. With the orchestra operating in the free-field sound room the level of the reflected sound is far below the level of the direct sound. The level of the reflected sound at the microphones for steady-state sound conditions is about 50 decibels below the direct sound. Therefore, it is impossible to detect any acoustics of the studio in the reproduced sound.

The listening room is the same as that used in the all-acoustic test. The

FIG. 9.75. Plan view of the schematic arrangement of the apparatus for frequency-range preference for stereophonically reproduced speech and music. (*After Olson, Acoustical Engineering, D. Van Nostrand Company, Inc., Princeton, 1957.*)

listening room is designed to be the acoustical equivalent of an average living room.

The reproducing system used in these tests employs two channels. Each channel consists of an RCA 44BX Velocity Microphone, an RCA OP-6 Amplifier, a laboratory developed triode-type push-pull power amplifier, laboratory developed high- and low-pass electrical filters, and an RCA LC1A Loudspeaker.

The over-all response-frequency characteristics depicting the ratio of the sound-pressure output from the loudspeaker in free space to the sound pressure at the microphone in free space, with and without the electrical filters, are shown in Fig. 9.76. In the restricted range condition there is

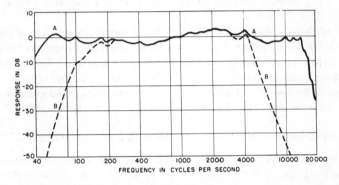

Fig. 9.76. Over-all response frequency characteristic of the two conditions used in the frequency-range preference for stereophonically reproduced speech and music. *A*. Wide frequency range. *B*. Restricted frequency range. (*After Olson, Acoustical Engineering, D. Van Nostrand Company, Inc., Princeton, 1957.*)

attenuation in both the low- and high-frequency ranges. The product of the low and high cutoff frequencies is 500,000 (cycles)^2. In this we have deviated from the original all-acoustic frequency-preference test in which a high-frequency cutoff alone was used. An argument in favor of the combination of high- and low-frequency cutoffs is that it approximates conventional radio and phonograph response-frequency characteristics.

The directivity pattern of the loudspeakers is very important where the listeners are located at relatively large angles with respect to the loudspeaker. In the loudspeakers used in these tests the variation in response at any frequency over a total angle of 90° is less than ±2 decibels.

The nonlinear distortion is another important factor in reproduced sound. The over-all nonlinear distortion was measured by supplying a distortionless signal to the input of the chain consisting of the voltage amplifier, power amplifier, and loudspeaker. The sound output of the loudspeaker was picked up by the microphone and fed to a harmonic analyzer. This method of measurement provides an over-all distortion characteristic from sound input to the microphone to sound output of the loudspeaker. The total nonlinear distortion measured at the peak level of the reproduced sound was less than 0.3 per cent. From the results reported in Sec. 9.20, it will be seen that this value of nonlinear distortion is sufficiently low to be practically imperceptible.

The level of the reproduced sound in the listening room is important in any subjective test. Tests have shown that a peak level of about 70 decibels to 80 decibels is most pleasing for serious listening in a small room. The average peak sound intensity level on a standard-level indicator was 75 decibels.

The same six-piece band was used in these tests as in the case of the all-acoustic frequency-range tests. The change from full-frequency range to restricted-frequency range was made every 30 seconds. The results of these tests indicate a preference for the full-frequency range. Similar tests were made for speech. The frequency-preference tests for speech also indicate a preference for the full-frequency range.

9.24 THE OBJECTIVE AND SUBJECTIVE ASPECTS OF SOUND REPRODUCTION[16]

The definition and connotation of the term sound reproduction imply that a facsimile of the original sound is achieved in the reproduction. Sound reproduction involves both objective and subjective considerations in the mechanism and phenomena required to achieve a high order of fidelity of performance. From an objective viewpoint, sound reproduction is a complex process because a large number of acoustic, mechanical and electronic elements and combinations thereof are involved. These elements must exhibit a high order of excellence of performance in order to provide a resemblance to a perfect transfer characteristic. From a subjective viewpoint, sound reproduction is a complex process because a large number of psychoacoustical effects are involved. These effects must be used in an appropriate manner in order to produce an ideal transfer characteristic which supplies a close artistic resemblance of the original live rendition.

The main purpose of sound reproduction is to provide the listener with the highest order of artistic and subjective resemblance to the condition of a live rendition. To achieve this objective requires the topmost degree of excellence of the physical performance of the equipment as directed by the psychological factors involved in the process.

[16] Olson, H. F., *Audio Engineering Society Preprint No. 348*, October 12–16, 1964.

In general, the state of the art in sound reproduction has advanced to a stage where a high order of physical performance can be obtained. Up until very recent times, the application of the psychological characteristics as related to sound reproduction has lagged behind the physical considerations. The commercialization of stereophonic sound in the consumer complex has hastened the work in the subjective aspects of sound reproduction.

Recent work in the field of psychological acoustics as related to the performance of a sound-reproducing system involves the following specific subjects: tolerable sound-pressure levels; sound-pressure level and ear-frequency response; loudness and dynamics; dynamics, noise and masking; auditory perspective; reverberation; quality and timbre.

A. Tolerable Sound-pressure Levels

The subjectively tolerable loudness of the sound reproduced in a small room in a home is an important factor involved in the reproduction of sound. Extensive subjective tests have been conducted on the stereophonic reproduction of sound at various sound-pressure levels under the acoustical conditions and environments of the average living room in the home. Studies have also been made of the reproduction level employed by consumers in their homes. These tests have shown that peak sound-pressure levels[17] of sound reproduction in the homes of consumers run from 70 to 90 decibels for 90 per cent of the listeners. The average listener in the home operates a sound-reproducing system at a peak sound-pressure level of 80 decibels.

The peak sound-pressure level of sound by a live performance in a large acoustic enclosure is about 100 decibels. Thus, it will be seen that the peak level of sound reproduction in the home is much lower than the level of a live performance.

The main reason why the average listener prefers a lower level of sound reproduction in the home as contrasted to the sound level in the concert hall, is that the tolerable peak sound-pressure level[18] in a small room is lower than the tolerable peak sound-pressure level in a large hall. The shorter mean free path and resultant faster growth and decay of sound in a small room appears to lead to a lower tolerable peak level in the small room. Subjective tests have indicated that the same results are obtained regardless of whether the sound program is live or reproduced.

[17] Peak sound-pressure level in these considerations is used to designate a level at which 95 per cent of the program lies below the peak level.

[18] The term tolerable peak sound-pressure level is used to designate the peak sound-pressure level of sound reproduction which the listener feels is acceptable and agreeable.

B. Sound-pressure Level and Ear-frequency Response

The frequency-response characteristic of the ear and the sound-pressure level of the reproduced program in a small room are two factors involved in the relatively low level of sound reproduction in the home.

The equal loudness frequency relations of hearing or the response-frequency characteristics of the human hearing mechanism have been determined by several investigators. The characteristics shown in Fig. 9.77 are an average of the investigations carried out by Fletcher and Munson,[19] Churcher and King,[20] and Robinson and Dadson.[21] As far as the compensation is concerned in going from one level to another for a change of 20 decibels or less—which is the point of interest—the individual data from each of the three different investigations yield very nearly the same results. Essentially the same results are obtained from the response-frequency characteristics of the hearing mechanism from the draft recommendation of the I.S.O.[22] Therefore, it seems logical to use the average data of the three investigations listed above.

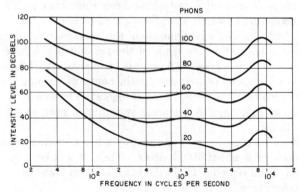

FIG. 9.77. Free-field contour lines of equal loudness for normal ears. Numbers on curves indicate loudness level. 0 dB = 10^{-6} watts per square centimeter. 0 dB = 0.0002 dyne per square centimeter.

The response-frequency characteristics of the human hearing mechanism as depicted in Fig. 9.77 indicate that certain of the frequency ranges must be increased or decreased in amplitude in order to maintain the quality balance of music when it is reproduced at a lower level than the original.

[19] Fletcher and Munson, *Jour. Acoust. Soc. Amer.*, Vol. 5, No. 1, p. 82, 1933.

[20] Churcher and King, *Jour. Inst. Elec. Eng.* (*London*), Vol. 81, No. 1, p. 57, 1937.

[21] Robinson and Dadson, *Brit. Jour. App. Phy.*, Vol. 7, No. 5, p. 116, 1956.

[22] International Organization for Standardization, Draft Recommendation, R 226, 1961.

Therefore, means must be provided to correct dynamically for the change in response of the ear with respect to frequency for a level drop of 20 decibels. There must be a continuous variation in the response-frequency characteristic as the level changes.

The frequency-response characteristics of the ear shown in Fig. 9.77 are essentially for free-field conditions. Therefore, the acoustic characteristics of the average room in a home and the loudspeaker operating in the room must be considered in establishing over-all frequency-response characteristics of the ear, the loudspeaker, and the room.

In the reproduction of sound in a room there are two sources of sound with respect to the listener, namely, the direct sound from the loudspeaker and the generally reflected sound. The acoustical characteristics of the average room in a residence accentuate the low-frequency response as can

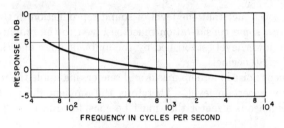

Fig. 9.78. Relative response derived from the dynamic average of the direct and reflected sound reproduced in a room in a residence.

be deduced from the reverberation characteristic of a typical living room in a home as shown in Fig. 8.14. The general run of direct-radiator loudspeakers exhibit increased directivity with increase of the frequency. The combination of the acoustical characteristics of the room and loudspeaker conspires to produce an accentuation in the low-frequency response as perceived by the listener. The relative response[23] at normal listening distances derived from the dynamic average of the direct and generally reflected sound for the case of music reproduced in a room in a residence is shown in Fig. 9.78.

C. Loudness and Dynamics

The loudness versus loudness level[24,25] of a performance must also be considered in the transition from the concert hall, auditorium, studio or any large room to the small room in the home. The loudness in sones is

[23] H. F. Olson, *Acoustical Engineering*, D. Van Nostrand Company, Princeton, 1957.
[24] Stephens, S. S., *Jour. Acoust. Soc. Amer.*, Vol. 27, No. 5, p. 815, 1955.
[25] Lochner and Burges, *Jour. Acoust. Soc. Amer.*, Vol. 34, No. 5, p. 576, 1962.

the average person's estimate of the loudness sensation. The loudness in sones is given by

$$S = 2^{(P-40)/10} \qquad (9.2)$$

where S = loudness in sones

P = loudness level, in phons

The loudness versus loudness level is another factor which must be considered in the transition from the relatively high-sound level in the large acoustic enclosure to the relatively low-sound level in a small room. Dynamics in the framework of sound reproduction is used to designate the relation between the subjective and objective aspects of the amplitude of the sound.

D. Dynamics, Noise and Masking

In view of the fact that the level of sound reproduction in the home is relatively low, some consideration must be given to the amplitude range between the lower level established by the ambient noise and the average upper level of 80 decibels.

The ambient noise in a room masks and renders inaudible the reproduced sounds below a certain level. Therefore, the ambient noise level in the average residence is a factor that must be considered in the reproduction of sound.

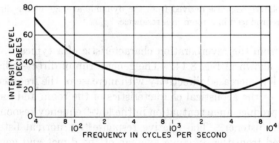

FIG. 9.79. Hearing limit for pure tones for the average listener in a typical residence.

The ambient noise level[26] for the entire audible range in 90 per cent of the residences falls somewhere between 33 and 52 decibels. The noise level in a room in an average residence is 43 decibels.

The spectrum of room noise is another factor that determines the level at which sounds disappear in the ambient as a function of frequency. The spectrum of typical room noise[27] decreases 5 decibels per octave with in-

[26] Seacord, D. F., *Jour. Acoust. Soc. Amer.*, Vol. 12, No. 1, p. 183, 1940.

[27] Hoth, D. F., *Jour. Acoust. Soc. Amer.*, Vol. 12, No. 4, p. 499, 1941.

crease in frequency as shown in Fig. 8.41. This type of characteristic appears to be the nature of practically all types of ambient noise.

From the average sound level and spectrum of the ambient noise it is possible to determine the lower hearing limit for pure tones.[28] The graph of Fig. 9.79 depicts the level below which a pure tone cannot be heard for a room in which the total sound level is 43 decibels. A large number of direct listening tests have been carried out to determine the level below which a pure tone disappears in the ambient. These data substantiate the characteristics of Fig. 9.79 within the usual limits of subjective tests. The threshold characteristic of Fig. 9.79 establishes the lower level of hearing in a room in an average residence.

E. Auditory Perspective[29]

Stereophonic sound reproduction as depicted in Fig. 9.25 provides auditory perspective of the reproduced sound.   Reproduction of sound in auditory perspective provides the illusion of the distribution of the reproduced sound in lateral directions as well as in depth in a geometrical configuration and correspondence which approximate the disposition of the original sound sources.

The microphone placement[30] with respect to the sound sources plays an important part in the auditory perspective aspects of the reproduced sound in the living room.   The sound sources and the microphone placement in the studio have been studied by means of listening tests in a typical living room.   In this work the most important result has been the observation that the subjective location of a sound source in stereophonic sound reproduction is determined by the phase and amplitude.   This has been illustrated by the experiment depicted in Fig. 9.26.   This experiment shows that the apparent position of the sound source can be shifted in a lateral direction from a point directly in front of either loudspeaker to any point between by varying the relative phases and/or relative intensities of the sound emanating from the loudspeakers of channels A and B.

F. Reverberation[31]

The growth, duration, decay and reverberation characteristics of the sound as perceived by the listener of Fig. 9.25 involve the combination of two rooms.   Here the objective is to provide an over-all growth, duration, decay and reverberation characteristic which provides the listener with a subjective impression of these properties which match those of the

[28] Fletcher and Munson, *Jour. Acoust. Soc. Amer.*, Vol. 9, No. 1, p. 1, 1937.
[29] Fletcher, H., *Jour. Soc. Mot. Pic. Tel. Eng.*, Vol. 61, No. 3, p. 415, 1953.
[30] Olson, H. F., *Jour. Audio Eng. Soc.*, Vol. 6, No. 2, p. 80, 1958.
[31] Olson, H. F., *Jour. Audio Eng. Soc.*, Vol. 12, No. 2, p. 98, 1964.

live rendition in the acoustic enclosure. In view of the fact that these characteristics in the living room play a minor part in the impression of the reverberant sound, the design of the acoustics of the studio[32] and the placement of microphones become the important considerations. In this connection, the growth characteristic plays the most important part in providing the desired artistic effects. The relative importance of the reflected sounds decreases with each reflection due to absorption at each encounter with boundaries. Therefore, the first reflections should be spaced in time so as to produce a smooth growth characteristic. In view of the fact that two rooms are involved in sound reproduction, the envelope of the transfer characteristic of the studio should be smooth with respect to frequency. The general reverberation of the sound which provides the blending of the reflected sounds with the direct sound imparts the artistic aspects associated with concert halls, auditoriums and theaters. The reverberation time of the acoustic enclosure in combination with the placement of the direct and reverberant microphones should be coordinated so that the over-all effect produced at the listener corresponds to the live condition. If the above general procedures are followed the general impression of the reverberant sound will be a subjective resemblance of the reverberant sound in the live condition.[33]

G. Quality and Timbre

Studies[34] have been carried out on the characteristics of musical instruments. These studies have included the frequency range, frequency spectrum, the directional pattern and the growth, duration and decay of the tones produced by the instruments.

The frequency range of musical instruments covers the entire audible frequency range. Therefore, to achieve faithful sound reproduction the reproducing system must cover the audible frequency range without frequency discrimination. The frequency response of the system must be free of variations otherwise the frequency spectrum of the instrument will be altered. The effect of nonuniform response with respect to frequency can be illustrated by referring to typical spectrums of musical instruments as shown in Sec. 6.3C. Obviously, if there are variations in the response the spectrum will be changed and as a result the character of the musical instrument will be changed.

[32] Bolle, Voldner, Pulley and Volkmann, *Jour. Audio Eng. Soc.*, Vol. 11, No. 1, p. 80, 1963.

[33] The reverberation of the live condition to be simulated in the room in a home may be that of a cathedral, church, concert hall, auditorium, theater, ballroom, restaurant, etc. In some popular music for example, the reverberation is almost nonexistent.

[34] See Chap. 6.

The directional pattern of voiced and musical instruments with respect to frequency and orientation are exceedingly complex as shown in Sec. 6.5. Therefore, the timbre which plays such an important part in the identification of the voice or instrument varies with the orientation. As a consequence great care must be taken in providing the orientation with respect to the microphone which will give the most realistic reproduced sound identified with the particular voice or instrument.

H. Implementation of Objectives

The main purpose of sound reproduction as manifested by the performance and experienced by the listener is to provide the listener with the highest order of artistic and subjective resemblance to the condition of the live rendition.

The psychological and psychophysical effects involved in achieving the objective have been described in the preceding sections. The purpose of this section is to delineate in abbreviated form the processes that must be implemented to attain the objective.

The growth, duration, decay, and reverberation characteristics as perceived by the listener involves the combination of two rooms. Here the objective is to provide an over-all growth, duration, decay and reverberation characteristic which provides the listener with a subjective impression of these properties which match those of the live rendition. In view of the fact that these characteristics in the average room in the home play a minor part in the impression of the reverberant sound, the design of the studio and the placement of the microphones are the important considerations.

The auditory perspective of the reproduced sound should approximate that of the best location in the standard live condition. To accomplish this result requires the correct placement of the microphones.

The quality of the reproduced sound must be preserved in reproduction. This means that all the characteristics of the musical instruments must be faithfully reproduced. The technical data relating to the characteristics of musical instruments must be employed as outlined in preceding sections in order to provide faithful reproduction of voices and instruments.

The dynamics of the sound which involves the relation between the subjective and objective aspects of the amplitude of the sound should simulate the live condition in the reproduced sound. One of the most important considerations relating to the dynamics is the subjective effect of the drop in sound level of approximately 20 decibels between the sound level of the live program and the tolerable sound level of the reproduced sound, the reason being that the frequency response of the human hearing mechanism is a function of the sound level.   Therefore, suitable dynamic frequency-response compensation as a function of the amplitude must be introduced by appropriate electronic means.

I. Dynamic Spectrum Equalizer

The preceding sections on the objective and subjective aspects of sound reproduction have been concerned with the relations involved in obtaining a simulation of the live condition in the reproduction of sound.   One of the main considerations relating to the dynamics is compensation for the 20 decibels difference in level between the live and reproduced sound.   An electronic system termed a Dynamic Spectrum Equalizer has been developed to provide the compensation for the difference in sound levels and other minor differences in the subjective aspects.

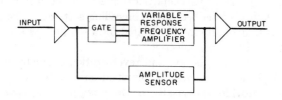

Fig. 9.80.   Schematic block diagram of the Dynamic Spectrum Equalizer.

A block diagram of the Dynamic Spectrum Equalizer is shown in Fig. 9.80.   The system operates in a continuous manner to change the response-frequency characteristic as a function of the amplitude.   Typical response-frequency characteristics for various levels are shown in Fig. 9.81.   There is a continuous variation in response from one level to another.   The response-frequency characteristics differ for different types of musical selections; in effect, when the levels are low the low-frequency components are accentuated.   For medium levels there are slight accentuations in the low-frequency region and the presence region of 2,000 to 6,000 cycles per second and a reduction in response in the region from 400 to 1,000 cycles per second.   For high sound levels there is accentuation in response in the presence region, and a reduction in the frequency range below 1,000 cycles per second.   When the sound level of the program is low, the objective is to raise the sound level of the appropriate frequency regions so that the music can be appreciated under the ambient noise and surrounding conditions of the average residence.   When the sound level of the program is high, the level of the presence region is raised and the level of the low-frequency range is lowered.   This procedure does not upset the dynamic balance but rather enhances this aspect of sound reproduction in a small room.

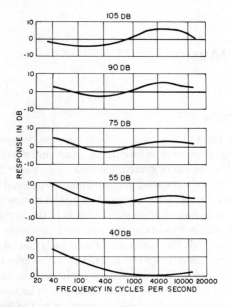

FIG. 9.81.   Response-frequency characteristics of the Dynamic Spectrum Equalizer for various sound levels of the program.

The Dynamic Spectrum Equalizer was designed to provide a dynamic alteration of the projection qualities of sound so that under conditions of playback, which differ from those in which the music was performed, the best perception of the qualities of the original performance would be obtained.

CHAPTER TEN

Electronic Music

10.1 INTRODUCTION

Music is both an art and a science. Music may be defined as the art of producing pleasing, expressive or intelligible combinations of tones. Most music is recorded by means of symbolic notations on paper for later rendition as sound. Music composed by humans is rendered (produced as sounds) by the voice or musician-actuated musical instruments. The ultimate objective destination of all musical sounds is the human ear. Thus the production of music consists of the following processes: the symbolic notation upon paper by the composer; the translation of the symbolic notation into musical sounds by the musician; the employment of either the human voice or a musical instrument or both; and the actuation of the human hearing mechanism by the musical sounds. The medium of transmission from the musician and musical instrument to the listener is sound waves. These sound waves carry the musical tones. The properties of a musical tone are frequency (pitch), intensity (loudness), growth, duration, decay, portamento, timbre, and vibrato and deviations as depicted in Fig. 10.1. Descriptions of the properties of a tone have been given in preceding chapters. Once a sound or a tone has been described by means of the characteristics of Fig. 10.1, it is possible to generate or produce this tone by electronic means. Thus it will be seen that it is possible to generate any tone produced by a voice or a musical instrument by employing an electronic system. In addition, it is possible to produce musical tones which cannot be produced by the voice or conventional instruments. In other words, the process of translating the musical notation on paper into the corresponding musical sounds can be accomplished by an electronic system. Furthermore, the electronic system can reproduce or create any sound or combinations of sounds, which have or have not been produced, that may have any possible musical significance.

New theories in the field of communication provide powerful tools for use by the composer in the composition of music. Random probability systems which operate under the guidance of certain rules and limits lead

to the composition of music.   Computers or specially designed computers
have been employed for the composition of music.

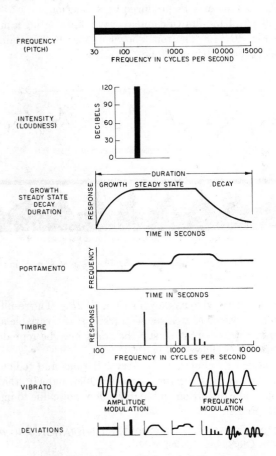

FIG. 10.1.   The fundamental characteristics of a tone.

The purpose of this chapter is to describe electronic systems for syn-
thesizing and composing music.

10.2   MUSIC SYNTHESIS BY ANIMATION[1,2]

The trace of any complex wave may be drawn by hand by following and
extending the exposition of Sec. 6.3A.   The trace of a wave includes all of

[1] Lewis and McLaren, *Jour. Soc. Motion Picture Eng.*, Vol. 50, No. 3, p. 233, 1948.
[2] McLaren, Norman, *Jour. Acoust. Soc. Amer.*, Vol. 31, No. 6, p. 839, 1959.

the information on the fundamental properties, namely, frequency, intensity, growth, duration, decay, portamento, timbre and vibrato. Therefore, any musical tone may be produced by animation.

An example of a hand-drawn complex wave representing a musical tone is shown in Fig. 10.2A. The tone shown in Fig. 10.2 contains many of the fundamental properties of a tone.

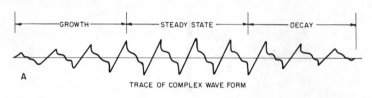

A TRACE OF COMPLEX WAVE FORM

B COMPLEX WAVE FORM FOR OPTICAL SOUND REPRODUCTION

Fig. 10.2 *A*. A hand-drawn complex wave. *B*. The wave of *A* inked in for reproduction by an optical sound reproducer.

The wave may be inked in as shown in Fig. 10.2B. The resultant drawing now becomes a variable area sound track. If the variable area sound track is drawn on transparent plastic the track may be reproduced by a system similar to that shown in Fig. 9.59.

Music synthesis by animation is a very laborious and tedious process. There are two factors that deteriorate the product, namely, the introduction of spurious wave distortion and high random noise due to inaccuracies in the drawing process.

10.3 ELECTRONIC MUSIC SYNTHESIS BY THE GENERATION AND MODIFICATION OF ORIGINAL SOUNDS[3,4,5,6]

Electronic synthesis of music may be carried out by the generation and modification of all manner of original sounds. A catalog of natural sounds may be augmented by the production of sound by all kinds of electronic generators. Modifying and shuffling these sounds creates a variety of new sounds which may be classified as discrete units. These units and combinations of units may be assembled to constitute musical productions.

[3] Douglas, *The Electrical Production of Music*, Philosophical Library, New York, 1957.

[4] Ussachevsky, Vladimir A., *Jour. Audio Eng. Soc.*, Vol. 6, No. 3, p. 202, 1958.

[5] Badings and deBruyn, *Philips Tech. Rev.*, Vol. 19, No. 6, p. 191, 1957/58.

[6] Searight, J., *Radio Electronics*, Vol. 36, No. 6, p. 36, 1965.

There are an infinite number of ways in which electronic synthesis of music may be carried out by the generation and modification of all manner of original sounds.   Electronic synthesis of music by the generation and the modification of original sounds has been termed Musique Concrete in France, Electronic Music in Germany, Italy, Holland and Japan, and Tape-Music at Columbia University.   The difference in the various schools exists in the specific type of approach that is employed to carry out the process.   For example, the source and nature of the original sounds and the modification processes are the areas in which the approaches of the various schools differ.   The purpose of this section is to describe the general process of the electronic synthesis by the generation and modification of original sounds.

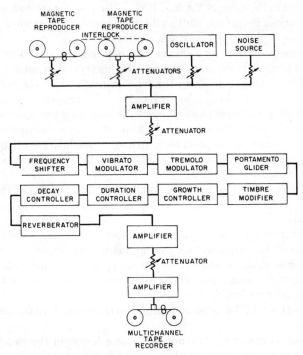

FIG. 10.3   Schematic diagram of the apparatus for the production of electronic music by mixing and modifying original sources of sound.

A generalized diagram depicting the main elements and the process of the production of electronic music by mixing and modifying original sources of sound is shown in Fig. 10.3.   The original sources of sound may be recordings on magnetic tape, oscillators and noise generators. The recordings on magnetic tape may be all manner of sounds such as the sounds of voice and musical instruments, the sounds of nature, etc. In some cases the output of more than one magnetic tape reproducer may be mixed as shown in Fig. 10.3. Oscillators with various wave shapes may be employed. For example, the output of an oscillator with a saw-tooth wave shape contains the fundamental and all the harmonics. The output of an oscillator with a triangular wave shape contains the fundamental and all the odd harmonics (see Sec. 6.3A). The noise generator is designed to produce white noise and other signals of a random nature. A mixing system is used to combine the output of the sound sources in the desired arrangement as determined by the musician. The output of the mixing system is fed to the input of the modifiers. The modifiers include elements for introducing frequency shift, vibrato, tremolo, portamento, timbre change, growth, duration, decay and reverberation.

The frequency shifter may change the frequency by a fixed amount as in the case of a single sideband system. If this type of frequency shifting is applied to a signal containing the fundamental and all the harmonics, the overtones will no longer be harmonics. The frequency shifter may change all the frequencies of all components by multiplication or division. In this case the frequency relations of the overtones are maintained.

The vibrato modulator provides a frequency modulation of the input frequency. The modulation frequency is usually somewhere below 10 cycles per second.

The tremolo modulator provides an amplitude modulation of the input frequency. The modulation frequency is usually below the audio-frequency range.

The portamento glider supplies a continuous frequency glide in going from a tone of one frequency to a tone of another frequency.

The timbre modifier consists of frequency selective networks which accentuates or attenuates the various components of the overtone structure of the complex tone input.

The growth controller determines the growth or attack characteristic of the tone.

The duration controller determines the time length of the steady state of the tone after the growth has been established.

The decay controller determines the decay characteristic of the tone. The decay follows the duration of the tone.

The growth, duration and decay controllers are interconnected because there is a sequence of events in the three characteristics. For example, the duration time may be zero which means that decay of the tone follows immediately after the growth of the tone.

Reverberation is an artistic embellishment which plays an important part in blending of a series of tones. Subjective considerations have established that for each type of music there is an optimum value of reverberation. A reverberator is included in the system of Fig. 10.3 to supply artificial reverberation to the synthesized sound (see Sec. 8.3K).

After all the characteristics of the modifiers are established the processed tone is recorded. The recording is made with a synchronized-multichannel magnetic tape recorder. In this way, several different programs may be recorded on a single magnetic tape. In most cases the magnetic tape recorder employs a tape with sprocket holes. Seven magnetic tracks can be recorded on the magnetic tape.

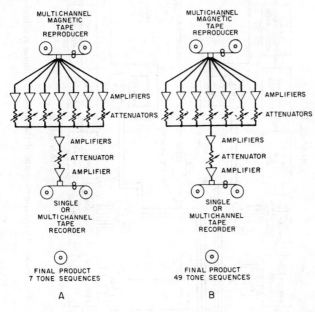

Fig. 10.4. *A*. Seven concurrent tone sequences recorded on seven separate tracks are converted to a single track. *B*. The seven tracks each containing seven concurrent tone sequences are converted to a single track comprising a combined total of 49 concurrent tone sequences.

The use of a synchronized-multichannel magnetic tape recorder makes it possible to record and synchronize seven tone sequences as shown in Fig. 10.4*A*. (See Sec. 9.12.) In Fig. 10.4*A*, the seven tone sequences are mixed and recorded on a single track. The program of the final product of Fig. 10.4*A* represents seven tone sequences. In Fig. 10.4*B*, seven sets of seven tones each are recorded on seven channel recorders. The record-

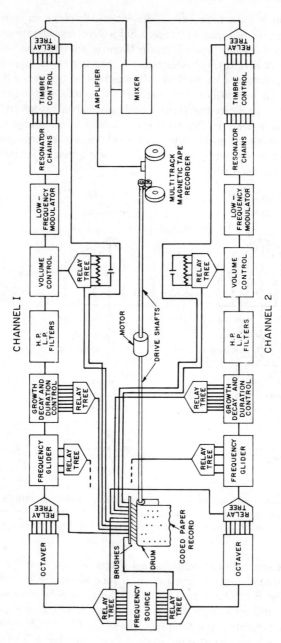

Fig. 10.5. Schematic diagram of the electronic music synthesizer.

ing of the seven tones is reproduced, mixed, combined, and recorded on a single track. The program of the final product of Fig. 10.4B represents the combination of 49 tone sequences. The process can be carried another step and the final product represents the combination of 343 sequences of tones which is equivalent to an orchestra of 343 pieces.

The electronic synthesis of music can be carried out to provide auditory perspective. The factors involved are phase and amplitude of the components in the two channels as outlined in Sec. 9.7.

The final product of the electronic music synthesis is recorded on magnetic tape or disc for rendition of the music in an auditorium or in the home.

A rather highly sophisticated system for modifying a tone has been described in the preceding exposition for the purpose of illustrating the process. All manner of changes can be made in the components and arrangement of the system depicted in Figs. 10.3 and 10.4 for the electronic synthesis of music by the generation and modification of original sounds depending upon the requirements of the composer.

10.4 RCA ELECTRONIC MUSIC SYNTHESIZER[7,8]

The properties of a tone are frequency, intensity, growth, duration, decay, portamento, timbre, vibrato, and deviations. When these properties of a tone as depicted in Fig. 10.1 are completely specified, the tone can be completely described. The RCA Electronic Music Synthesizer is based upon the construction of a tone in terms of the properties as depicted in Fig. 10.1.

The use of an electronic music synthesizer for the production of musical sounds opens an entirely new field for the production of recorded music. For example, there is the possibility of entirely new tone complexes and combinations which cannot be achieved with conventional instruments. Furthermore, in the case of conventional instruments, the musician is limited to the use of lips, mouth, ten fingers, two hands and two feet to perform the different operations. This limitation does not exist in the electronic music synthesizer. Conventional instruments produce various noises such as the rushing of wind in wind instruments, bow scratch in the viol family, various clatters and rattles in plucked and struck-string instruments, and mechanism rattle in any instrument in which keys, valves, levers and shafts are used. These undesirable noises do not exist in the electronic music synthesizer. With the advent of the electronic method for the production of musical tones, new musical compositions can be written which take advantage of the superior characteristics of the electronic music synthesizer.

From the preceding discussions and Fig. 10.1, it is evident that, in order

[7] Olson and Belar, *Jour. Acoust. Soc. Amer.*, Vol. 27, No. 3, p. 595, 1955.
[8] Olson, Belar and Timmens, *Jour. Acoust. Soc. Amer.*, Vol. 32, No. 3, p. 311, 1960.

to synthesize any musical tone whatsoever, the electronic music synthesizer must provide the following facilities: means for producing a tone with any fundamental frequency within the audio-frequency range; means for producing a tone with any overtone structure; means for producing a tone of any growth, duration, or decay characteristic; means for introducing a vibrato; means for changing the intensity of the tone; means for providing a portamento or glide from a tone of one frequency to a tone of a different frequency; means for introducing, in some instances, various deviations in these characteristics.

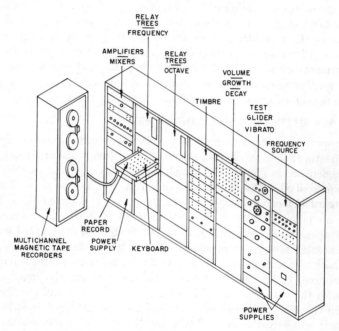

FIG. 10.6. Perspective view of the electronic music synthesizer.

A schematic block diagram of an electronic music synthesizer with means for producing all the characteristics of a musical tone outlined above is shown in Fig. 10.5. The coded paper record controls all the functions of the electronic music synthesizer. The information is recorded and stored in the paper record in the form of perforations. When the paper record is run through the machine, brushes slide over the paper record. An electrical circuit is closed when a brush passes over a perforation. The brushes which pass over the paper record actuate electrical circuits in the relay trees. The use of a relay tree makes it possible to record the information on the paper record in the binary code system. There is a separate relay tree for each of the characteristics of a tone. In this way any of the characteristics of a tone can be obtained or changed at any instant. Thus it will be seen that any tone whatsoever can be produced by providing the proper information in the coded paper record. The output of the synthesizer is recorded on a magnetic tape recorder. The coded paper record and the magnetic tape recorder are driven to synchronism by an interconnecting cable drive. In the complete electronic music synthesizer two complete channels, as shown in Fig. 10.5, are used and operated from the single coded paper record. This makes it possible for the coded paper record to set up one channel while the other channel is in operation and producing a tone. Furthermore, one channel can start playing a tone before the other channel stops playing a tone. Everything is duplicated in the second channel except the 12 tuning fork oscillators which supply the tones in one octave.

A schematic-perspective view of the complete electronic music synthesizer is shown in Fig. 10.6. The seven racks contain all of the electronic equipment for performing the different functions. The information

Fig. 10.7. Photograph of the punched paper record, the keyboard puncher, the brushes, and the paper drive mechanism.

contained in the perforated paper record in conjunction with the brushes provides the means for actuating the relays. The relays in turn activate the appropriate electronic elements. The electrical output is recorded on a magnetic tape recorder.

The paper record is punched by means of a keyboard punching system shown in Fig. 10.7. The keys are colored to facilitate the operation of punching the codes. The note-selecting group of 1, 2, 4 and 8 are red. The octave group is yellow. The growth, duration and decay group is blue. The timbre group is gray. The volume control group is black.

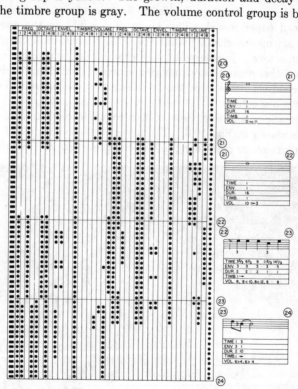

Fig. 10.8.   A copy of the paper record containing a phrase of *Obelin*.

Referring to Fig. 10.7, it will be seen that the punched record consists of rows of holes. Each row of holes passes under a brush. When the brush passes over a hole, the brush makes contact with the drum, and as a result closes the actuating circuit in the relay tree. Each brush is equipped with several springs arranged so that the brush never breaks contact with the drum before making contact at the adjacent hole. Thus a row of holes will provide continuous contact and at the same time give the same result as a slot in the paper. Slots cut in the paper will result in a very weak paper record that can be easily torn and with poor lateral rigidity.

A simple punched record for playing a phrase of *Obelin* is shown in Fig. 10.8. The record is drawn to scale and has the length indicated in inches. A paper speed of 4 inches per second was chosen for this selection. The corresponding measures in conventional musical notation are also shown in Fig. 10.8. The rows of holes are numbered in the binary code numbering. Referring to Fig. 10.8, it will be seen that the coding of the left half is for one synthesizer channel and the coding on the right for the other synthesizer channel. Referring to the growth, duration, and decay coding, it will be seen that the notes are executed alternately by the first and second channel. Fig. 10.8 shows the changes in growth, decay, and volume which are the synthesist's interpretation of this musical selection as called for in detail by the coded information shown below the conventional music notation.

When the paper record has been punched and the settings of the various elements of the synthesizer have been established, the next step is the recording of the output of the synthesizer. The output is recorded on a synchronized seven-channel magnetic tape recorder. The outputs of seven series of tones are converted to a single track on a seven-channel magnetic tape recorder as shown in Fig. 10.4*A*. This can be carried out seven times and the outputs combined as shown in Fig. 10.4*B*. The combined output now represents 49 series of tones. This can be extended another step to 343 series of tones which is equivalent to an orchestra of 343 pieces.

The first step in the electronic synthesis of music is the evolution of the musical score by the composer or arranger. The composer or arranger determines the various tone structures which will be employed to produce the first product. He writes the musical score for each of these series of tones. The musical score carries the specific information on the scale, frequency, and duration of the tone. The musician indicates on the musical score the growth and decay and the timbre. He also indicates glides and vibrato if these are used. When the musical score with the specific notations has been completed, as shown in Fig. 10.9, the next step is the transfer from the musical score to the code.

Any musical score is itself a coded presentation of the composer's work which, before it can become music, requires further interpretation. Many factors including artistic considerations enter into the problem of interpretation making this the most difficult part of synthesis, but also the most rewarding from the viewpoint of the unlimited possibilities offered. Once the interpretation is made, the rest of the work can be done sys-

Fig. 10.9. Excerpt from the musical score for *Obelin* composed by Jim Timmens.

tematically as will be shown.   Returning to the example, of which score an excerpt is shown in Fig. 10.9, it was decided to synthesize a rhythm section in the Latin American style which would sound plausible, but not a copy of any existing instrument exactly.   The aim was to perform an experiment in creating music which would be commercial in today's

Fig. 10.10. General classification of nine sound categories called for in the musical score for *Obelin* by Jim Timmens as arranged for the electronic music synthesizer.

	RAPID GROWTH	SLOW GROWTH	VARIOUS GROWTH	IMMEDIATE DECAY (LOG)	VARIOUS DECAYS	NOISE SOURCES	FREQUENCY SOURCES	RESONATORS	FILTERS	FREQUENCY GLIDERS
CLAVIS	x			x		x		x		
CONGA	x			x		x		x		
BONGO	x			x		x		x		
BONGO PNEUMATICO		x			x	x		x		
BELL	x				x		x		x	
BASS	x		x				x		x	
SCRATCHER			x		x	x		x		
PLUCKED STRINGS	x			x			x		x	
MELODY			x		x		x		x	x

market, not too different yet still possessing novelty. Before the score had been completed experiments were made with short sections trying different envelopes of sounds in the percussion category, that is, envelopes with rapid growth followed immediately by a period of decay. After selecting resonators and filters which gave the desired result, the next task was the determination of an envelope normally associated with a lip reed instrument but still using the sound source and timbre quality selected for the percussive sound. The result is a nonexistent instrument which was

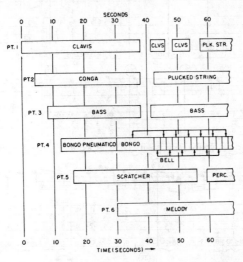

Fig. 10.11.   Organization of the parts for *Obelin*.

dubbed Bongo Pneumatico. The resultant sound suited the genre of the selection, and so it was decided to use it. Having established in similar manner various other interpretations, a schedule was made showing the general classification of the sound categories by which the score was interpreted. The classification schedule is depicted in Fig. 10.10. The number of setups and parts are determined from the schedule of Fig. 10.10. The procedure is carried out by considering the number of different sounds which must appear simultaneously in the finished selection and the resultant setups required. The organization of the parts for the sample selection is shown in Fig. 10.11. The writing of the individual parts is the next step.

Techniques were developed to set forth the specifications in both conventional musical and coded technical terms. The latter includes the desired interpretation and follows in a more detailed manner the percept of the performance score developed by Seashore.[9]   Fig. 10.12 shows page one for part 6 which carries the melody for *Obelin*.

Referring to Fig. 10.12, the bars denoting the measures in conventional notation are numbered.   The distance between bars represents 16 rows of holes in the paper record into which the information is to be punched.

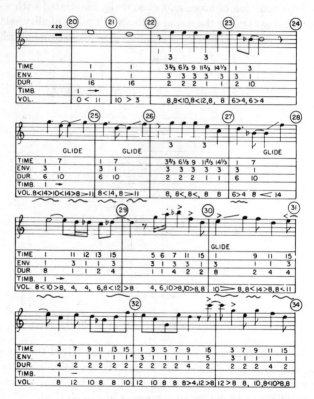

FIG. 10.12.   Synthesized code sheet for page 1, part 6, of *Obelin*.   MM: quarter note = 240.

At a paper speed of four inches per second and with a normal hole spacing of four to the inch, this means 16 holes per second will pass under the brushes in the playing of this part.   The normal length for a quarter note is then four holes or one inch in order to play at a metronome speed of 240 quarter notes to the minute.   The time at which the growth of a note is to begin is interpreted in the first tabulation under the staff.   Thus, any

[9] Seashore, C. E., *Psychology of Music*, McGraw-Hill Book Co., New York, 1938.

deviation desired can be specified. The next code specified is the envelope. In the example given, it was decided that this part should be of sustained tones capable of being varied in loudness during the sounding of a note and have different degrees of attack. Accordingly, a setup was made providing for different rates of growth according to the number called for in the schedule. The decay inherent in the instrument was arranged to be be always about the same in time constant but the duration between growth and decay was made unlimited. The duration itself was specified in the third row of the code tabulation. It may be noted that the duration of similar notes is not always specified to be of the same length. These deviations are one of the aspects of interpretation.

For the duration of the four measures shown in Fig. 10.12, the timbre was the same and specified as code 1. The actual timbre for that portion was produced with a harmonic spectrum containing all harmonics diminishing with their order but with the higher orders reduced more drastically. The last tabulation of codes concerns volume. The volume is nearly always changing as it would with a lip reed or bowed instrument.

Figure 10.8 shows how the directions outlined on the part of Figs. 10.11 and 10.12 were carried out in punching out the paper tape record. Time now becomes a longitudinal dimension measured in inches or in holes. The frequency of each note is determined directly from the music notation which is easily learned. The only difference between the notation and the suggested *American Standard for Subscripts* is that in order to make all relay trees have the same numbering, $C_0 = 16.352$ cycles is noted by C_1, but since octaves can be switched at will manually, this is no great handicap. Other parts in addition to the one described were recorded on the multiple magnetic tape recorder as depicted in Fig. 10.4 and the accompanying text.

Synthetic reverberation was also added by means of an electronic reverberator as described in Sec. 8.3K.

The preceding example has outlined the process employed in the production of an original composition by means of the electronic music synthesizer. To produce a new composition is difficult. To make a facsimile of an existing recording is also difficult. The latter process was also carried out and will be described in the text which follows.

Two different piano selections, namely, *Polonaise in A Flat Op. 53* (Chopin) and *Clair de Lune* (Debussy) and a violin and piano selection, *Old Refrain* (Kreisler), all played by famous artists and reproduced from commercial disc records were compared with synthesized versions of the same selections. The piano selections were recorded by Iturbi, Rubinstein, and Horowitz, and the violin in *Old Refrain* was recorded by Kreisler. The synthesizer versions, completed on August 5, 1953, and the identical passages from the commercial phonograph records were recorded on magnetic tape and intermixed and played to various people. They were asked to tell which was synthesized and which was not. Interpreting the results by standard statistical methods, it can be said with 70 per cent certainty that only one out of four persons can tell which is which.

The preceding experiment demonstrates that the electronic music synthesizer possesses the inherent capabilities of producing great music. The analysis and synthesis was carried out by an acoustical engineer. Furthermore, it should be noted that this work was done with only the help of conventional equipment to aid in the analysis which is not as complete as demanded for synthesis; thus much of the work was done by cut and try. This points to an important feature of the synthesizer. Manual dexterity is not required. A synthesis once learned can be added to the fund of knowledge without further practice to be able to perform it. The ability of the synthesist is, therefore, always increasing.

If a composer has in mind what he wants to achieve, the effects can be obtained by means of the electronic music synthesizer, regardless of whether he can play a musical instrument or not. The composer or musician can produce the sound of any existing musical instrument as well as other sounds, regardless of whether they have ever existed. The results which the composer and musician wish to achieve can be obtained and demonstrated as the music is being composed and played. Once a particular result has been obtained, it can be retained forever. Thus it will be seen that the electronic music provides a powerful tool for the composer or musician because he can reproduce or create any sound or combination of sounds which have or have not been produced, that may have any musical significance.

The above experiment demonstrates the potential capabilities of the synthesizer in the ability to copy existing selections. For an acoustical engineer to copy the performance of famous artists in the matter of a few weeks shows the tremendous possibilities of the synthesizer. For example, it would be impossible for even a genius with no previous experience in playing a musical instrument to imitate all of these artists on these two instruments with the order of fidelity that was achieved. Thus, it seems quite obvious that a trained or professional musician could produce great musical renditions by means of the electronic music synthesizer.

The above experiment illustrates another use for the synthesizer, in addition to those which have been listed, namely, the rejuvenation of old recordings where the master is in poor condition. That is, a new record without distortion and noise can be made by the synthesizer.

One use envisioned for the music synthesizer is the production of music for sale in the form of phonograph records. In order to demonstrate further the potentialities of the music synthesizer, complete musical selections were synthesized. A partial list of these selections indicating the style of music and the date on which the synthesis was completed was as follows:

Blue Skies (Berlin) April 1, 1952	In the style of a dance band.
Nola (Arndt) May 28, 1952	In the style of the piano.
Stephen Foster Medley, December 12, 1952	*Oh Susanna, De Camptown Races, My Old Kentucky Home, Old Black Joe, Old Folks at Home,* and *Hard Times Come Again No Mo.* In the style of bowed, plucked, and struck instruments; air, mechanical, and lip reed instruments; and percussion instruments.
Holy Night (Adams) December 15, 1952	In the style of the organ.
"Fugue No. 2" from *Well Tempered Clavichord* (Bach) July 15, 1953	In the style of ancient struck and plucked strings in several variations.
Hungarian Dance No. 1 (Brahms) September 3, 1953	In the gypsy style without copying any particular instrument, but varying colors adapted for easy synthesis.
Sweet and Low (Tennyson-Barnby) January 15, 1954	Voice and instrumental accompaniment to show that voice can be synthesized.
Spoken Voice March 26, 1954	A few spoken sentences were synthesized to show the versatility of the synthesizer.

The results obtained with the electronic music synthesizer as exemplified by the musical selections outlined above demonstrate that excellent musical performance can be produced by means of this new system; in fact, the performance of the musical synthesizer speaks for itself.

A phonograph record, RCA Victor LM-1922, entitled "The Sounds and Music of the RCA Electronic Music Synthesizer" illustrates the characteristics of a tone and music produced by the synthesizer.

Following the work described in the preceding text several composers have carried out the production of music by means of the electronic music synthesizer.

Richard Maltby, the well-known band leader, composer, and recording artist while with RCA Victor contributed original compositions and arrangements on the RCA Electronic Music Synthesizer.

Jim Timmens, a renowned composer and arranger, produced several original compositions on the RCA Electronic Music Synthesizer. These included *Obelin*, the original compositions used in the above exposition as an example and *Quito*, and unique arrangements of existing compositions.

RCA Electronic Music Synthesizer Mark II has been in operation at the

Columbia-Princeton Electronic Music Center since 1959. Milton Babbitt,
Vladimir Ussachevsky and Otto Leuming and other composers have synthesized music and given several public concerts of synthesized music. The
plans are to continue this program of the production of synthesized music
without interruption.

10.5   ELECTRONIC MUSIC SYNTHESIS BY DIGITAL COMPUTER[10,11]

Music may be synthesized by means of a digital computer. In this process
there is the key process involving a digital to acoustic converter. A schematic block diagram of the conversion process is shown in Fig. 10.13. A
magnetic tape is prepared by the computer on which is recorded successive
digitized samples of the acoustic output. The numbers are then converted
to pulses the amplitudes of which are proportional to the numbers. The
pulses are smoothed by a low-pass filter to obtain the input for an audio
loudspeaker. If the sampling rate is 20,000 per second the top frequency
from the loudspeaker will be 10,000 cycles per second. Each sample is
produced from a four-decimal number. The signal-to-noise ratio will be of
the order of 60 decibels. Within the limits of the frequency range and the
signal-to-noise ratio, the converter can produce any sound whatsoever,
provided the appropriate sequence of digital samples can be generated.

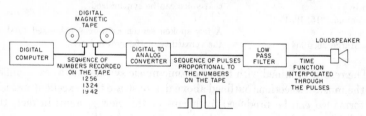

FIG. 10.13.   Digital to acoustic converter.

The basic procedure of the listing of 20,000 numbers per second by the
composer does provide a high order of generality. However, such a procedure is impossibly tedious and out of the question. Furthermore, such
a procedure does not lead to a practical solution of the effective control of
the parameters involved in the synthesis of music by means of a computer.
Therefore, as in the case of the electronic synthesis of music outlined in the
preceding section, the synthesis by a computer must also be based upon
the fundamental properties of a musical tone as depicted in Fig. 10.1.

The basic form of the generating program is a scheme for producing a
sequence of sounds representing individual "instruments." The "instru-

[10] Matthews, M. V., *Bell Syst. Tech. Jour.*, Vol. 40, No. 3, p. 677, 1961.

[11] Matthews, Pierce and Gutman, *Gravesaner Blatter*, No. 23/24, p. 119, 1962.

ments" are formed by combining a set of basic building blocks termed unit generators. Appropriate combinations of these unit generators can produce sounds of almost any desired complexity.

The compiling program is greatly simplified by the use of macro instructions which specify a sequence of computer instructions by a single statement. In this way each unit generator can be specified by a single macro statement.

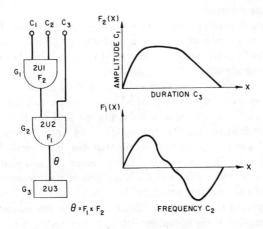

Fig. 10.14. An "instrument" with predetermined frequency, amplitude, growth, duration and decay.

The first step in the production of a musical selection is to punch a set of cards which specify the "instruments" of the "orchestra." These "instruments" are fed into the computer together with the computing program and the computer punches a card deck which is the music-generating program or "orchestra." Next a series of note cards or score must be prepared. These provide the properties of music such as frequency, duration and amplitude of the notes to be generated from the punched cards. These equivalents are termed compilers and produce a sequence of numbers equivalent to the sound of the particular "instrument." An "instrument" with a predetermined frequency, amplitude, growth, duration and decay is depicted in Fig. 10.14. The output of the generator G_3 provides a signal of a specified frequency, amplitude, growth, duration and decay from input θ digits. The magnitudes represented by these digits are determined by the products of two functions $F_1(X)$ and $F_2(X)$ shown in Fig.

10.14. The waveform in Fig. 10.14 is periodic but it can be of any desired form. The control signal C_2 determines frequency at which the function $F_1(X)$ is repeated. The control signal C_2 need not be constant. $F_2(X)$ is an envelope control for some given growth and decay characteristic as depicted in Fig. 10.14. The maximum amplitude is determined by the number C_1 and the total duration by C_3. In this single situation the composer called for a given musical rate of five numbers. Three are C_1, C_2 and C_3. The other two numbers select instruments to play and at what particular time interval.

Instruments of greater versatility and complexity may be developed by the addition of vibrato generators, a portamento glider, and other components. The "instruments" are compiled by punching the cards according to logic and operation of the computer.

All the "instruments" are compiled and inserted in the computer.

Each note is specified by one card which gives the duration of the note, the "instrument" on which it will be played and all the fundamental quantities required by the "instrument." The note sequences for each "instrument" are treated separately so that the note cards for different "instruments" can be interleaved. If two notes are sounded together it is only necessary that the sum of the durations of the preceding notes and sets of each of the two "instruments" be equal.

As in the case of the synthesizer described in the preceding section, the process is simplified if the composer listens to one series of notes at a time and makes suitable modifications and then groups the combination of tones after each one has been approved. Minor modifications can then be made in the group if such procedure appears to be desirable.

The exposition in the preceding text of this section has described one process for the electronic synthesis of music employing a digital computer. As the programing for computers is simplified, the general procedure will be simplified. However, the composer operating the synthesizer, regardless of the method employed for the electronic synthesis of music, is the final judge of the rendition. The analog and digital synthesizers can produce any tone whatsoever regardless of whether it has ever been produced before or not. Therefore, the method which will be selected is the one which provides the greatest ease of translating what the composer has in mind into the final sound product.

10.6 ELECTRONIC COMPOSITION OF MUSIC[12,13,14]

Some of the processes in the art of music composition can be acquired and learned by following the work of composers and mastering the fundamentals

[12] Pinkerton, Richard C., *Sci. Amer.*, Vol. 194, No. 2, p. 77, 1956.

[13] Hiller and Isaacson, *Experimental Music*, McGraw-Hill Book Company, New York, 1959.

[14] Olson and Belar, *Jour. Acoust. Soc. Amer.*, Vol. 33, No. 9, p. 1163, 1961.

of music. Composition can be stimulated by a variety of experiences and may be aided by research. However, the creative process of the composer is not fully understood because the ability to create is a gift.

Scientists have given considerable attention to investigations in music. The work of Helmholtz[15] on the pitch and timbre of musical tones is a classical contribution to musical acoustics. The outstanding investigations of Seashore[16] established the fundamentals of both the objective and subjective aspects of physics and psychology applied to the field of music. Pioneering work in the application of science to the field of music has stimulated both scientists and musicians to develop new means for the production of music. Tremendous advances made in electronics have stimulated investigations in the development of new musical instruments employing electronic systems. The electronic music synthesizer is an example of the application of modern communication principles to the field of music. The synthesizer produces musical tones from the electronically generated, fundamental, physical properties of a tone and a coded record.

For decades, composers have employed various aids for assistance in the composition of music. The new information theories[17] in the field of communications provide powerful new tools for use in the composition of music. Employing information theories, the art of composing music may be considered as the development of order from a chaotic state. That is to say, the subject of musical composition can be studied and implemented by applying certain general principles of information theory. In this theory, the information contained in a message is characterized as being dependent upon the number of available choices. As one specific example, random probability systems allow for the generation of sequences of notes in a series which are not completely random nor completely ordered but each is selected in random fashion with a probability which depends upon preceding notes. This coincides with the conventional process of musical composition as described by many composers, namely, a procedure involving a series of choices of musical elements from an essentially limitless variety of raw materials.

A. Music and Modern Communication Theories

Music may be considered as a form of communication. On this assumption modern communication principles may be employed as an aid in the composition of music. An important element of the theory of modern communication is the concept of entropy. Entropy may be considered as a measure of the degree of randomness in any system. In a condition in which there is a high order of uncertainty, the entropy is high. On the

[15] Helmholtz, H., *The Sensations of Tone*, Dover Publications, Inc., New York, 1954.

[16] Seashore, C. A., *Psychology of Music*, McGraw-Hill Book Co., New York, 1938.

[17] Shannon and Weaver, *The Mathematical Theory of Communication*, The University of Illinois Press, Urbana, 1949.

other hand, in a condition in which there is a high order of similarity or symmetry, the entropy is low. Considerations of the ramifications of entropy have led investigators to associate entropy with beauty and melody. Melody may be described as an agreeable succession or arrangement of sounds. That is to say, melody contains certain basic qualities of similarity, regularity or symmetry in the arrangement of the musical tones. From the standpoint of modern communication theory, the entropy of the melody must be sufficiently low so that a definite pattern is established. On the other hand, the entropy must be high enough to incorporate sufficient complexity to provide a degree of sophistication.

From the foregoing, the general conclusion may be drawn that the musical tones of a melody possess a relatively low value of entropy. This conclusion may be substantiated by a statistical analysis of music as described in the section which follows.

B. Statistical Analysis of Musical Compositions

Elementary statistical analysis has been carried out on simple musical selections that are known to have enjoyed age-old popularity. For this purpose, a statistical study of Stephen Foster music will be described. Specifically, the following Stephen Foster songs were selected: *Old Black Joe; Old Folks at Home; Massa's in the Cold Cold Ground; Hard Times, Come Again No More; Uncle Ned; My Old Kentucky Home; Oh Susannah; Camptown Races; Oh Boys, Carry Me 'long; Ring, Ring de Banjo; Under the Willow She's Sleeping.*

Fig. 10.15.   The notes of the musical scale used in the analysis of Stephen Foster songs.

The first order of approximation is the selection of the number of times each note occurs. The 12 notes of the musical scale used in the example in this paper for analyzing Stephen Foster songs are shown in Fig. 10.15. An analysis showed that these 12 notes were sufficient for composing the desired type of music. All songs were transposed[18] to the key of D in the analysis. The frequency of occurrence of the 12 notes in the 11 songs of Stephen Foster is shown in Table 10.1.

[18] G# is an accidental note.

TABLE 10.1. RELATIVE FREQUENCY OF THE NOTES IN ELEVEN STEPHEN FOSTER SONGS

Note	B_3	$C_4^\#$	D_4	E_4	$F_4^\#$	G_4	$G_4^\#$	A_4	B_4	$C_5^\#$	D_5	E_5
Relative frequency	17	18	58	26	38	23	17	67	42	29	30	17

A more complicated structure is obtained if successive notes are chosen not independently, but when their probabilities depend upon the preceding notes. In the simplest case of this type, a choice depends upon the preceding note only and not the one before that. The structure can be specified by the dinote probabilities; that is, the relative frequency of the dinote A, B, etc. The dinote probabilities for the eleven Stephen Foster songs are shown in Table 10.2. The base of the probability was chosen as sixteen because the relay tree in the random unit of the composing machine has 16 output contacts. The last note is one of the 12 notes shown at the top of Table 10.2. Under the last note and opposite the first note of the dinote there is a numeral indicating how many chances there are that the last note of the dinote will follow said note. For example, in Table 10.2, the first dinote tabulated is B_3, D_4. There are 16 chances in 16 (a certainty) that the note D_4 will follow the note B_3. On the third line of Table 10.2, nine dinotes are indicated with probabilities of 1/16, 1/16, 1/8, 5/16, 3/16, 1/16, 1/16, 1/16, and 1/16 respectively. Specifically, there is 1 chance in 16 that B_3, $C_4^\#$, G_4, A_4, $C_5^\#$ or D_5 will follow D_4. There is 1 chance in 8 that D_4 will follow D_4. There are 5 chances in 16 that E_4 will follow D_4. There are 3 chances in 16 that $F_4^\#$ will follow D_4.

TABLE 10.2. TWO-NOTE SEQUENCES OF ELEVEN STEPHEN FOSTER SONGS
Probability of following note

Note	B_3	$C_4^\#$	D_4	E_4	$F_4^\#$	G_4	$G_4^\#$	A_4	B_4	$C_5^\#$	D_5	E_5
B_3			16									
$C_4^\#$			16									
D_4	1	1	2	5	3	1		1		1	1	
E_4		1	6	3	4			1			1	
$F_4^\#$			2	4	5	2		2	1			
G_4				4	3			6	3			
$G_4^\#$								16				
A_4		1		5	1	1		4	3		1	
B_4		1		1	1			9	2		2	
$C_5^\#$									8		8	
D_5								4	7	3	1	1
E_5								6		10		

Probability of note following the preceding note expressed in sixteenths.

TABLE 10.3. THREE-NOTE SEQUENCES OF ELEVEN STEPHEN FOSTER SONGS

Dinote	B₃	C♯₄	D₄	E₄	F♯₄	G₄	G♯₄	A₄	B₄	C♯₅	D₅	E₅
B₃D₄			16									
C♯₄D₄			5	6				5				
D₄B₃			16									
D₄C♯₄			16									
D₄D₄		2	2	9	2	1						
D₄E₄			3	4	8			1				
D₄F♯₄				7	3	2		4				
D₄G₄					11				5			
D₄A₄					4			12				
D₄C♯₄											16	
D₄D₅								2	11	3		
E₄C♯₄			16									
E₄D₄	1		1	4	5			1		1	3	
E₄E₄		1	12	1	2							
E₄F♯₄			1	3	6	4		1	1			
E₄A₄								13	3			
E₄D₄										16		
F♯₄D₄				12	3	1						
F♯₄E₄		2	7	3	2			1			1	
F♯₄F♯₄			3	4	6	2		1				
F♯₄G₄					4	3		6	3			
F♯₄A₄					2			10	3		1	
F♯₄B₄								16				
G₄F♯₄				8		8						
G₄G₄						8		8				
G₄A₄			2					10			4	
G₄B₄								16				
G♯₄A₄									16			
A₄D₄			11	5								
A₄F♯₄			5	4	3	1		2	1			
A₄G₄				16								
A₄G♯₄								16				
A₄A₄					4	1	1	5	5			
A₄B₄			1		1			12	1		1	
A₄D₅								6	5	3	2	
B₄D₄			16									
B₄F♯₄				11	5							
B₄G₄									16			
B₄A₄			1		9	1		2	1		2	
B₄B₄					2			12			2	
B₄D₅								9	2	5		
C♯₅B₄								16				
C♯₅D₅									6			10
D₅A₄					14			2				
D₅B₄						1		5	6		4	
D₅C♯₅									12		4	
D₅D₅									16			
D₅E₅								5	11			
E₅A₄								16				
E₅C♯₅											16	

Probability of note following a dinote expressed in sixteenths.

The next increase in complexity involves trinote frequencies. The choice of a note depends upon the preceding two notes but not on any before them. The structure can be specified by the trinote probabilities; that is, the relative frequency of the trinote A B C , etc. Continuing with the trinote probabilities, a more complicated stochastic process is obtained. The trinote probabilities for the 11 Stephen Foster songs are shown in Table 10.3. The last note of the trinote is one of the 12 notes shown at the top of the table. Under the last note, and opposite the first two notes of said trinote (a dinote), there is a numeral indicating how many chances there are that the last note of the trinote will follow said dinote. For example, in Table 10.3 the first trinote tabulated is $B_3D_4D_4$. There are 16 chances in 16 (a certainty) that the note D_4 will follow the dinote B_3D_4. On the second line of the tabulation, three trinotes are indicated, namely, $C_4^\sharp D_4 D_4$, $C_4^\sharp D_4 E_4$, and $C_4^\sharp D_4 A_4$. Their probabilities are 5/16, 6/16, and 5/16, respectively. Specifically, there are 5 chances in 16 that the note A_4 will follow the sequence $C_4^\sharp D_4$. On the third line, only one trinote is indicated, namely, $D_4B_3D_4$, with a probability of 16/16, that is, 16 chances in 16 (a certainty) that note D_4 will follow the dinote D_4B_3.

The way the trinote probabilities are determined and expressed in sixteenths will be understood from the following example. Suppose there are three trinotes based on the dinote A,B and it is found that these trinotes A B A, A B B, and A B C occur 7, 14, and 13 times, respectively. The total occurrences are 34. Thus, there are 7 chances in 34 that the note A will follow the dinote A B. Expressed to the nearest sixteenth, there are 3 chances in 16 that note A will follow the dinote A B. The other trinote probabilities are determined similarly.

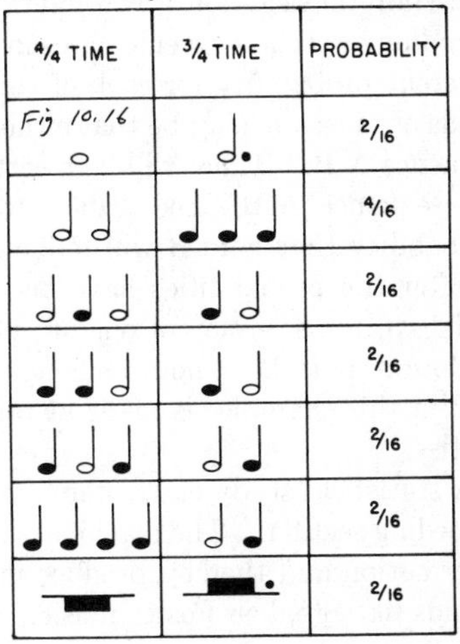

4/4 TIME	3/4 TIME	PROBABILITY
Fig 10.16		2/16
		4/16
		2/16
		2/16
		2/16
		2/16
		2/16

FIG. 10.16. The rhythm probability for 4/4 and 3/4 time.

A series of sound pulses, with varied spacing in time and different intensities, possess a subjective quality termed rhythm. (See Secs. 2.4 and 7.4J.) Rhythm in music is a regular occurrence of stressed and relaxed sound pulses. The term rhythm means a repetition of groups of sounds at regular intervals. There must be at least two similar groups in order to establish rhythm. Rhythm must be incorporated into a succession of notes in order to make a satisfactory melody. The probabilities for rhythm may be established in the same way as they are for notes. A rhythm probability for 3/4 and 4/4 time is shown in Fig. 10.16.

A consideration of the Tables 10.1, 10.2, 10.3 and Fig. 10.16 shows that there is considerable redundancy and, as a consequence, a definite pattern in the music of Stephen Foster. These results indicate that a composing machine employing a random probability system based upon Stephen Foster songs may be developed, which will produce music which will sound like Stephen Foster songs but will be new. Furthermore, the electronic music composing machine will be of value to the composer in the production of new music of his type.

C. Philosophy of the Electronic Music Composing Machine Employing a Random Probability System

The electronic music composing machine, which has been developed as an aid to music composition, depends upon a random selection of notes weighed by a probability based upon preceding events. In developing the composing machine it was first determined, estimated, or arbitrarily decided what the probabilities are that certain events, such as the sounding of certain musical notes, will follow a preceding event as, for example, the sounding of certain specific notes in succession. In general, there will be a different probability for each of certain events. For example, for certain kinds of music, it may be that either note A or note C will follow the note-sequence A B. There will be a certain probability that note A will follow the sequence A B. See Table 10.3. Likewise, there will be a certain probability that note B will follow A B and that note C will follow A B.

After the probabilities have been determined, as given in Tables 10.2 and 10.3, the selection of one of said certain events (such as the selection of note A note B or note C) following the preceding events (sounding of A B in this example) is made at random and as a function of said probabilities.

A statistical study of 11 Stephen Foster songs has been described in the preceding section. The machine is designed to operate from this statistical information and thereby produce music. The resulting music while new, sounds like Stephen Foster music.

The electronic music composing machine described in this section is particularly suitable for composing music since music follows certain general rules and patterns with many possible good answers. The machine, however, is not limited to this particular use. The machine may be employed in any field in which the probability of a certain event occurring depends on the kind of event or events that preceded and wherein the probability is known at least approximately or assumed.

D. Description of an Electronic Music Composing Machine Employing a Random Probability System[19],[20]

A block diagram of the electronic music composing machine is shown in Fig. 10.17. The system shown in Fig. 10.17 is designed to provide means for selecting an event at random and, also as a function of the probability that said event will follow a preceding event.

One of the important elements of the electronic music composing machine is the random number generator shown in Fig. 10.18. There are

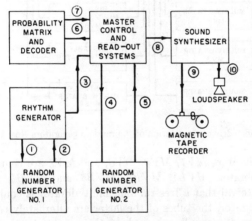

FIG. 10.17. Schematic diagram of the electronic music composing machine employing the random-probability tone-series generating system.

[19] A question may be raised: Why not employ a computer instead of the machine described in this section? The answers are as follows: First, the machine described in this section is indeed a species of computer developed for a particular application. Second, the computer must be modified to perform the functions of the machine described in this section. Third, for composing music, it is desirable to provide sound output. To do this with a computer presents a problem almost as great as the actual construction of the machine described in this section. Fourth, the cost of employing a computer to carry out the work in the manner in which the machine described in this report is employed would be prohibitive.

[20] Olson and Belar, *Jour. Acoust. Soc. Amer.*, Vol. 33, No. 9, p. 1163, 1961.

four units, identical except as to the frequency of operation. The multi-vibrators *MV1*, *MV2*, *MV3*, and *MV4* are free-running units. The output of the free-running multivibrators are connected to the bistable multi-vibrators through a control system described later in this section.

The multivibrators *MV1*, *MV2*, *MV3*, and *MV4* oscillate continuously, each at a slightly different frequency. The particular frequencies are not important. In the actual machine the frequencies were between 1,000 and 1,500 cycles.

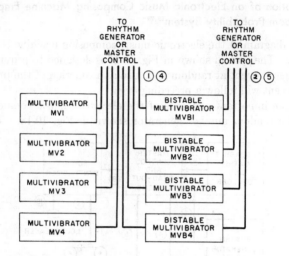

Fig. 10.18.   Schematic diagram of the random generators Nos. 1 and 2.

The multivibrators *MV1*, *MV2*, *MV3*, and *MV4* are connected to bi-stable multivibrators *MVB1*, *MVB2*, *MVB3*, and *MVB4*, respectively. During the interval that a free-running multivibrator is coupled to a bi-stable multivibrator, the tubes of the latter are alternately rendered con-ducting. When the coupling is broken, the tube of the bistable multi-vibrator that is conducting at that instant remains conducting. This selection process of a conducting bistable multivibrator is random because of the high frequency of the oscillation of the free-running multivibrator, as compared to the frequency of the operation of the switch of the control system for making and breaking the coupling, occurring only a few times a second.

Two random generators are used in the composing machine, one for the rhythm generator designated as No. 1, and one for the note selection through the master control system designated as No. 2 in Fig. 10.17.

The rhythm generator of the composing machine operates to produce variations in rhythm by random choice within the defined framework and along predetermined probabilities. The rhythm generator of Fig. 10.19 includes a bank of motor-driven switches which connect the multivibrators to the bistable multivibrators of the random generator of Fig. 10.18. The motor-driven rotary switches may be connected to provide different rhythmic patterns. The notes and the probabilities for 3/4 and 4/4 time are shown in Fig. 10.16. The right-hand column of the table shows the probability that a particular switch connection will be selected to control the system for one measure. These probabilities are employed in connecting the wafers of the rotary switch to the input of the relay tree of Fig. 10.19.

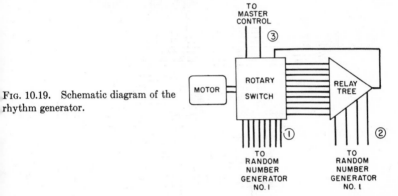

FIG. 10.19. Schematic diagram of the rhythm generator.

The motor-driven switch and relay tree of Fig. 10.19 may be connected to provide either 3/4 or 4/4 time. Four wafers of the rotary switch are connected to the random generator and provide the connection between free-running multivibrators and bistable multivibrators. The four terminals of the relay tree of Fig. 10.19 are connected to the output of the random generator of Fig. 10.18. After the sounding of the first note of a measure initiated by the motor-driven switch in the rhythm generator, that note will continue to sound until the next note is sounded. However, the sounding of one note is actually stopped an instant before the next note is sounded. The sequence of events will be described in greater detail in connection with the master control system.

The output of the rhythm generator is connected to the master control unit. When the connections to the wafer switch and the relay tree of Fig. 10.19 are made in accordance with the probability of Fig. 10.16, the output of the rhythm generator will, over a long period of time, provide rhythms with the probability shown in Fig. 10.16. The output of the rhythm generator is connected to the master control system and provides the control on the sounding of each note.

The master control system is depicted in Fig. 10.20.   One of its basic elements is a rotary stepping switch with eight sections.   The rotary stepping switch is started in operation by the rhythm generator and makes one rotation; that is, it goes through the sequence of six positions for each note selected.   One complete rotation requires about 0.12 second.   Four

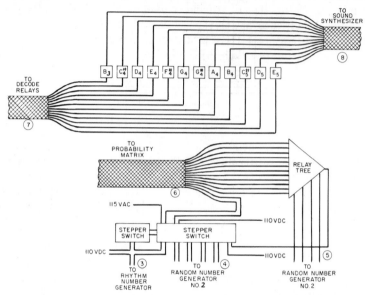

Fig. 10.20.   Schematic diagram of the master control and read-out systems.

sections of the rotary stepping switch are connected to the random number generator No. 2 and provide the means for completing the circuits between the free-running multivibrators and the bistable multivibrators.   The four outputs of the random generator are connected to the four inputs of the relay tree in the master control system.   The 12 outputs of the relay tree are connected to the probability matrix and decode switch.   The 12 outputs represent the notes of the scale shown in Fig. 10.15.   Over a long period of time, the 12 outputs are activated the same number of times, but in a random fashion, because the activation depends upon the output of random generator No. 2.

The matrix and decode system consists of a stepper switch and decode relays as shown in Fig. 10.21.   Twelve sections of the switch are employed and represent the 12 notes of the scale of Fig. 10.15.   Fifty positions of the

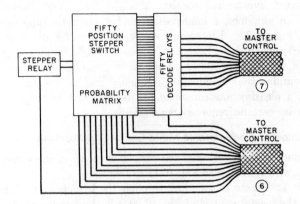

FIG. 10.21. Schematic diagram of the probability matrix and decode system.

switch are used to correspond to the dinotes of the trinotes of Table 10.3. The input to the probability matrix system from the master control system is random. Therefore, over a long period of time the output of the decode system will correspond to the probabilities of Table 10.3. The output of the decode system is connected to 12 relays in the master control unit of Fig. 10.20. These 12 relays represent the 12 notes of Fig. 10.15, and control the activation of the 12 tones in the sound synthesizer.

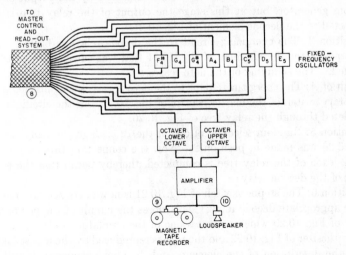

FIG. 10.22. Schematic diagram of the electronic sound synthesizer, loudspeaker and magnetic tape recorder.

The sound synthesizer consists of eight frequency generators, two octavers, an amplifier, a loudspeaker, and a magnetic tape recorder, as shown in Fig. 10.22. Employing octavers, that is, frequency dividers, it is possible to generate the 12 sounds by means of eight frequency generators. The wave shape of the electrical output is a saw-tooth. Therefore, the output contains the fundamental frequency and all the harmonics. In this way, a pleasing musical sound can be obtained by employing filters to emphasize certain frequency regions.

E. Operation of the Electronic Music Composing Machine

The sequence of operation of the various elements of the system will be described, with reference to Fig. 10.17 as follows:

Position 1: A note is sounding through the loudspeaker. The rotary switch of the master control of Fig. 10.20 is in the homing position. The motor of the rhythm control generator of Fig. 10.19 is driving the rotary switch which in turn selects a rhythm from the random multivibrator of Fig. 10.18 in combination with relay tree of Fig. 10.19. The output of the rhythm generator activates the stepper of the rotary switch of the master control of Fig. 10.20, and the switch moves to position 2 thereby initiating another cycle of operation.

Position 2: In this position, the random generator No. 2 of Fig. 10.17 is actuated. A block diagram of the random generator is shown in Fig. 10.18. The relay tree of Fig. 10.20 is also actuated by the output of the random generator, but at this stage the output of the relay tree is not connected to the decode element of Fig. 10.21.

Position 3: The energizing circuit to the decode relays is broken, thus deenergizing the decode relay of Fig. 10.21 that was causing a note to be sounded. The sounding of the note ceases.

Position 4: The energizing circuit to the decode relays is again closed by the rotary switch of the master control of Fig. 10.20, but the circuit is not completed through the relay tree of Fig. 10.20.

Position 5: Random selection of one of the 12 leads of the relay tree of Fig. 10.20 was made in position 2. Now the connection through one of the 12 leads of the relay tree is completed, thereby permitting the energizing of the decode relays.

Position 6: The stepper switch of Fig. 10.21 is now energized and moves to the appropriate decode relay. This closes the circuit of the one of the relays of Fig. 10.20 which activates one of the oscillators and octavers of the synthesizer of Fig. 10.22 and the tone is reproduced by the loudspeaker. From the description of the elements and the above sequence, it will be seen that the note selection is made in accordance with both the random selection and the probability wiring of the system of Fig. 10.21.

Position 1: The rotary switch of the control system of Fig. 10.20 stops on this, the homing position, and remains in this position until reactivated by a signal from the rhythm generator. The note selected continues to sound until another sequence is started by the rhythm generator. In this way, a continuous series of notes is generated. These are recorded by the magnetic tape recorder. A selection of the most melodic or desirable series can be obtained from the tremendous number produced by the machine.

The composing machine was arranged to produce music which will be new but which will sound like Stephen Foster songs by charging the memory machine with the information as outlined in Secs. 10.6B and C. The output of the composing machine was recorded on a magnetic recorder. The most pleasing measures were recorded. The yield of the machine was quite good. For example, in a typical run, 29 measures were obtained from an output of 44 measures. As expected the synthesized output of the machine sounds like Stephen Foster music.[21]

The electronic music composing machine is not limited to the application of the type described in the preceding exposition. For example, the machine may be charged with musical compositions or the style of music writing by a composer. By listening to the sound output, the composer can select from the hundreds, thousands or more tone sequences and select those which appeal to him. This, in turn, may be added to or substituted for the charge in the electronic music composing machine. In this method of operation the electronic music composing machine becomes an aid to composition.

F. Musical Composition by Means of a Digital Computer[22,23]

A process and technique has been developed for writing music by means of a digital computer. In this process random integers, considered to be equivalent to musical notes, are first generated and then selected and screened by mathematical operations which express the rules of musical composition. The control over the musical output is limited solely by the input instructions and factors not specifically accounted for are left entirely to chance. Studies of the problems of strict counterpoint of certain modern compositional procedures such as dissonant chromatic writing

[21] Two selections each about three minutes in length were recorded on magnetic tape. These selections were played before meetings of the National Academy of Sciences, the Acoustical Society of America and the Audio Engineering Society. The general opinions of those in attendance were that the music sounded like Stephen Foster music but not like any particular Stephen Foster song.

[22] Hiller and Isaacson, *Experimental Music*, McGraw-Hill Book Company, New York, 1959.

[23] Hiller and Isaacson, *Jour. Audio Eng. Soc.*, Vol. 6, No. 3, p. 154, 1958.

and tone nongeneration and of writing music by more abstract procedure
based upon certain techniques of probability theory have been carried out
The purpose of this section is to describe the electronic composition o
music by means of the digital computer.

**1. The Monto Carlo Method of Melody Composition Employing the
Digital Computer.**   The composition of the melody by the digital computer employs the Monto Carlo method and therefore operates on the
laws of chance or upon the laws of random probability.   As in the case o
the electronic composing machine described in the preceding section, there
must be restrictions or screening applied to the random probability process
otherwise the result would indeed be nothing but random and meaningless
The first objective was the generation of a random sequence of integers
which provided for the composition of a simple but recognizable melody
the second objective was to achieve simple polyphonic writing.   Strict
counterpoint was selected as the musical medium for Experiment One of
Table 10.4 which consisted of three computer programs designed to produce successively monody, two-part writing and four part writing.   The
notes used in the composition are shown in Fig. 10.23.

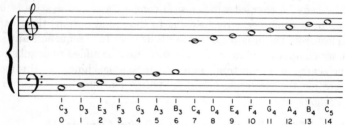

Fig. 10.23.   The notes of the musical scale as used in Experiments One and Two.
(*After Hiller and Isaacson.*)

In Experiment Two the procedure followed the block diagram of Fig.
10.24.   The block diagram depicts the rules in the computer which depend
upon the transmission of musical concepts into arithmetic operations.
Fig. 10.24 is a condensed version of the routine for strict counterpoint.
The diagram groups the rules of counterpoint shown in Table 10.4 into
three basic categories.   The first category consists of the melodic rules
governing linear relationships between successive notes in a given melodic
line.   The second category consists of harmonic rules controlling vertical
relationships between melodic lines.   The last category is made up of what
may be called combined rules which express more complicated interactions.
The block diagram of Fig. 10.24 also depicts the "try again" technique for
building the total composition not only for efficiency but also for simulating
more closely the normal processes in musical composition.

The main routine for chromatic writing used in Experiment Three is shown in Fig. 10.25. Three principal rules were applied here. The first was a chromatic jump-stepwise rule used in strict counterpoint. The second was a complex rule for resolving tritones by inward or outward chromatic progression. The third was an octave range rule in which the melodic lines in an octave were equated backwards to the 15 notes in the key of C in the major scale from C_3 to C_5 inclusive as shown in Fig. 10.23. The

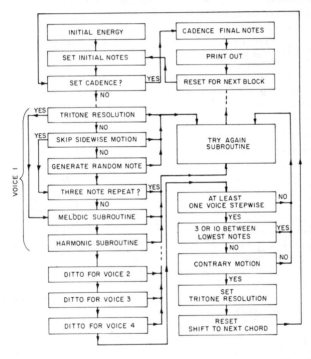

FIG. 10.24. Block diagram for the subroutine for Experiment Two. (*After Hiller and Isaacson.*)

randomly generated notes were processed by a sorting process accepting some and rejecting others. The accepted notes were stored in the memory of the computer and assembled step by step into a machine representation of the finished musical composition. The latter composition was converted into a printed representation in number notation which was then transcribed by hand into a musical score. Following this the following rules were employed for successive intervals, namely: 1, no tritones are permitted; 2, no sevenths are permitted; 3, the melody must start and end on C_4; and 4, the range of the melody from the highest to the lowest note must not exceed one octave. Melodies varying in length from 3 to 12 notes were generated in which the notes were chosen at random. When-

TABLE 10.4. "ILLIAC SUITE" EXPERIMENTS SUMMARIZED (*After Hiller and Isaacson*)

Experiment One: Monody, Two-part and Four-part Writing

Only a limited selection of first species counterpoint rules for controlling the musical output.

(a) Monody: *cantus firmi* 3 to 12 notes in length
(b) Two-part *cantus firmus* settings
(c) Four-part *cantus firmus* settings

Experiment Two: Four-part First Species Counterpoint

Counterpoint rules were added successively to random white-note music as follows:

(a) Random white-note music
(b) Skip-stepwise rule; no more than one successive repeat
(c) Opening C chord; *cantus firmus* begins and ends on C; cadence on C; B-F tritone only in VII_6 chord; tritone resolves to C-E except leading into cadence
(d) Octave range rule
(e) Consonant harmonies only except for $\frac{6}{4}$ chords
(f) Dissonant melodic intervals (seconds, sevenths, tritones) forbidden
(g) No parallel unisons, octaves, fifths
(h) No parallel fourths; no $\frac{6}{4}$ chords, no repeat of climax in highest voice

Experiment Three: Experimental Music

Rhythm, dynamics, playing instructions and simple chromatic writing.

(a) Basic rhythm, dynamics and playing instructions code
(b) Random chromatic music
(c) Random chromatic music combined with modified rhythm, dynamics and playing instructions code
(d) Chromatic music controlled by an octave range rule, a tritone resolution rule and a skip-stepwise rule
(e) Controlled chromatic music combined with modified rhythm, dynamics and playing instructions code
(f) Interval rows, tone rows and restricted tone rows

Experiment Four: "Markoff Chain" Music

(a) Variation of zeroth-order harmonic probability function from complete tonal restriction to "average" distribution
(b) Variation of zeroth-order harmonic probability function from random to "average" distribution
(c) Zeroth-order harmonic and proximity probability functions and functions combined additively
(d) First-order harmonic and proximity probability functions and functions combined additively
(e) Zeroth-order harmonic and proximity functions on strong and weak beats, respectively, and vice-versa
(f) First-order harmonic and proximity functions on strong and weak beats, respectively, and vice-versa
(g) i^{th}-order harmonic function on strong beats, first-order proximity function on weak beats; extended cadence; simple closed form

ever a rule was violated the melody attempt was terminated and the whole process started over again.

2. Experimental Process of Composition by Means of the Digital Computer. The main portion of the work on the electronic composition of music by means of the digital computer except for the note-selection process described in the preceding section involved four main experiments as shown in Table 10.4.

In Experiment One the main objectives were to develop a technique for the composition of a simple but recognizable melody and, to achieve simple polyphonic writing. Strict counterpoint was selected as the musical medium for Experiment One of Table 10.4 which consisted of three-computer programs designed to produce successively monody, two-part writing and four-part writing. The notes used in the composition are shown in Fig. 10.23.

In Experiment Two the procedure followed the block diagram of Fig. 10.24. The block diagram depicts the rules in the computer which depend

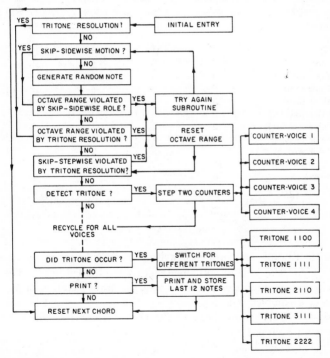

FIG. 10.25. Block diagram for the subroutine for Experiment Three. (*After Hiller and Isaacson.*)

upon the transmission of musical concepts into arithmetic operations. Fig. 10.24 is a condensed version of the routine for strict counterpoint. The diagram groups the rules of counterpoint shown in Table 10.4 into three basic categories. The first category consists of the melodic rules governing linear relationships between successive notes in a given melodic line. The second category consists of harmonic rules controlling vertical relationships between melodic lines. The last category is made up of what may be called combined rules which express more complicated interactions. The block diagram of Fig. 10.24 also depicts the "try again" technique for building the total composition not only for efficiency but also because it simulates more closely the normal processes in musical composition.

TABLE 10.5.   TABLE OF FUNCTIONS FOR THE GENERATION OF "MARKOFF-CHAIN" MUSIC IN EXPERIMENT 4.   (*After Hiller and Isaacson*)

Interval	Stochastic variable, v_j	Harmonic function, x_j	Proximity function, y_j	Combined function, $z_j = x_j + y_j$
Unison	0	13	13	26
Octave	12	12	1	13
Fifth	7	11	6	17
Fourth	5	10	8	18
Major third	4	9	9	18
Minor sixth	8	8	5	13
Minor third	3	7	10	17
Major sixth	9	6	4	10
Major second	2	5	11	16
Minor seventh	10	4	3	7
Minor second	1	3	12	15
Major seventh	11	2	2	4
Tritone	6	1	7	8

$$\sum_{j=0}^{12} x_j = 91 \qquad \sum_{j=0}^{12} y_j = 91 \qquad \sum_{j=0}^{12} z_j = 182 = 2 \times 91$$

$$[x_j = x(v_j)][y_j = y(v_j)]$$

The main routine for chromatic writing used in Experiment Three is shown in Fig. 10.25. Three principal rules were applied here. The first was a chromatic jump-stepwise rule used in strict counterpoint. The second was a complex rule for resolving tritones by inward or outward chromatic progression while the last was an octave range rule which applied backwards for each particular melodic line as far as 12 notes.

Experiment Four illustrates transition probability functions related to harmonic series and to rules of melodic writing which were devised for successive note selection. Table 10.5 depicts the generation of "Markoff chain" music in Experiment Four.

In the programing rhythm, random binary numbers were generated to simulate rhythm permitting ones to represent holds or rests depending on playing instructions produced separately by another part of this same particular program. There are 16 possible quarter-note rhythms in a common-time measure or eight basic patterns in a ternary rhythm, the other basic metrical pattern. In generating rhythms, the binary numbers were permitted to be applied to all possible combinations of four voices and also for as many measures as desired depending upon the rules in effect. The coding of simple rhythmic patterns is shown in Fig. 10.26.

Among the results of the musical composition by means of the digital computer is the *Illiac Suite* by Hiller and Isaacson which has been published by the New Music Editions, New York.

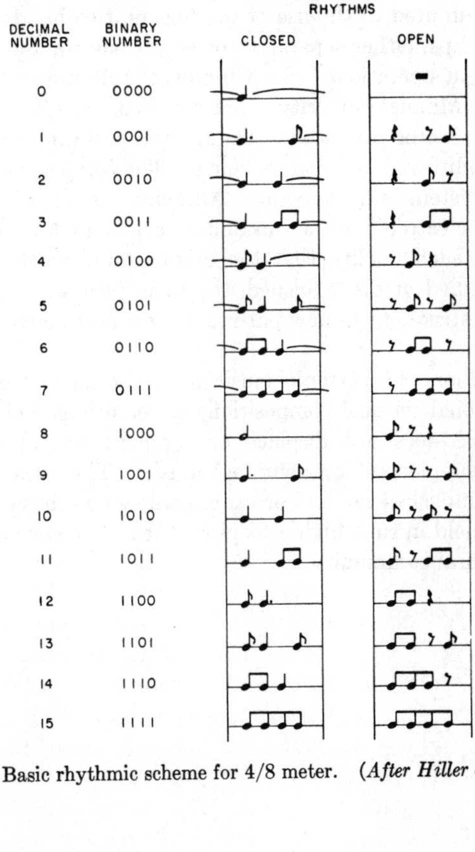

FIG. 10.26. Basic rhythmic scheme for 4/8 meter. (*After Hiller and Isaacson.*)

10.7 POTENTIAL OF ELECTRONIC MUSIC

The preceding subject matter of this chapter provides a brief exposition on electronic music. Electronic music involves the composition and production of music by various electronic means. The ultimate product is a record, either tape or disc, for reproduction through sound-reproducing systems in the concert hall or the home.

Electronic music is not confined to a particular or singular style for special uses, places or audiences. As indicated in the preceding sections, electronic music is based upon the very fundamentals of a tone. Therefore, it is revolutionary in nature in that there are no inherent limitations. The electronic system can produce or create any sound or combination of sounds that may have any musical significance. There is the possibility of entirely new tone complexes and combinations which cannot be achieved in conventional instruments. In the case of conventional instruments, the musician is limited to the use of ten fingers, two hands, two feet and the mouth and lips, either separately or in various combinations, to perform the different operations. Such limitations do not exist in the electronic system. Manual dexterity is not required. Anything in the mind of the composer can be produced. The limitation is only the depth range and other capabilities of the creative being. The composer is the master— the electronic system is his servant. Although, as described in the preceding sections, electronic music can imitate existing conventional musical renditions with such fidelity that the original and synthesized versions cannot be identified or distinguished, the main purpose is not to imitate the past but to strike out in new pathways with new materials and structures.

The introduction of electronic music does not mean that existing or future conventional musical compositions or renditions will be displaced. Electronic music does not displace or supplant anything or anyone. The idea is to supplement conventional music. This idea is rapidly becoming an accomplished fact. For example, in some music centers, electronic music is held in such high esteem that it has become an established part of the cultural community.

Index

CATALOGUE OF DOVER BOOKS

PHYSICS

General physics

FOUNDATIONS OF PHYSICS, R. B. Lindsay & H. Margenau. Excellent bridge between semi-popular works & technical treatises. A discussion of methods of physical description, construction of theory; valuable for physicist with elementary calculus who is interested in ideas that give meaning to data, tools of modern physics. Contents include symbolism, mathematical equations; space & time foundations of mechanics; probability; physics & continua; electron theory; special & general relativity; quantum mechanics; causality. "Thorough and yet not overdetailed. Unreservedly recommended," NATURE (London). Unabridged, corrected edition. List of recommended readings. 35 illustrations. xi + 537pp. 5⅜ x 8.
S377 Paperbound **$2.75**

FUNDAMENTAL FORMULAS OF PHYSICS, ed. by D. H. Menzel. Highly useful, fully inexpensive reference and study text, ranging from simple to highly sophisticated operations. Mathematics integrated into text—each chapter stands as short textbook of field represented. Vol. 1: Statistics, Physical Constants, Special Theory of Relativity, Hydrodynamics, Aerodynamics, Boundary Value Problems in Math. Physics; Viscosity, Electromagnetic Theory, etc. Vol. 2: Sound, Acoustics, Geometrical Optics, Electron Optics, High-Energy Phenomena, Magnetism, Biophysics, much more. Index. Total of 800pp. 5⅜ x 8.
Vol. 1 S595 Paperbound **$2.00**
Vol. 2 S596 Paperbound **$2.00**

MATHEMATICAL PHYSICS, D. H. Menzel. Thorough one-volume treatment of the mathematical techniques vital for classic mechanics, electromagnetic theory, quantum theory, and relativity. Written by the Harvard Professor of Astrophysics for junior, senior, and graduate courses, it gives clear explanations of all those aspects of function theory, vectors, matrices, dyadics, tensors, partial differential equations, etc., necessary for the understanding of the various physical theories. Electron theory, relativity, and other topics seldom presented appear here in considerable detail. Scores of definitions, conversion factors, dimensional constants, etc. "More detailed than normal for an advanced text . . . excellent set of sections on Dyadics, Matrices, and Tensors," JOURNAL OF THE FRANKLIN INSTITUTE. Index. 193 problems, with answers. x + 412pp. 5⅜ x 8.
S56 Paperbound **$2.00**

THE SCIENTIFIC PAPERS OF J. WILLARD GIBBS. All the published papers of America's outstanding theoretical scientist (except for "Statistical Mechanics" and "Vector Analysis"). Vol I (thermodynamics) contains one of the most brilliant of all 19th-century scientific papers—the 300-page "On the Equilibrium of Heterogeneous Substances," which founded the science of physical chemistry, and clearly stated a number of highly important natural laws for the first time; 8 other papers complete the first volume. Vol II includes 2 papers on dynamics, 8 on vector analysis and multiple algebra, 5 on the electromagnetic theory of light, and 6 miscellaneous papers. Biographical sketch by H. A. Bumstead. Total of xxxvi + 718pp. 5⅝ x 8⅜.
S721 Vol I Paperbound **$2.50**
S722 Vol II Paperbound **$2.00**
The set **$4.50**

BASIC THEORIES OF PHYSICS, Peter Gabriel Bergmann. Two-volume set which presents a critical examination of important topics in the major subdivisions of classical and modern physics. The first volume is concerned with classical mechanics and electrodynamics: mechanics of mass points, analytical mechanics, matter in bulk, electrostatics and magnetostatics, electromagnetic interaction, the field waves, special relativity, and waves. The second volume (Heat and Quanta) contains discussions of the kinetic hypothesis, physics and statistics, stationary ensembles, laws of thermodynamics, early quantum theories, atomic spectra, probability waves, quantization in wave mechanics, approximation methods, and abstract quantum theory. A valuable supplement to any thorough course or text.
Heat and Quanta: Index. 8 figures. x + 300pp. 5⅜ x 8½.   S968 Paperbound **$2.00**
Mechanics and Electrodynamics: Index. 14 figures. vii + 280pp. 5⅜ x 8½.
S969 Paperbound **$1.75**

THEORETICAL PHYSICS, A. S. Kompaneyets. One of the very few thorough studies of the subject in this price range. Provides advanced students with a comprehensive theoretical background. Especially strong on recent experimentation and developments in quantum theory. Contents: Mechanics (Generalized Coordinates, Lagrange's Equation, Collision of Particles, etc.), Electrodynamics (Vector Analysis, Maxwell's equations, Transmission of Signals, Theory of Relativity, etc.), Quantum Mechanics (the Inadequacy of Classical Mechanics, the Wave Equation, Motion in a Central Field, Quantum Theory of Radiation, Quantum Theories of Dispersion and Scattering, etc.), and Statistical Physics (Equilibrium Distribution of Molecules in an Ideal Gas, Boltzmann statistics, Bose and Fermi Distribution, Thermodynamic Quantities, etc.). Revised to 1961. Translated by George Yankovsky, authorized by Kompaneyets. 137 exercises. 56 figures. 529pp. 5⅜ x 8½. S972 Paperbound **$2.50**

ANALYTICAL AND CANONICAL FORMALISM IN PHYSICS, André Mercier. A survey, in one volume, of the variational principles (the key principles—in mathematical form—from which the basic laws of any one branch of physics can be derived) of the several branches of physical theory, together with an examination of the relationships among them. Contents: the Lagrangian Formalism, Lagrangian Densities, Canonical Formalism, Canonical Form of Electrodynamics, Hamiltonian Densities, Transformations, and Canonical Form with Vanishing Jacobian Determinant. Numerous examples and exercises. For advanced students, teachers, etc. 6 figures. Index. viii + 222pp. 5⅜ x 8½.
S1077 Paperbound **$1.75**

Acoustics, optics, electricity and magnetism, electromagnetics, magneto-hydrodynamics

THE THEORY OF SOUND, Lord Rayleigh. Most vibrating systems likely to be encountered in practice can be tackled successfully by the methods set forth by the great Nobel laureate, Lord Rayleigh. Complete coverage of experimental, mathematical aspects of sound theory. Partial contents: Harmonic motions, vibrating systems in general, lateral vibrations of bars, curved plates or shells, applications of Laplace's functions to acoustical problems, fluid friction, plane vortex-sheet, vibrations of solid bodies, etc. This is the first inexpensive edition of this great reference and study work. Bibliography. Historical introduction by R. B. Lindsay. Total of 1040pp. 97 figures. 5⅜ x 8.
S292, S293, Two volume set, paperbound, **$4.70**

THE DYNAMICAL THEORY OF SOUND, H. Lamb. Comprehensive mathematical treatment of the physical aspects of sound, covering the theory of vibrations, the general theory of sound, and the equations of motion of strings, bars, membranes, pipes, and resonators. Includes chapters on plane, spherical, and simple harmonic waves, and the Helmholtz Theory of Audition. Complete and self-contained development for student and specialist; all fundamental differential equations solved completely. Specific mathematical details for such important phenomena as harmonics, normal modes, forced vibrations of strings, theory of reed pipes, etc. Index. Bibliography. 86 diagrams. viii + 307pp. 5⅜ x 8.
S655 Paperbound **$2.00**

WAVE PROPAGATION IN PERIODIC STRUCTURES, L. Brillouin. A general method and application to different problems: pure physics, such as scattering of X-rays of crystals, thermal vibration in crystal lattices, electronic motion in metals; and also problems of electrical engineering. Partial contents: elastic waves in 1-dimensional lattices of point masses. Propagation of waves along 1-dimensional lattices. Energy flow. 2 dimensional, 3 dimensional lattices. Mathieu's equation. Matrices and propagation of waves along an electric line. Continuous electric lines. 131 illustrations. Bibliography. Index. xii + 253pp. 5⅜ x 8.
S34 Paperbound **$2.00**

THEORY OF VIBRATIONS, N. W. McLachlan. Based on an exceptionally successful graduate course given at Brown University, this discusses linear systems having 1 degree of freedom, forced vibrations of simple linear systems, vibration of flexible strings, transverse vibrations of bars and tubes, transverse vibration of circular plate, sound waves of finite amplitude, etc. Index. 99 diagrams. 160pp. 5⅜ x 8.
S190 Paperbound **$1.50**

LIGHT: PRINCIPLES AND EXPERIMENTS, George S. Monk. Covers theory, experimentation, and research. Intended for students with some background in general physics and elementary calculus. Three main divisions: 1) Eight chapters on geometrical optics—fundamental concepts (the ray and its optical length, Fermat's principle, etc.), laws of image formation, apertures in optical systems, photometry, optical instruments etc.; 2) 9 chapters on physical optics—interference, diffraction, polarization, spectra, the Rayleigh refractometer, the wave theory of light, etc.; 3) 23 instructive experiments based directly on the theoretical text. "Probably the best intermediate textbook on light in the English language. Certainly, it is the best book which includes both geometrical and physical optics," J. Rud Nielson, PHYSICS FORUM. Revised edition. 102 problems and answers. 12 appendices. 6 tables. Index. 270 illustrations. xi +489pp. 5⅜ x 8½.
S341 Paperbound **$2.50**

PHOTOMETRY, John W. T. Walsh. The best treatment of both "bench" and "illumination" photometry in English by one of Britain's foremost experts in the field (President of the International Commission on Illumination). Limited to those matters, theoretical and practical, which affect the measurement of light flux, candlepower, illumination, etc., and excludes treatment of the use to which such measurements may be put after they have been made. Chapters on Radiation, The Eye and Vision, Photo-Electric Cells, The Principles of Photometry, The Measurement of Luminous Intensity, Colorimetry, Spectrophotometry, Stellar Photometry, The Photometric Laboratory, etc. Third revised (1958) edition. 281 illustrations. 10 appendices. xxiv + 544pp. 5½ x 9¼.
S319 Paperbound **$3.00**

EXPERIMENTAL SPECTROSCOPY, R. A. Sawyer. Clear discussion of prism and grating spectrographs and the techniques of their use in research, with emphasis on those principles and techniques that are fundamental to practically all uses of spectroscopic equipment. Beginning with a brief history of spectroscopy, the author covers such topics as light sources, spectroscopic apparatus, prism spectroscopes and graphs, diffraction grating, the photographic process, determination of wave length, spectral intensity, infrared spectroscopy, spectrochemical analysis, etc. This revised edition contains new material on the production of replica gratings, solar spectroscopy from rockets, new standard of wave length, etc. Index. Bibliography. 111 illustrations. x + 358pp. 5⅜ x 8½.
S1045 Paperbound **$2.25**

FUNDAMENTALS OF ELECTRICITY AND MAGNETISM, L. B. Loeb. For students of physics, chemistry, or engineering who want an introduction to electricity and magnetism on a higher level and in more detail than general elementary physics texts provide. Only elementary differential and integral calculus is assumed. Physical laws developed logically, from magnetism to electric currents, Ohm's law, electrolysis, and on to static electricity, induction, etc. Covers an unusual amount of material; one third of book on modern material: solution of wave equation, photoelectric and thermionic effects, etc. Complete statement of the various electrical systems of units and interrelations. 2 Indexes. 75 pages of problems with answers stated. Over 300 figures and diagrams. xix +669pp. 5⅜ x 8.
S745 Paperbound **$3.50**

MATHEMATICAL ANALYSIS OF ELECTRICAL AND OPTICAL WAVE-MOTION, Harry Bateman. Written by one of this century's most distinguished mathematical physicists, this is a practical introduction to those developments of Maxwell's electromagnetic theory which are directly connected with the solution of the partial differential equation of wave motion. Methods of solving wave-equation, polar-cylindrical coordinates, diffraction, transformation of coordinates, homogeneous solutions, electromagnetic fields with moving singularities, etc. Index. 168pp. 5⅜ x 8.
S14 Paperbound **$1.75**

PRINCIPLES OF PHYSICAL OPTICS, Ernst Mach. This classical examination of the propagation of light, color, polarization, etc. offers an historical and philosophical treatment that has never been surpassed for breadth and easy readability. Contents: Rectilinear propagation of light. Reflection, refraction. Early knowledge of vision. Dioptrics. Composition of light. Theory of color and dispersion. Periodicity. Theory of interference. Polarization. Mathematical representation of properties of light. Propagation of waves, etc. 279 illustrations, 10 portraits. Appendix. Indexes. 324pp. 5⅜ x 8.
S178 Paperbound **$2.00**

THE THEORY OF OPTICS, Paul Drude. One of finest fundamental texts in physical optics, classic offers thorough coverage, complete mathematical treatment of basic ideas. Includes fullest treatment of application of thermodynamics to optics; sine law in formation of images, transparent crystals, magnetically active substances, velocity of light, apertures, effects depending upon them, polarization, optical instruments, etc. Introduction by A. A. Michelson. Index. 110 illus. 567pp. 5⅜ x 8.
S532 Paperbound **$2.45**

ELECTRICAL THEORY ON THE GIORGI SYSTEM, P. Cornelius. A new clarification of the fundamental concepts of electricity and magnetism, advocating the convenient m.k.s. system of units that is steadily gaining followers in the sciences. Illustrating the use and effectiveness of his terminology with numerous applications to concrete technical problems, the author here expounds the famous Giorgi system of electrical physics. His lucid presentation and well-reasoned, cogent argument for the universal adoption of this system form one of the finest pieces of scientific exposition in recent years. 28 figures. Index. Conversion tables for translating earlier data into modern units. Translated from 3rd Dutch edition by L. J. Jolley. x + 187pp. 5½ x 8¾.
S909 Clothbound **$6.00**

ELECTRIC WAVES: BEING RESEARCHES ON THE PROPAGATION OF ELECTRIC ACTION WITH FINITE VELOCITY THROUGH SPACE, Heinrich Hertz. This classic work brings together the original papers in which Hertz—Helmholtz's protegé and one of the most brilliant figures in 19th-century research—probed the existence of electromagnetic waves and showed experimentally that their velocity equalled that of light, research that helped lay the groundwork for the development of radio, television, telephone, telegraph, and other modern technological marvels. Unabridged republication of original edition. Authorized translation by D. E. Jones. Preface by Lord Kelvin. Index of names. 40 illustrations. xvii + 278pp. 5⅜ x 8½.
S57 Paperbound **$1.75**

PIEZOELECTRICITY: AN INTRODUCTION TO THE THEORY AND APPLICATIONS OF ELECTRO-MECHANICAL PHENOMENA IN CRYSTALS, Walter G. Cady. This is the most complete and systematic coverage of this important field in print—now regarded as something of scientific classic. This republication, revised and corrected by Prof. Cady—one of the foremost contributors in this area—contains a sketch of recent progress and new material on Ferroelectrics. Time Standards, etc. The first 7 chapters deal with fundamental theory of crystal electricity. 5 important chapters cover basic concepts of piezoelectricity, including comparisons of various competing theories in the field. Also discussed: piezoelectric resonators (theory, methods of manufacture, influences of air-gaps, etc.); the piezo oscillator; the properties, history, and observations relating to Rochelle salt; ferroelectric crystals; miscellaneous applications of piezoelectricity; pyroelectricity; etc. "A great work," W. A. Wooster, NATURE. Revised (1963) and corrected edition. New preface by Prof. Cady. 2 Appendices. Indices. Illustrations. 62 tables. Bibliography. Problems. Total of 1 + 822pp. 5⅜ x 8½.
S1094 Vol. I Paperbound **$2.50**
S1095 Vol. II Paperbound **$2.50**
Two volume set Paperbound **$5.00**

MAGNETISM AND VERY LOW TEMPERATURES, H. B. G. Casimir. A basic work in the literature of low temperature physics. Presents a concise survey of fundamental theoretical principles, and also points out promising lines of investigation. Contents: Classical Theory and Experimental Methods, Quantum Theory of Paramagnetism, Experiments on Adiabatic Demagnetization. Theoretical Discussion of Paramagnetism at Very Low Temperatures, Some Experimental Results, Relaxation Phenomena. Index. 89-item bibliography. ix + 95pp. 5⅜ x 8.
S943 Paperbound **$1.25**

SELECTED PAPERS ON NEW TECHNIQUES FOR ENERGY CONVERSION: THERMOELECTRIC METHODS; THERMIONIC; PHOTOVOLTAIC AND ELECTRICAL EFFECTS; FUSION, Edited by Sumner N. Levine. Brings together in one volume the most important papers (1954-1961) in modern energy technology. Included among the 37 papers are general and qualitative descriptions of the field as a whole, indicating promising lines of research. Also: 15 papers on thermoelectric methods, 7 on thermionic, 5 on photovoltaic, 4 on electrochemical effect, and 2 on controlled fusion research. Among the contributors are: Joffe, Maria Telkes, Herold, Herring, Douglas, Jaumot, Post, Austin, Wilson, Pfann, Rappaport, Morehouse, Domenicali, Moss, Bowers, Harman, Von Doenhoef. Preface and introduction by the editor. Bibliographies. xxviii + 451pp. 6⅛ x 9¼.
S37 Paperbound **$3.00**

SUPERFLUIDS: MACROSCOPIC THEORY OF SUPERCONDUCTIVITY, Vol. I, Fritz London. The major work by one of the founders and great theoreticians of modern quantum physics. Consolidates the researches that led to the present understanding of the nature of superconductivity. Prof. London here reveals that quantum mechanics is operative on the macroscopic plane as well as the submolecular level. Contents: Properties of Superconductors and Their Thermodynamical Correlation; Electrodynamics of the Pure Superconducting State; Relation between Current and Field; Measurements of the Penetration Depth; Non-Viscous Flow vs. Superconductivity; Micro-waves in Superconductors; Reality of the Domain Structure; and many other related topics. A new epilogue by M. J. Buckingham discusses developments in the field up to 1960. Corrected and expanded edition. An appreciation of the author's life and work by L. W. Nordheim. Biography by Edith London. Bibliography of his publications. 45 figures. 2 Indices. xviii + 173pp. 5⅝ x 8⅜. S44 Paperbound **$1.75**

SELECTED PAPERS ON PHYSICAL PROCESSES IN IONIZED PLASMAS, Edited by Donald H. Menzel, Director, Harvard College Observatory. 30 important papers relating to the study of highly ionized gases or plasmas selected by a foremost contributor in the field, with the assistance of Dr. L. H. Aller. The essays include 18 on the physical processes in gaseous nebulae, covering problems of radiation and radiative transfer, the Balmer decrement, electron temperatures, spectrophotometry, etc. 10 papers deal with the interpretation of nebular spectra, by Bohm, Van Vleck, Aller, Minkowski, etc. There is also a discussion of the intensities of "forbidden" spectral lines by George Shortley and a paper concerning the theory of hydrogenic spectra by Menzel and Pekeris. Other contributors: Goldberg, Hebb, Baker, Bowen, Ufford, Liller, etc. viii + 374pp. 6⅛ x 9¼. S60 Paperbound **$2.95**

THE ELECTROMAGNETIC FIELD, Max Mason & Warren Weaver. Used constantly by graduate engineers. Vector methods exclusively: detailed treatment of electrostatics, expansion methods, with tables converting any quantity into absolute electromagnetic, absolute electrostatic, practical units. Discrete charges, ponderable bodies, Maxwell field equations, etc. Introduction. Indexes. 416pp. 5⅜ x 8. S185 Paperbound **$2.25**

THEORY OF ELECTRONS AND ITS APPLICATION TO THE PHENOMENA OF LIGHT AND RADIANT HEAT, H. Lorentz. Lectures delivered at Columbia University by Nobel laureate Lorentz. Unabridged, they form a historical coverage of the theory of free electrons, motion, absorption of heat, Zeeman effect, propagation of light in molecular bodies, inverse Zeeman effect, optical phenomena in moving bodies, etc. 109 pages of notes explain the more advanced sections. Index. 9 figures. 352pp. 5⅜ x 8. S173 Paperbound **$2.00**

FUNDAMENTAL ELECTROMAGNETIC THEORY, Ronold P. King, Professor Applied Physics, Harvard University. Original and valuable introduction to electromagnetic theory and to circuit theory from the standpoint of electromagnetic theory. Contents: Mathematical Description of Matter—stationary and nonstationary states; Mathematical Description of Space and of Simple Media—Field Equations, Integral Forms of Field Equations, Electromagnetic Force, etc.; Transformation of Field and Force Equations; Electromagnetic Waves in Unbounded Regions; Skin Effect and Internal Impedance—in a solid cylindrical conductor, etc.; and Electrical Circuits—Analytical Foundations, Near-zone and quasi-near zone circuits, Balanced two-wire and four-wire transmission lines. Revised and enlarged version. New preface by the author. 5 appendices (Differential operators: Vector Formulas and Identities, etc.). Problems. Indexes. Bibliography. xvi + 580pp. 5⅜ x 8½. S1023 Paperbound **$3.00**

Hydrodynamics

A TREATISE ON HYDRODYNAMICS, A. B. Basset. Favorite text on hydrodynamics for 2 generations of physicists, hydrodynamical engineers, oceanographers, ship designers, etc. Clear enough for the beginning student, and thorough source for graduate students and engineers on the work of d'Alembert, Euler, Laplace, Lagrange, Poisson, Green, Clebsch, Stokes, Cauchy, Helmholtz, J. J. Thomson, Love, Hicks, Greenhill, Besant, Lamb, etc. Great amount of documentation on entire theory of classical hydrodynamics. Vol I: theory of motion of frictionless liquids, vortex, and cyclic irrotational motion, etc. 132 exercises. Bibliography. 3 Appendixes. xii + 264pp. Vol II: motion in viscous liquids, harmonic analysis, theory of tides, etc. 112 exercises, Bibliography. 4 Appendixes. xv + 328pp. Two volume set. 5⅜ x 8.
S724 Vol I Paperbound **$1.75**
S725 Vol II Paperbound **$1.75**
The set **$3.50**

HYDRODYNAMICS, Horace Lamb. Internationally famous complete coverage of standard reference work on dynamics of liquids & gases. Fundamental theorems, equations, methods, solutions, background, for classical hydrodynamics. Chapters include Equations of Motion, Integration of Equations in Special Gases, Irrotational Motion, Motion of Liquid in 2 Dimensions, Motion of Solids through Liquid-Dynamical Theory, Vortex Motion, Tidal Waves, Surface Waves, Waves of Expansion, Viscosity, Rotating Masses of liquids. Excellently planned, arranged; clear, lucid presentation. 6th enlarged, revised edition. Index. Over 900 footnotes, mostly bibliographical. 119 figures. xv + 738pp. 6⅛ x 9¼. S256 Paperbound **$3.75**

HYDRODYNAMICS, H. Dryden, F. Murnaghan, Harry Bateman. Published by the National Research Council in 1932 this enormous volume offers a complete coverage of classical hydrodynamics. Encyclopedic in quality. Partial contents: physics of fluids, motion, turbulent flow, compressible fluids, motion in 1, 2, 3 dimensions; viscous fluids rotating, laminar motion, resistance of motion through viscous fluid, eddy viscosity, hydraulic flow in channels of various shapes, discharge of gases, flow past obstacles, etc. Bibliography of over 2,900 items. Indexes. 23 figures. 634pp. 5⅜ x 8.　　　　　　　　　S303 Paperbound **$2.75**

Mechanics, dynamics, thermodynamics, elasticity

MECHANICS, J. P. Den Hartog. Already a classic among introductory texts, the M.I.T. professor's lively and discursive presentation is equally valuable as a beginner's text, an engineering student's refresher, or a practicing engineer's reference. Emphasis in this highly readable text is on illuminating fundamental principles and showing how they are embodied in a great number of real engineering and design problems: trusses, loaded cables, beams, jacks, hoists, etc. Provides advanced material on relative motion and gyroscopes not usual in introductory texts. "Very thoroughly recommended to all those anxious to improve their real understanding of the principles of mechanics." MECHANICAL WORLD. Index. List of equations. 334 problems, all with answers. Over 550 diagrams and drawings. ix + 462pp. 5⅜ x 8.
　　　　　　　　　S754 Paperbound **$2.00**

THEORETICAL MECHANICS: AN INTRODUCTION TO MATHEMATICAL PHYSICS, J. S. Ames, F. D. Murnaghan. A mathematically rigorous development of theoretical mechanics for the advanced student, with constant practical applications. Used in hundreds of advanced courses. An unusually thorough coverage of gyroscopic and baryscopic material, detailed analyses of the Coriolis acceleration, applications of Lagrange's equations, motion of the double pendulum, Hamilton-Jacobi partial differential equations, group velocity and dispersion, etc. Special relativity is also included. 159 problems. 44 figures. ix + 462pp. 5⅜ x 8.
　　　　　　　　　S461 Paperbound **$2.25**

THEORETICAL MECHANICS: STATICS AND THE DYNAMICS OF A PARTICLE, W. D. MacMillan. Used for over 3 decades as a self-contained and extremely comprehensive advanced undergraduate text in mathematical physics, physics, astronomy, and deeper foundations of engineering. Early sections require only a knowledge of geometry; later, a working knowledge of calculus. Hundreds of basic problems, including projectiles to the moon, escape velocity, harmonic motion, ballistics, falling bodies, transmission of power, stress and strain, elasticity, astronomical problems. 340 practice problems plus many fully worked out examples make it possible to test and extend principles developed in the text. 200 figures. xvii + 430pp. 5⅜ x 8.　　　　　　　　　S467 Paperbound **$2.25**

THEORETICAL MECHANICS: THE THEORY OF THE POTENTIAL, W. D. MacMillan. A comprehensive, well balanced presentation of potential theory, serving both as an introduction and a reference work with regard to specific problems, for physicists and mathematicians. No prior knowledge of integral relations is assumed, and all mathematical material is developed as it becomes necessary. Includes: Attraction of Finite Bodies; Newtonian Potential Function; Vector Fields, Green and Gauss Theorems; Attractions of Surfaces and Lines; Surface Distribution of Matter; Two-Layer Surfaces; Spherical Harmonics; Ellipsoidal Harmonics; etc. "The great number of particular cases . . . should make the book valuable to geophysicists and others actively engaged in practical applications of the potential theory," Review of Scientific Instruments. Index. Bibliography. xiii + 469pp. 5⅜ x 8.　　　　　S486 Paperbound **$2.50**

THEORETICAL MECHANICS: DYNAMICS OF RIGID BODIES, W. D. MacMillan. Theory of dynamics of a rigid body is developed, using both the geometrical and analytical methods of instruction. Begins with exposition of algebra of vectors, it goes through momentum principles, motion in space, use of differential equations and infinite series to solve more sophisticated dynamics problems. Partial contents: moments of inertia, systems of free particles, motion parallel to a fixed plane, rolling motion, method of periodic solutions, much more. 82 figs. 199 problems. Bibliography. Indexes. xii + 476pp. 5⅜ x 8.　　　　S641 Paperbound **$2.50**

MATHEMATICAL FOUNDATIONS OF STATISTICAL MECHANICS, A. I. Khinchin. Offering a precise and rigorous formulation of problems, this book supplies a thorough and up-to-date exposition. It provides analytical tools needed to replace cumbersome concepts, and furnishes for the first time a logical step-by-step introduction to the subject. Partial contents: geometry & kinematics of the phase space, ergodic problem, reduction to theory of probability, application of central limit problem, ideal monatomic gas, foundation of thermo-dynamics, dispersion and distribution of sum functions. Key to notations. Index. viii + 179pp. 5⅜ x 8.
　　　　　　　　　S147 Paperbound **$1.50**

ELEMENTARY PRINCIPLES IN STATISTICAL MECHANICS, J. W. Gibbs. Last work of the great Yale mathematical physicist, still one of the most fundamental treatments available for advanced students and workers in the field. Covers the basic principle of conservation of probability of phase, theory of errors in the calculated phases of a system, the contributions of Clausius, Maxwell, Boltzmann, and Gibbs himself, and much more. Includes valuable comparison of statistical mechanics with thermodynamics: Carnot's cycle, mechanical definitions of entropy, etc. xvi + 208pp. 5⅜ x 8.　　　　　　　S707 Paperbound **$1.45**

Technological, historical

A DIDEROT PICTORIAL ENCYCLOPEDIA OF TRADES AND INDUSTRY, Manufacturing and the Technical Arts in Plates Selected from "L'Encyclopédie ou Dictionnaire Raisonné des Sciences, des Arts, et des Métiers" of Denis Diderot. Edited with text by C. Gillispie. This first modern selection of plates from the high point of 18th century French engraving is a storehouse of valuable technological information to the historian of arts and science. Over 2000 illustrations on 485 full-page plates, most of them original size, show the trades and industries of a fascinating era in such great detail that the processes and shops might very well be reconstructed from them. The plates teem with life, with men, women, and children performing all of the thousands of operations necessary to the trades before and during the early stages of the industrial revolution. Plates are in sequence, and show general operations, closeups of difficult operations, and details of complex machinery. Such important and interesting trades and industries are illustrated as sowing, harvesting, beekeeping, cheesemaking, operating windmills, milling flour, charcoal burning, tobacco processing, indigo, fishing, arts of war, salt extraction, mining, smelting, casting iron, steel, extracting mercury, zinc, sulphur, copper, etc., slating, tinning, silverplating, gilding, making gunpowder, cannons, bells, shoeing horses, tanning, papermaking, printing, dyeing, and more than 40 other categories. Professor Gillispie, of Princeton, supplies a full commentary on all the plates, identifying operations, tools, processes, etc. This material, presented in a lively and lucid fashion, is of great interest to the reader interested in history of science and technology. Heavy library cloth. 920pp. 9 x 12. T421 Two volume set **$18.50**

CHARLES BABBAGE AND HIS CALCULATING ENGINES, edited by P. Morrison and E. Morrison. Babbage, leading 19th century pioneer in mathematical machines and herald of modern operational research, was the true father of Harvard's relay computer Mark I. His Difference Engine and Analytical Engine were the first machines in the field. This volume contains a valuable introduction on his life and work; major excerpts from his autobiography, revealing his eccentric and unusual personality; and extensive selections from "Babbage's Calculating Engines," a compilation of hard-to-find journal articles by Babbage, the Countess of Lovelace, L. F. Menabrea, and Dionysius Lardner. 8 illustrations, Appendix of miscellaneous papers. Index. Bibliography. xxxviii + 400pp. 5⅜ x 8. T12 Paperbound **$2.00**

HISTORY OF HYDRAULICS, Hunter Rouse and Simon Ince. First history of hydraulics and hydrodynamics available in English. Presented in readable, non-mathematical form, the text is made especially easy to follow by the many supplementary photographs, diagrams, drawings, etc. Covers the great discoveries and developments from Archimedes and Galileo to modern giants— von Mises, Prandtl, von Karman, etc. Interesting browsing for the specialist; excellent introduction for teachers and students. Discusses such milestones as the two-piston pump of Ctesibius, the aqueducts of Frontius, the anticipations of da Vinci, Stevin and the first book on hydrodynamics, experimental hydraulics of the 18th century, the 19th-century expansion of practical hydraulics and classical and applied hydrodynamics, the rise of fluid mechanics in our time, etc. 200 illustrations. Bibliographies. Index. xii + 270pp. 5¾ x 8.
S1131 Paperbound **$2.00**

BRIDGES AND THEIR BUILDERS, David Steinman and Sara Ruth Watson. Engineers, historians, everyone who has ever been fascinated by great spans will find this book an endless source of information and interest. Dr. Steinman, recipient of the Louis Levy medal, was one of the great bridge architects and engineers of all time, and his analysis of the great bridges of history is both authoritative and easily followed. Greek and Roman bridges, medieval bridges, Oriental bridges, modern works such as the Brooklyn Bridge and the Golden Gate Bridge, and many others are described in terms of history, constructional principles, artistry, and function. All in all this book is the most comprehensive and accurate semipopular history of bridges in print in English. New, greatly revised, enlarged edition. 23 photographs, 26 line drawings. Index. xvii + 401pp. 5⅜ x 8. T431 Paperbound **$2.00**

Prices subject to change without notice.

Dover publishes books on art, music, philosophy, literature, languages, history, social sciences, psychology, handcrafts, orientalia, puzzles and entertainments, chess, pets and gardens, books explaining science, intermediate and higher mathematics, mathematical physics, engineering, biological sciences, earth sciences, classics of science, etc. Write to:

Dept. catrr.
Dover Publications, Inc.
180 Varick Street, N.Y. 14, N.Y.